IEE POWER AND ENERGY SERIES 39

Series Editors: Professor A. T. Johns
D. F. Warne

Power systems electromagnetic transients simulation

Other volumes in this series:

Volume 1   **Power circuits breaker theory and design** C. H. Flurscheim (Editor)
Volume 2   **Electric Fuses** A. Wright and P. G. Newbery
Volume 3   **Z-transform electromagnetic transient analysis in high-voltage networks** W. Derek Humpage
Volume 4   **Industrial microwave heating** A. C. Metaxas and R. J. Meredith
Volume 5   **Power system economics** T. W. Berrie
Volume 6   **High voltage direct current transmission** J. Arrillaga
Volume 7   **Insulators for high voltages** J. S. T. Looms
Volume 8   **Variable frequency AC motor drive systems** D. Finney
Volume 9   **Electricity distribution network design** E. Lakervi and E. J. Holmes
Volume 10 **SF_6 switchgear** H. M. Ryan and G. R. Jones
Volume 11 **Conduction and induction heating** E. J. Davies
Volume 12 **Overvoltage protection of low-voltage systems** P. Hasse
Volume 13 **Statistical techniques for high-voltage engineering** W. Hauschild and W. Mosch
Volume 14 **Uninterruptible power supplies** J. D. St. Aubyn and J. Platts (Editors)
Volume 15 **Digital protection for power systems** A. T. Johns and S. K. Salman
Volume 16 **Electricity economics and planning** T. W. Berrie
Volume 17 **High voltage engineering and testing** H. M. Ryan (Editor)
Volume 18 **Vacuum switchgear** A. Greenwood
Volume 19 **Electrical safety: a guide to the causes and prevention of electrical hazards** J. Maxwell Adams
Volume 20 **Electric fuses, 2nd Edn.** A. Wright and P. G. Newbery
Volume 21 **Electricity distribution network design, 2nd Edn.** E. Lakervi and E. J. Holmes
Volume 22 **Artificial intelligence techniques in power systems** K. Warwick, A. Ekwue and R. Aggarwal (Editors)
Volume 23 **Financial and economic evaluation of projects in the electricity supply industry** H. Khatib
Volume 24 **Power system commissioning and maintenance practice** K. Harker
Volume 25 **Engineers' handbook of industrial microwave heating** R. J. Meredith
Volume 26 **Small electric motors** H. Moczala
Volume 27 **AC-DC power system analysis** J. Arrillaga and B. C. Smith
Volume 28 **Protection of electricity distribution networks** J. Gers and E. J. Holmes
Volume 29 **High voltage direct current transmission** J. Arrillaga
Volume 30 **Flexible AC transmission systems (FACTS)** Y.-H. Song and A. T. Johns (Editors)
Volume 31 **Embedded generation** N. Jenkins, R. Allan, P. Crossley, D. Kirschen and G. Strbac
Volume 32 **High voltage engineering and testing, 2nd Edn.** H. M. Ryan (Editor)
Volume 33 **Overvoltage protection of low voltage systems** P. Hasse
Volume 34 **The lightning flash** G. V. Cooray
Volume 35 **Contol Techniques drives and controls handbook** W. Drury (Editor)
Volume 36 **Voltage quality in electrical systems** J. Schlabbach, D. Blume and T. Stephanblome
Volume 38 **The electric car: development and future of battery, hybrid and fuel-cell cars** M. H. Westbrook

Power systems electromagnetic transients simulation

Neville Watson
and Jos Arrillaga

The Institution of Electrical Engineers

Published by: The Institution of Electrical Engineers, London,
United Kingdom

© 2003: The Institution of Electrical Engineers

This publication is copyright under the Berne Convention and the
Universal Copyright Convention. All rights reserved. Apart from any fair
dealing for the purposes of research or private study, or criticism or
review, as permitted under the Copyright, Designs and Patents Act, 1988,
this publication may be reproduced, stored or transmitted, in any forms or
by any means, only with the prior permission in writing of the publishers,
or in the case of reprographic reproduction in accordance with the terms
of licences issued by the Copyright Licensing Agency. Inquiries
concerning reproduction outside those terms should be sent to the
publishers at the undermentioned address:

The Institution of Electrical Engineers,
Michael Faraday House,
Six Hills Way, Stevenage,
Herts. SG1 2AY, United Kingdom

While the authors and the publishers believe that the information and
guidance given in this work are correct, all parties must rely upon their
own skill and judgment when making use of them. Neither the authors
nor the publishers assume any liability to anyone for any loss or damage
caused by any error or omission in the work, whether such error or
omission is the result of negligence or any other cause. Any and all such
liability is disclaimed.

The moral right of the authors to be identified as author of this work has
been asserted by them in accordance with the Copyright, Designs and
Patents Act 1988.

British Library Cataloguing in Publication Data

Arrillaga, J.
Power systems electromagnetic transients simulation –
(IEE power and energy series; no. 39)
1. Electrical power systems 2. Transients (Electricity)
I. Title II. Watson, N. R. III. Institution of Electrical Engineers
621.3'191

ISBN 0 85296 106 5

Typeset in India by Newgen Imaging Systems
Printed in the UK by MPG Books Limited, Bodmin, Cornwall

Contents

List of figures		xiii
List of tables		xxi
Preface		xxiii
Acronyms and constants		xxv

1 Definitions, objectives and background — 1
- 1.1 Introduction — 1
- 1.2 Classification of electromagnetic transients — 3
- 1.3 Transient simulators — 4
- 1.4 Digital simulation — 5
  - 1.4.1 State variable analysis — 5
  - 1.4.2 Method of difference equations — 5
- 1.5 Historical perspective — 6
- 1.6 Range of applications — 9
- 1.7 References — 9

2 Analysis of continuous and discrete systems — 11
- 2.1 Introduction — 11
- 2.2 Continuous systems — 11
  - 2.2.1 State variable formulations — 13
    - 2.2.1.1 Successive differentiation — 13
    - 2.2.1.2 Controller canonical form — 14
    - 2.2.1.3 Observer canonical form — 16
    - 2.2.1.4 Diagonal canonical form — 18
    - 2.2.1.5 Uniqueness of formulation — 19
    - 2.2.1.6 Example — 20
  - 2.2.2 Time domain solution of state equations — 20
  - 2.2.3 Digital simulation of continuous systems — 22
    - 2.2.3.1 Example — 27
- 2.3 Discrete systems — 30

vi *Contents*

	2.4	Relationship of continuous and discrete domains	32
	2.5	Summary	34
	2.6	References	34

3 State variable analysis — 35

	3.1	Introduction	35
	3.2	Choice of state variables	35
	3.3	Formation of the state equations	37
		3.3.1 The transform method	37
		3.3.2 The graph method	40
	3.4	Solution procedure	43
	3.5	Transient converter simulation (TCS)	44
		3.5.1 Per unit system	45
		3.5.2 Network equations	46
		3.5.3 Structure of TCS	49
		3.5.4 Valve switchings	51
		3.5.5 Effect of automatic time step adjustments	53
		3.5.6 TCS converter control	55
	3.6	Example	59
	3.7	Summary	64
	3.8	References	65

4 Numerical integrator substitution — 67

	4.1	Introduction	67
	4.2	Discretisation of R, L, C elements	68
		4.2.1 Resistance	68
		4.2.2 Inductance	68
		4.2.3 Capacitance	70
		4.2.4 Components reduction	71
	4.3	Dual Norton model of the transmission line	73
	4.4	Network solution	76
		4.4.1 Network solution with switches	79
		4.4.2 Example: voltage step applied to RL load	80
	4.5	Non-linear or time varying parameters	88
		4.5.1 Current source representation	89
		4.5.2 Compensation method	89
		4.5.3 Piecewise linear method	91
	4.6	Subsystems	92
	4.7	Sparsity and optimal ordering	95
	4.8	Numerical errors and instabilities	97
	4.9	Summary	97
	4.10	References	98

5 The root-matching method — 99

	5.1	Introduction	99
	5.2	Exponential form of the difference equation	99

5.3		z-domain representation of difference equations	102
5.4		Implementation in EMTP algorithm	105
5.5		Family of exponential forms of the difference equation	112
	5.5.1	Step response	114
	5.5.2	Steady-state response	116
	5.5.3	Frequency response	117
5.6		Example	118
5.7		Summary	120
5.8		References	121

6 Transmission lines and cables — 123

6.1		Introduction	123
6.2		Bergeron's model	124
	6.2.1	Multiconductor transmission lines	126
6.3		Frequency-dependent transmission lines	130
	6.3.1	Frequency to time domain transformation	132
	6.3.2	Phase domain model	136
6.4		Overhead transmission line parameters	137
	6.4.1	Bundled subconductors	140
	6.4.2	Earth wires	142
6.5		Underground cable parameters	142
6.6		Example	146
6.7		Summary	156
6.8		References	156

7 Transformers and rotating plant — 159

7.1		Introduction	159
7.2		Basic transformer model	160
	7.2.1	Numerical implementation	161
	7.2.2	Parameters derivation	162
	7.2.3	Modelling of non-linearities	164
7.3		Advanced transformer models	165
	7.3.1	Single-phase UMEC model	166
	7.3.1.1	UMEC Norton equivalent	169
	7.3.2	UMEC implementation in PSCAD/EMTDC	171
	7.3.3	Three-limb three-phase UMEC	172
	7.3.4	Fast transient models	176
7.4		The synchronous machine	176
	7.4.1	Electromagnetic model	177
	7.4.2	Electromechanical model	183
	7.4.2.1	Per unit system	184
	7.4.2.2	Multimass representation	184
	7.4.3	Interfacing machine to network	185
	7.4.4	Types of rotating machine available	189
7.5		Summary	190
7.6		References	191

viii Contents

8	**Control and protection**	**193**
8.1	Introduction	193
8.2	Transient analysis of control systems (TACS)	194
8.3	Control modelling in PSCAD/EMTDC	195
	8.3.1 Example	198
8.4	Modelling of protective systems	205
	8.4.1 Transducers	205
	8.4.2 Electromechanical relays	208
	8.4.3 Electronic relays	209
	8.4.4 Microprocessor-based relays	209
	8.4.5 Circuit breakers	210
	8.4.6 Surge arresters	211
8.5	Summary	213
8.6	References	214
9	**Power electronic systems**	**217**
9.1	Introduction	217
9.2	Valve representation in EMTDC	217
9.3	Placement and location of switching instants	219
9.4	Spikes and numerical oscillations (chatter)	220
	9.4.1 Interpolation and chatter removal	222
9.5	HVDC converters	230
9.6	Example of HVDC simulation	233
9.7	FACTS devices	233
	9.7.1 The static VAr compensator	233
	9.7.2 The static compensator (STATCOM)	241
9.8	State variable models	243
	9.8.1 EMTDC/TCS interface implementation	244
	9.8.2 Control system representation	248
9.9	Summary	248
9.10	References	249
10	**Frequency dependent network equivalents**	**251**
10.1	Introduction	251
10.2	Position of FDNE	252
10.3	Extent of system to be reduced	252
10.4	Frequency range	253
10.5	System frequency response	253
	10.5.1 Frequency domain identification	253
	10.5.1.1 Time domain analysis	255
	10.5.1.2 Frequency domain analysis	257
	10.5.2 Time domain identification	262
10.6	Fitting of model parameters	262
	10.6.1 RLC networks	262
	10.6.2 Rational function	263
	10.6.2.1 Error and figure of merit	265

10.7	Model implementation	266
10.8	Examples	267
10.9	Summary	275
10.10	References	275

11 Steady state applications — 277

11.1	Introduction	277
11.2	Initialisation	278
11.3	Harmonic assessment	278
11.4	Phase-dependent impedance of non-linear device	279
11.5	The time domain in an ancillary capacity	281
	11.5.1 Iterative solution for time invariant non-linear components	282
	11.5.2 Iterative solution for general non-linear components	284
	11.5.3 Acceleration techniques	285
11.6	The time domain in the primary role	286
	11.6.1 Basic time domain algorithm	286
	11.6.2 Time step	286
	11.6.3 DC system representation	287
	11.6.4 AC system representation	287
11.7	Voltage sags	288
	11.7.1 Examples	290
11.8	Voltage fluctuations	292
	11.8.1 Modelling of flicker penetration	294
11.9	Voltage notching	296
	11.9.1 Example	297
11.10	Discussion	297
11.11	References	300

12 Mixed time-frame simulation — 303

12.1	Introduction	303
12.2	Description of the hybrid algorithm	304
	12.2.1 Individual program modifications	307
	12.2.2 Data flow	307
12.3	TS/EMTDC interface	307
	12.3.1 Equivalent impedances	308
	12.3.2 Equivalent sources	310
	12.3.3 Phase and sequence data conversions	310
	12.3.4 Interface variables derivation	311
12.4	EMTDC to TS data transfer	313
	12.4.1 Data extraction from converter waveforms	313
12.5	Interaction protocol	313
12.6	Interface location	316
12.7	Test system and results	317
12.8	Discussion	319
12.9	References	319

13 Transient simulation in real time — 321
- 13.1 Introduction — 321
- 13.2 Simulation with dedicated architectures — 322
  - 13.2.1 Hardware — 323
  - 13.2.2 RTDS applications — 325
- 13.3 Real-time implementation on standard computers — 327
  - 13.3.1 Example of real-time test — 329
- 13.4 Summary — 330
- 13.5 References — 331

A Structure of the PSCAD/EMTDC program — 333
- A.1 References — 340

B System identification techniques — 341
- B.1 s-domain identification (frequency domain) — 341
- B.2 z-domain identification (frequency domain) — 343
- B.3 z-domain identification (time domain) — 345
- B.4 Prony analysis — 346
- B.5 Recursive least-squares curve-fitting algorithm — 348
- B.6 References — 350

C Numerical integration — 351
- C.1 Review of classical methods — 351
- C.2 Truncation error of integration formulae — 354
- C.3 Stability of integration methods — 356
- C.4 References — 357

D Test systems data — 359
- D.1 CIGRE HVDC benchmark model — 359
- D.2 Lower South Island (New Zealand) system — 359
- D.3 Reference — 365

E Developing difference equations — 367
- E.1 Root-matching technique applied to a first order lag function — 367
- E.2 Root-matching technique applied to a first order differential pole function — 368
- E.3 Difference equation by bilinear transformation for RL series branch — 369
- E.4 Difference equation by numerical integrator substitution for RL series branch — 369

		Contents	xi

F MATLAB code examples — **373**
- F.1 Voltage step on RL branch — 373
- F.2 Diode fed RL branch — 374
- F.3 General version of example F.2 — 376
- F.4 Frequency response of difference equations — 384

G FORTRAN code for state variable analysis — **389**
- G.1 State variable analysis program — 389

H FORTRAN code for EMT simulation — **395**
- H.1 DC source, switch and RL load — 395
- H.2 General EMT program for d.c. source, switch and RL load — 397
- H.3 AC source diode and RL load — 400
- H.4 Simple lossless transmission line — 402
- H.5 Bergeron transmission line — 404
- H.6 Frequency-dependent transmission line — 407
- H.7 Utility subroutines for transmission line programs — 413

Index — 417

List of figures

1.1	Time frame of various transient phenomena	2
1.2	Transient network analyser	4
2.1	Impulse response associated with s-plane pole locations	23
2.2	Step response of lead–lag function	29
2.3	Norton of a rational function in z-domain	31
2.4	Data sequence associated with z-plane pole locations	32
2.5	Relationship between the domains	33
3.1	Non-trivial dependent state variables	36
3.2	Capacitive loop	38
3.3	(a) Capacitor with no connection to ground, (b) small capacitor added to give a connection to ground	39
3.4	K matrix partition	41
3.5	Row echelon form	41
3.6	Modified state variable equations	42
3.7	Flow chart for state variable analysis	43
3.8	Tee equivalent circuit	45
3.9	TCS branch types	47
3.10	TCS flow chart	50
3.11	Switching in state variable program	51
3.12	Interpolation of time upon valve current reversal	52
3.13	NETOMAC simulation responses	54
3.14	TCS simulation with 1 ms time step	55
3.15	Steady state responses from TCS	56
3.16	Transient simulation with TCS for a d.c. short-circuit at 0.5 s	57
3.17	Firing control mechanism based on the phase-locked oscillator	58
3.18	Synchronising error in firing pulse	58
3.19	Constant $\alpha_{\text{order}}(15°)$ operation with a step change in the d.c. current	60
3.20	RLC test circuit	60
3.21	State variable analysis with 50 μs step length	61
3.22	State variable analysis with 50 μs step length	62
3.23	State variable analysis with 50 μs step length and $\dot{x}$ check	62

xiv List of figures

3.24	State variable with 50 μs step length and step length optimisation	63
3.25	Both $\dot{x}$ check and step length optimisation	63
3.26	Error comparison	64
4.1	Resistor	68
4.2	Inductor	68
4.3	Norton equivalent of the inductor	69
4.4	Capacitor	70
4.5	Norton equivalent of the capacitor	71
4.6	Reduction of RL branch	73
4.7	Reduction of RLC branch	74
4.8	Propagation of a wave on a transmission line	74
4.9	Equivalent two-port network for a lossless line	76
4.10	Node 1 of an interconnected circuit	77
4.11	Example using conversion of voltage source to current source	78
4.12	Network solution with voltage sources	80
4.13	Network solution with switches	81
4.14	Block diagonal structure	81
4.15	Flow chart of EMT algorithm	82
4.16	Simple switched RL load	83
4.17	Equivalent circuit for simple switched RL load	83
4.18	Step response of an RL branch for step lengths of $\Delta t = \tau/10$ and $\Delta t = \tau$	86
4.19	Step response of an RL branch for step lengths of $\Delta t = 5\tau$ and $\Delta t = 10\tau$	87
4.20	Piecewise linear inductor represented by current source	89
4.21	Pictorial view of simultaneous solution of two equations	91
4.22	Artificial negative damping	92
4.23	Piecewise linear inductor	92
4.24	Separation of two coupled subsystems by means of linearised equivalent sources	93
4.25	Interfacing for HVDC link	94
4.26	Example of sparse network	96
5.1	Norton equivalent for RL branch	106
5.2	Switching test system	107
5.3	Step response of switching test system for $\Delta t = \tau$	107
5.4	Step response of switching test system for $\Delta t = 5\tau$	108
5.5	Step response of switching test system for $\Delta t = 10\tau$	108
5.6	Resonance test system	109
5.7	Comparison between exponential form and Dommel's method to a 5 kHz excitation for resonance test system. $\Delta t = 25\,\mu s$	109
5.8	Comparison between exponential form and Dommel's method to a 5 kHz excitation for resonance test system. $\Delta t = 10\,\mu s$	110
5.9	Comparison between exponential form and Dommel's method to 10 kHz excitation for resonance test system	110

List of figures xv

5.10	Response of resonance test system to 10 kHz excitation, blow-up of exponential form's response	111
5.11	Diode test system	111
5.12	Response to diode test system (a) Voltage (b) Current	112
5.13	Input as function of time	113
5.14	Control or electrical system as first order lag	113
5.15	Comparison step response of switching test system for $\Delta t = \tau$	114
5.16	Comparison step response of switching test system for $\Delta t = 5\tau$	115
5.17	Comparison of step response of switching test system for $\Delta t = 10\tau$	115
5.18	Root-matching type (d) approximation to a step	116
5.19	Comparison with a.c. excitation (5 kHz) ($\Delta t = \tau$)	116
5.20	Comparison with a.c. excitation (10 kHz) ($\Delta t = \tau$)	117
5.21	Frequency response for various simulation methods	118
6.1	Decision tree for transmission line model selection	124
6.2	Nominal PI section	124
6.3	Equivalent two-port network for line with lumped losses	125
6.4	Equivalent two-port network for half-line section	125
6.5	Bergeron transmission line model	126
6.6	Schematic of frequency-dependent line	129
6.7	Thevenin equivalent for frequency-dependent transmission line	132
6.8	Norton equivalent for frequency-dependent transmission line	132
6.9	Magnitude and phase angle of propagation function	134
6.10	Fitted propagation function	135
6.11	Magnitude and phase angle of characteristic impedance	137
6.12	Transmission line geometry	138
6.13	Matrix elimination of subconductors	141
6.14	Cable cross-section	142
6.15	Step response of a lossless line terminated by its characteristic impedance	147
6.16	Step response of a lossless line with a loading of double characteristic impedance	148
6.17	Step response of a lossless line with a loading of half its characteristic impedance	149
6.18	Step response of Bergeron line model for characteristic impedance termination	149
6.19	Step response of Bergeron line model for a loading of half its characteristic impedance	150
6.20	Step response of Bergeron line model for a loading of double characteristic impedance	150
6.21	Comparison of attenuation (or propagation) constant	151
6.22	Error in fitted attenuation constant	151
6.23	Comparison of surge impedance	152
6.24	Error in fitted surge impedance	152

xvi List of figures

6.25	Step response of frequency-dependent transmission line model (load = 100 Ω)	153
6.26	Step response of frequency-dependent transmission line model (load = 1000 Ω)	154
6.27	Step response of frequency-dependent transmission line model (load = 50 Ω)	154
7.1	Equivalent circuit of the two-winding transformer	160
7.2	Equivalent circuit of the two-winding transformer, without the magnetising branch	161
7.3	Transformer example	161
7.4	Transformer equivalent after discretisation	163
7.5	Transformer test system	163
7.6	Non-linear transformer	164
7.7	Non-linear transformer model with in-rush	165
7.8	Star–delta three-phase transformer	165
7.9	UMEC single-phase transformer model	166
7.10	Magnetic equivalent circuit for branch	167
7.11	Incremental and actual permeance	168
7.12	UMEC Norton equivalent	170
7.13	UMEC implementation in PSCAD/EMTDC	171
7.14	UMEC PSCAD/EMTDC three-limb three-phase transformer model	173
7.15	UMEC three-limb three-phase Norton equivalent for blue phase (Y-g/Y-g)	175
7.16	Cross-section of a salient pole machine	177
7.17	Equivalent circuit for synchronous machine equations	180
7.18	The a.c. machine equivalent circuit	182
7.19	d-axis flux paths	183
7.20	Multimass model	184
7.21	Interfacing electrical machines	186
7.22	Electrical machine solution procedure	187
7.23	The a.c. machine system	188
7.24	Block diagram synchronous machine model	189
8.1	Interface between network and TACS solution	194
8.2	Continuous system model function library (PSCAD/EMTDC)	196/7
8.3	First-order lag	198
8.4	Simulation results for a time step of 5 μs	201
8.5	Simulation results for a time step of 50 μs	202
8.6	Simulation results for a time step of 500 μs	202
8.7	Simple bipolar PWM inverter	204
8.8	Simple bipolar PWM inverter with interpolated turn ON and OFF	204
8.9	Detailed model of a current transformer	206
8.10	Comparison of EMTP simulation (solid line) and laboratory data (dotted line) with high secondary burden	207
8.11	Detailed model of a capacitive voltage transformer	208

List of figures xvii

8.12	Diagram of relay model showing the combination of electrical, magnetic and mechanical parts	209
8.13	Main components of digital relay	210
8.14	Voltage–time characteristic of a gap	211
8.15	Voltage–time characteristic of silicon carbide arrestor	212
8.16	Voltage–time characteristic of metal oxide arrestor	213
8.17	Frequency-dependent model of metal oxide arrestor	213
9.1	Equivalencing and reduction of a converter valve	218
9.2	Current chopping	221
9.3	Illustration of numerical chatter	222
9.4	Numerical chatter in a diode-fed RL load $\left(R_{\mathrm{ON}} = 10^{-10},\ R_{\mathrm{OFF}} = 10^{10}\right)$	223
9.5	Forced commutation benchmark system	223
9.6	Interpolation for GTO turn-OFF (switching and integration in one step)	224
9.7	Interpolation for GTO turn-OFF (using instantaneous solution)	224
9.8	Interpolating to point of switching	226
9.9	Jumps in variables	226
9.10	Double interpolation method (interpolating back to the switching instant)	227
9.11	Chatter removal by interpolation	228
9.12	Combined zero-crossing and chatter removal by interpolation	229
9.13	Interpolated/extrapolated source values due to chatter removal algorithm	230
9.14	(a) The six-pulse group converter, (b) thyristor and snubber equivalent circuit	231
9.15	Phase-vector phase-locked oscillator	231
9.16	Firing control for the PSCAD/EMTDC valve group model	232
9.17	Classic V–I converter control characteristic	232
9.18	CIGRE benchmark model as entered into the PSCAD draft software	234
9.19	Controller for the PSCAD/EMTDC simulation of the CIGRE benchmark model	235
9.20	Response of the CIGRE model to five-cycle three-phase fault at the inverter bus	236
9.21	SVC circuit diagram	237
9.22	Thyristor switch-OFF with variable time step	238
9.23	Interfacing between the SVC model and the EMTDC program	239
9.24	SVC controls	240
9.25	Basic STATCOM circuit	241
9.26	Basic STATCOM controller	242
9.27	Pulse width modulation	243
9.28	Division of a network	244
9.29	The converter system to be divided	245
9.30	The divided HVDC system	246

xviii *List of figures*

9.31	Timing synchronisation	246
9.32	Control systems in EMTDC	247
10.1	Curve-fitting options	254
10.2	Current injection	254
10.3	Voltage injection	255
10.4	PSCAD/EMTDC schematic with current injection	256
10.5	Voltage waveform from time domain simulation	257
10.6	Typical frequency response of a system	258
10.7	Reduction of admittance matrices	259
10.8	Multifrequency admittance matrix	260
10.9	Frequency response	261
10.10	Two-port frequency dependent network equivalent (admittance implementation)	261
10.11	Three-phase frequency dependent network equivalent (impedance implementation)	262
10.12	Ladder circuit of Hingorani and Burbery	263
10.13	Ladder circuit of Morched and Brandwajn	264
10.14	Magnitude and phase response of a rational function	268
10.15	Comparison of methods for the fitting of a rational function	269
10.16	Error for various fitted methods	269
10.17	Small passive network	270
10.18	Magnitude and phase fit for the test system	271
10.19	Comparison of full and a passive FDNE for an energisation transient	272
10.20	Active FDNE	272
10.21	Comparison of active FDNE response	273
10.22	Energisation	273
10.23	Fault inception and removal	274
10.24	Fault inception and removal with current chopping	274
11.1	Norton equivalent circuit	282
11.2	Description of the iterative algorithm	283
11.3	Test system at the rectifier end of a d.c. link	288
11.4	Frequency dependent network equivalent of the test system	288
11.5	Impedance/frequency of the frequency dependent equivalent	289
11.6	Voltage sag at a plant bus due to a three-phase fault	290
11.7	Test circuit for transfer switch	291
11.8	Transfer for a 30 per cent sag at 0.8 power factor with a 3325 kVA load	292
11.9	EAF system single line diagram	293
11.10	EAF without compensation	293
11.11	EAF with SVC compensation	294
11.12	EAF with STATCOM compensation	294
11.13	Test system for flicker penetration (the circles indicate busbars and the squares transmission lines)	295

List of figures xix

11.14	Comparison of P_{st} indices resulting from a positive sequence current injection	296
11.15	Test system for the simulation of voltage notching	298
11.16	Impedance/frequency spectrum at the 25 kV bus	299
11.17	Simulated 25 kV system voltage with drive in operation	299
11.18	Simulated waveform at the 4.16 kV bus (surge capacitor location)	300
12.1	The hybrid concept	304
12.2	Example of interfacing procedure	305
12.3	Modified TS steering routine	306
12.4	Hybrid interface	308
12.5	Representative circuit	308
12.6	Derivation of Thevenin equivalent circuit	309
12.7	Comparison of total r.m.s. power, fundamental frequency power and fundamental frequency positive sequence power	314
12.8	Normal interaction protocol	315
12.9	Interaction protocol around a disturbance	315
12.10	Rectifier terminal d.c. voltage comparisons	318
12.11	Real and reactive power across interface	318
12.12	Machine variables – TSE (TS variables)	319
13.1	Schematic of real-time digital simulator	321
13.2	Prototype real-time digital simulator	323
13.3	Basic RTDS rack	324
13.4	RTDS relay set-up	326
13.5	Phase distance relay results	327
13.6	HVDC control system testing	327
13.7	Typical output waveforms from an HVDC control study	328
13.8	General structure of the DTNA system	328
13.9	Test system	329
13.10	Current and voltage waveforms following a single-phase short-circuit	330
A.1	The PSCAD/EMTDC Version 2 suite	333
A.2	DRAFT program	334
A.3	RUNTIME program	335
A.4	RUNTIME program showing controls and metering available	335
A.5	MULTIPLOT program	336
A.6	Interaction in PSCAD/EMTDC Version 2	337
A.7	PSCAD/EMTDC flow chart	338
A.8	PSCAD Version 3 interface	339
C.1	Numerical integration from the sampled data viewpoint	353
D.1	CIGRE HVDC benchmark test system	359
D.2	Frequency scan of the CIGRE rectifier a.c. system impedance	361
D.3	Frequency scan of the CIGRE inverter a.c. system impedance	361
D.4	Frequency scan of the CIGRE d.c. system impedance	362
D.5	Lower South Island of New Zealand test system	363

List of tables

1.1	EMTP-type programs	8
1.2	Other transient simulation programs	8
2.1	First eight steps for simulation of lead–lag function	29
3.1	State variable analysis error	61
4.1	Norton components for different integration formulae	72
4.2	Step response of RL circuit to various step lengths	85
5.1	Integrator characteristics	101
5.2	Exponential form of difference equation	104
5.3	Response for $\Delta t = \tau = 50\,\mu s$	119
5.4	Response for $\Delta t = 5\tau = 250\,\mu s$	119
5.5	Response for $\Delta t = 10\tau = 500\,\mu s$	120
6.1	Parameters for transmission line example	146
6.2	Single phase test transmission line	146
6.3	s-domain fitting of characteristic impedance	153
6.4	Partial fraction expansion of characteristic admittance	153
6.5	Fitted attenuation function (s-domain)	155
6.6	Partial fraction expansion of fitted attenuation function (s-domain)	155
6.7	Pole/zero information from PSCAD V2 (characteristic impedance)	155
6.8	Pole/zero information from PSCAD V2 (attenuation function)	156
9.1	Overheads associated with repeated conductance matrix refactorisation	219
10.1	Numerator and denominator coefficients	268
10.2	Poles and zeros	268
10.3	Coefficients of z^{-1} (no weighting factors)	270
10.4	Coefficients of z^{-1} (weighting-factor)	271
11.1	Frequency dependent equivalent circuit parameters	289
C.1	Classical integration formulae as special cases of the tunable integrator	353
C.2	Integrator formulae	354
C.3	Linear inductor	354
C.4	Linear capacitor	355
C.5	Comparison of numerical integration algorithms ($\Delta T = \tau/10$)	356

C.6	Comparison of numerical integration algorithms ($\Delta T = \tau$)	356
C.7	Stability region	357
D.1	CIGRE model main parameters	360
D.2	CIGRE model extra information	360
D.3	Converter information for the Lower South Island test system	362
D.4	Transmission line parameters for Lower South Island test system	362
D.5	Conductor geometry for Lower South Island transmission lines (in metres)	363
D.6	Generator information for Lower South Island test system	363
D.7	Transformer information for the Lower South Island test system	364
D.8	System loads for Lower South Island test system (MW, MVar)	364
D.9	Filters at the Tiwai-033 busbar	364
E.1	Coefficients of a rational function in the z-domain for admittance	370
E.2	Coefficients of a rational function in the z-domain for impedance	371
E.3	Summary of difference equations	372

Preface

The analysis of electromagnetic transients has traditionally been discussed under the umbrella of circuit theory, the main core course in the electrical engineering curriculum, and therefore the subject of very many textbooks. However, some of the special characteristics of power plant components, such as machine non-linearities and transmission line frequency dependence, have not been adequately covered in conventional circuit theory. Among the specialist books written to try and remedy the situation are H. A. Peterson's *Transient performance in power systems* (1951) and A. Greenwood's *Electric transients in power systems* (1971). The former described the use of the transient network analyser to study the behaviour of linear and non-linear power networks. The latter described the fundamental concepts of the subject and provided many examples of transient simulation based on the Laplace transform.

By the mid-1960s the digital computer began to determine the future pattern of power system transients simulation. In 1976 the IEE published an important monograph, *Computation of power system transients*, based on pioneering computer simulation work carried out in the UK by engineers and mathematicians.

However, it was the IEEE classic paper by H. W. Dommel *Digital computer solution of electromagnetic transients in single and multiphase networks* (1969), that set up the permanent basic framework for the simulation of power system electromagnetic transients in digital computers. Electromagnetic transient programs based on Dommel's algorithm, commonly known as the EMTP method, have now become an essential part of the design of power apparatus and systems. They are also being gradually introduced in the power curriculum of electrical engineering courses and play an increasing role in their research and development programs.

Applications of the EMTP method are constantly reported in the IEE, IEEE and other international journals, as well as in the proceedings of many conferences, some of them specifically devoted to the subject, like the International Conference on Power System Transients (IPST) and the International Conference on Digital Power System Simulators (ICDS). In 1997 the IEEE published a volume entitled *Computer analysis of electric power system transients*, which contained a comprehensive selection of papers considered as important contributions in this area. This was followed in 1998 by the special publication TP-133-0 *Modeling and analysis of system transients using digital programs*, a collection of published guidelines produced by various IEEE taskforces.

Although there are well documented manuals to introduce the user to the various existing electromagnetic transients simulation packages, there is a need for a book with cohesive technical information to help students and professional engineers to understand the topic better and minimise the effort normally required to become effective users of the EMT programs. Hopefully this book will fill that gap.

Basic knowledge of power system theory, matrix analysis and numerical techniques is presumed, but many references are given to help the readers to fill the gaps in their understanding of the relevant material.

The authors would like to acknowledge the considerable help received from many experts in the field, prior to and during the preparation of the book. In particular they want to single out Hermann Dommel himself, who, during his study leave in Canterbury during 1983, directed our early attempts to contribute to the topic. They also acknowledge the continuous help received from the Manitoba HVDC Research Centre, specially the former director Dennis Woodford, as well as Garth Irwin, now both with Electranix Corporation. Also, thanks are due to Ani Gole of the University of Manitoba for his help and for providing some of the material covered in this book. The providing of the paper by K. Strunz is also appreciated. The authors also wish to thank the contributions made by a number of their colleagues, early on at UMIST (Manchester) and later at the University of Canterbury (New Zealand), such as J. G. Campos Barros, H. Al Kashali, Chris Arnold, Pat Bodger, M. D. Heffernan, K. S. Turner, Mohammed Zavahir, Wade Enright, Glenn Anderson and Y.-P. Wang. Finally J. Arrillaga wishes to thank the Royal Society of New Zealand for the financial support received during the preparation of the book, in the form of the James Cook Senior Research Fellowship.

Acronyms and constants

Acronyms

APSCOM	Advances in Power System Control, Operation and Management
ATP	Alternative Transient Program
BPA	Bonneville Power Administration (USA)
CIGRE	Conference Internationale des Grands Reseaux Electriques (International Conference on Large High Voltage Electric Systems)
DCG	Development Coordination Group
EMT	Electromagnetic Transient
EMTP	Electromagnetic Transients Program
EMTDC[1]	Electromagnetic Transients Program for DC
EPRI	Electric Power Research Institute (USA)
FACTS	Flexible AC Transmission Systems
ICDS	International Conference on Digital Power System Simulators
ICHQP	International Conference on Harmonics and Quality of Power
IEE	The Institution of Electrical Engineers
IEC	International Electrotechnical Commission
IEEE	Institute of Electrical and Electronics Engineers
IREQ	Laboratoire Simulation de Reseaux, Institut de Recherche d'Hydro-Quebec
NIS	Numerical Integration Substitution
MMF	Magneto-Motive Force
PES	Power Engineering Society
PSCAD[2]	Power System Computer Aided Design
RTDS[3]	Real-Time Digital Simulator
SSTS	Solid State Transfer Switch
TACS	Transient Analysis of Control Systems

[1] EMTDC is a registered trademark of the Manitoba Hydro
[2] PSCAD is a registered trademark of the Manitoba HVDC Research Centre
[3] RTDS is a registered trademark of the Manitoba HVDC Research Centre

TCS Transient Converter Simulation (state variable analysis program)
TRV Transient Recovery Voltage
UIE Union International d'Electrothermie/International Union of Electroheat

Constants

ε_0 permittivity of free space ($8.85 \times 10^{-12}\,\text{C}^2\,\text{N}^{-1}\text{m}^{-2}$ or $\text{F}\,\text{m}^{-1}$)

μ_0 permeability of free space ($4\pi \times 10^{-7}\,\text{Wb}\,\text{A}^{-1}\,\text{m}^{-1}$ or $\text{H}\,\text{m}^{-1}$)

π 3.1415926535

c Speed of light ($2.99793 \times 10^8\,\text{m}\,\text{s}^{-1}$)

Chapter 1

Definitions, objectives and background

1.1 Introduction

The operation of an electrical power system involves continuous electromechanical and electromagnetic distribution of energy among the system components. During normal operation, under constant load and topology, these energy exchanges are not modelled explicitly and the system behaviour can be represented by voltage and current phasors in the frequency domain.

However, following switching events and system disturbances the energy exchanges subject the circuit components to higher stresses, resulting from excessive currents or voltage variations, the prediction of which is the main objective of power system transient simulation.

Figure 1.1 shows typical time frames for a full range of power system transients. The transients on the left of the figure involve predominantly interactions between the magnetic fields of inductances and the electric fields of capacitances in the system; they are referred to as *electromagnetic transients*. The transients on the right of the figure are mainly affected by interactions between the mechanical energy stored in the rotating machines and the electrical energy stored in the network; they are accordingly referred to as *electromechanical transients*. There is a grey area in the middle, namely the transient stability region, where both effects play a part and may need adequate representation.

In general the lightning stroke produces the highest voltage surges and thus determines the insulation levels. However at operating voltages of 400 kV and above, system generated overvoltages, such as those caused by the energisation of transmission lines, can often be the determining factor for insulation coordination.

From the analysis point of view the electromagnetic transients solution involves a set of first order differential equations based on Kirchhoff's laws, that describe the behaviour of *RLC* circuits when excited by specified stimuli. This is a well documented subject in electrical engineering texts and it is therefore assumed that the reader is familiar with the terminology and concepts involved, as well as their physical interpretation.

2 Power systems electromagnetic transients simulation

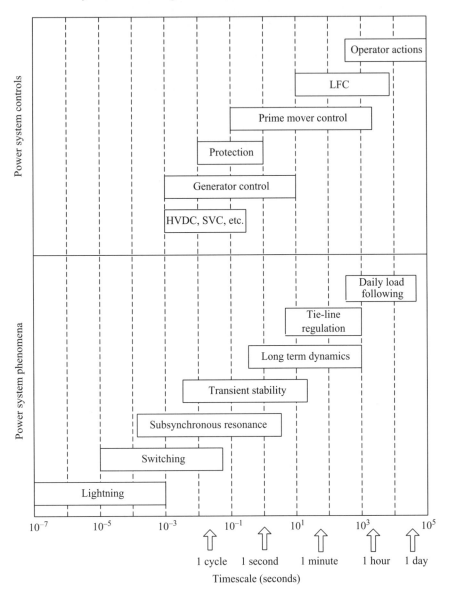

Figure 1.1 *Time frame of various transient phenomena*

It is the primary object of this book to describe the application of efficient computational techniques to the solution of electromagnetic transient problems in systems of any size and topology involving linear and non-linear components. This is an essential part in power system design to ensure satisfactory operation, derive the component ratings and optimise controller and protection settings. It is also an important diagnostic tool to provide post-mortem information following system incidents.

1.2 Classification of electromagnetic transients

Transient waveforms contain one or more oscillatory components and can thus be characterised by the natural frequencies of these oscillations. However in the simulation process, the accurate determination of these oscillations is closely related to the equivalent circuits used to represent the system components. No component model is appropriate for all types of transient analysis and must be tailored to the scope of the study.

From the modelling viewpoint, therefore, it is more appropriate to classify transients by the time range of the study, which is itself related to the phenomena under investigation. The key issue in transient analysis is the selection of a model for each component that realistically represents the physical system over the time frame of interest.

Lightning, the fastest-acting disturbance, requires simulation in the region of nano to micro-seconds. Of course in this time frame the variation of the power frequency voltage and current levels will be negligible and the electronic controllers will not respond; on the other hand the stray capacitance and inductance of the system components will exercise the greatest influence in the response.

The time frame for switching events is in micro to milliseconds, as far as insulation coordination is concerned, although the simulation time can go into cycles, if system recovery from the disturbance is to be investigated. Thus, depending on the information sought, switching phenomena may require simulations on different time frames with corresponding component models, i.e. either a fast transient model using stray parameters or one based on simpler equivalent circuits but including the dynamics of power electronic controllers. In each case, the simulation step size will need to be at least one tenth of the smallest time constant of the system represented.

Power system components are of two types, i.e. those with essentially lumped parameters, such as electrical machines and capacitor or reactor banks, and those with distributed parameters, including overhead lines and underground or submarine cables. Following a switching event these circuit elements are subjected to voltages and currents involving frequencies between 50 Hz and 100 kHz. Obviously within such a vast range the values of the component parameters and of the earth path will vary greatly with frequency. The simulation process therefore must be capable of reproducing adequately the frequency variations of both the lumped and distributed parameters. The simulation must also represent such non-linearities as magnetic saturation, surge diverter characteristics and circuit-breaker arcs. Of course, as important, if not more, as the method of solution is the availability of reliable data and the variation of the system components with frequency, i.e. a fast transient model including stray parameters followed by one based on simpler equivalent circuits.

1.3 Transient simulators

Among the tools used in the past for the simulation of power system transients are the electronic analogue computer, the transient network analyser (TNA) and the HVDC simulator.

The electronic analogue computer basically solved ordinary differential equations by means of several units designed to perform specific functions, such as adders, multipliers and integrators as well as signal generators and a multichannel cathode ray oscilloscope.

Greater versatility was achieved with the use of scaled down models and in particular the TNA [1], shown in Figure 1.2, is capable of emulating the behaviour of the actual power system components using only low voltage and current levels. Early limitations included the use of lumped parameters to represent transmission lines, unrealistic modelling of losses, ground mode of transmission lines and magnetic non-linearities. However all these were largely overcome [2] and TNAs are still in use for their advantage of operating in real time, thus allowing many runs to be performed quickly and statistical data obtained, by varying the instants of switching. The real-time nature of the TNA permits the connection of actual control hardware and its performance validated, prior to their commissioning in the actual power system. In particular, the TNA is ideal for testing the control hardware and software associated with FACTS and HVDC transmission. However, due to their cost and maintenance requirements TNAs and HVDC models are being gradually displaced by real-time digital simulators, and a special chapter of the book is devoted to the latter.

Figure 1.2   Transient network analyser

1.4 Digital simulation

Owing to the complexity of modern power systems, the simulators described above could only be relied upon to solve relatively simple problems. The advent of the digital computer provided the stimulus to the development of more accurate and general solutions. A very good description of the early digital methods can be found in a previous monograph of this series [3].

While the electrical power system variables are continuous, digital simulation is by its nature discrete. The main task in digital simulation has therefore been the development of suitable methods for the solution of the differential and algebraic equations at discrete points.

The two broad classes of methods used in the digital simulation of the differential equations representing continuous systems are numerical integration and difference equations. Although the numerical integration method does not produce an explicit difference equation to be simulated, each step of the solution can be characterised by a difference equation.

1.4.1 State variable analysis

State variable analysis is the most popular technique for the numerical integration of differential equations [4]. This technique uses an indefinite numerical integration of the system variables in conjunction with the differential equation (to obtain the derivatives of the states).

The differential equation is expressed in implicit form. Instead of rearranging it into an explicit form, the state variable approach uses a predictor–corrector solution, such that the state equation predicts the state variable derivative and the trapezoidal rule corrects the estimates of the state variables.

The main advantages of this method are its simplicity and lack of overhead when changing step size, an important property in the presence of power electronic devices to ensure that the steps are made to coincide with the switching instants. Thus the numerical oscillations inherent in the numerical integration substitution technique do not occur; in fact the state variable method will fail to converge rather than give erroneous answers. Moreover, non-linearities are easier to represent in state variable analysis. The main disadvantages are greater solution time, extra code complexity and greater difficulty to model distributed parameters.

1.4.2 Method of difference equations

In the late 1960s H. W. Dommel of BPA (Bonneville Power Administration) developed a digital computer algorithm for the efficient analysis of power system electromagnetic transients [5]. The method, referred to as EMTP (ElectroMagnetic Transients Program), is based on the difference equations model and was developed around the transmission system proposed by Bergeron [6].

Bergeron's method uses linear relationships (characteristics) between the current and the voltage, which are invariant from the point of view of an observer travelling with the wave. However, the time intervals or discrete steps required by the digital solution generate truncation errors which can lead to numerical instability. The use of the trapezoidal rule to discretise the ordinary differential equations has improved the situation considerably in this respect.

Dommel's EMTP method combines the method of characteristics and the trapezoidal rule into a generalised algorithm which permits the accurate simulation of transients in networks involving distributed as well as lumped parameters.

To reflect its main technical characteristics, Dommel's method is often referred to by other names, the main one being numerical integration substitution. Other less common names are the method of companion circuits (to emphasise the fact that the difference equation can be viewed as a Norton equivalent, or companion, for each element in the circuit) and the nodal conductance approach (to emphasise the use of the nodal formulation).

There are alternative ways to obtain a discrete representation of a continuous function to form a difference equation. For example the root-matching technique, which develops difference equations such that the poles of its corresponding rational function match those of the system being simulated, results in a very accurate and stable difference equation. Complementary filtering is another technique of the numerical integration substitution type to form difference equations that is inherently more stable and accurate. In the control area the widely used bilinear transform method (or Trustin's method) is the same as numerical integration substitution developed by Dommel in the power system area.

1.5 Historical perspective

The EMTP has become an industrial standard and many people have contributed to enhance its capability. With the rapid increase in size and complexity, documentation, maintenance and support became a problem and in 1982 the EMTP Development Coordination Group (DCG) was formed to address it.

In 1984 EPRI (Electric Power Research Institute) reached agreement with DCG to take charge of documentation, conduct EMTP validation tests and add a more user-friendly input processor. The development of new technical features remained the primary task of DCG. DCG/EPRI version 1.0 of EMTP was released in 1987 and version 2.0 in 1989.

In order to make EMTP accessible to the worldwide community, the Alternative Transient Program (ATP) was developed, with W.S. Meyer (of BPA) acting as coordinator to provide support. Major contributions were made, among them TACS (Transient Analysis of Control Systems) by L. Dube in 1976, multi-phase untransposed transmission lines with constant parameters by C. P. Lee, a frequency-dependent transmission line model and new line constants program by J. R. Marti, three-phase transformer models by H. W. and I. I. Dommel, a synchronous machine model by V. Brandwajn, an underground cable model by L. Marti and synchronous machine data conversion by H. W. Dommel.

Inspired by the work of Dr. Dommel and motivated by the need to solve the problems of frequently switching components (specifically HVDC converters) through the 1970s D. A. Woodford (of Manitoba Hydro) helped by A. Gole and R. Menzies developed a new program still using the EMTP concept but designed around a.c.–d.c. converters. This program, called EMTDC (Electromagnetic Transients Program for DC), originally ran on mainframe computers.

With the development and universal availability of personal computers (PCs) EMTDC version 1 was released in the late 1980s. A data driven program can only model components coded by the programmer, but, with the rapid technological developments in power systems, it is impractical to anticipate all future needs. Therefore, to ensure that users are not limited to preprogrammed component models, EMTDC required the user to write two FORTRAN files, i.e. DSDYN (Digital Simulator DYNamic subroutines) and DSOUT (Digital Simulator OUTput subroutines). These files are compiled and linked with the program object libraries to form the program. A BASIC program was used to plot the output waveforms from the files created.

The Manitoba HVDC Research Centre developed a comprehensive graphical user interface called PSCAD (Power System Computer Aided Design) to simplify and speed up the simulation task. PSCAD/EMTDC version 2 was released in the early 1990s for UNIX workstations. PSCAD comprised a number of programs that communicated via TCP/IP sockets. DRAFT for example allowed the circuit to be drawn graphically, and automatically generated the FORTRAN files needed to simulate the system. Other modules were TLINE, CABLE, RUNTIME, UNIPLOT and MULTIPLOT.

Following the emergence of the Windows operating system on PCs as the dominant system, the Manitoba HVDC Research Centre rewrote PSCAD/EMTDC for this system. The Windows/PC based PSCAD/EMTDC version was released in 1998.

The other EMTP-type programs have also faced the same challenges with numerous graphical interfaces being developed, such as ATP_Draw for ATP. A more recent trend has been to increase the functionality by allowing integration with other programs. For instance, considering the variety of specialised toolboxes of MATLAB, it makes sense to allow the interface with MATLAB to benefit from the use of such facilities in the transient simulation program.

Data entry is always a time-consuming exercise, which the use of graphical interfaces and component libraries alleviates. In this respect the requirements of universities and research organisations differ from those of electric power companies. In the latter case the trend has been towards the use of database systems rather than files using a vendor-specific format for power system analysis programs. This also helps the integration with SCADA information and datamining. An example of database usage is PowerFactory (produced by DIgSILENT). University research, on the other hand, involves new systems for which no database exists and thus a graphical entry such as that provided by PSCAD is the ideal tool.

A selection, not exhaustive, of EMTP-type programs and their corresponding Websites is shown in Table 1.1. Other transient simulation programs in current use are listed in Table 1.2. A good description of some of these programs is given in reference [7].

Table 1.1 EMTP-type programs

Program	Organisation	Website address
EPRI/DCG EMTP	EPRI	www.emtp96.com/
ATP program		www.emtp.org/
MicroTran	Microtran Power Systems Analysis Corporation	www.microtran.com/
PSCAD/EMTDC	Manitoba HVDC Research Centre	www.hvdc.ca/
NETOMAC	Siemens	www.ev.siemens.de/en/pages/
NPLAN	BCP Busarello + Cott + Partner Inc.	
EMTAP	EDSA	www.edsa.com/
PowerFactory	DIgSILENT	www.digsilent.de/
Arene	Anhelco	www.anhelco.com/
Hypersim	IREQ (Real-time simulator)	www.ireq.ca/
RTDS	RTDS Technologies	rtds.ca
Transient Performance Advisor (TPA)	MPR (MATLAB based)	www.mpr.com
Power System Toolbox	Cherry Tree (MATLAB based)	www.eagle.ca/ cherry/

Table 1.2 Other transient simulation programs

Program	Organisation	Website address
ATOSEC5	University of Quebec at Trios Rivieres	cpee.uqtr.uquebec.ca/dctodc/ato5_1htm
Xtrans	Delft University of Technology	eps.et.tudelft.nl
KREAN	The Norwegian University of Science and Technology	www.elkraft.ntnu.no/sie10aj/Krean1990.pdf
Power Systems Blockset	MATHworks (MATLAB based)	www.mathworks.com/products/
	TransEnergie Technologies	www.transenergie-tech.com/en/
SABER	Avant (formerly Analogy Inc.)	www.analogy.com/
SIMSEN	Swiss Federal Institute of Technology	simsen.epfl.ch/

1.6 Range of applications

Dommel's introduction to his classical paper [5] started with the following statement: 'This paper describes a general solution method for finding the time response of electromagnetic transients in arbitrary single or multi-phase networks with lumped and distributed parameters'.

The popularity of the EMTP method has surpassed all expectations, and three decades later it is being applied in practically every problem requiring time domain simulation. Typical examples of application are:

- Insulation coordination, i.e. overvoltage studies caused by fast transients with the purpose of determining surge arrestor ratings and characteristics.
- Overvoltages due to switching surges caused by circuit breaker operation.
- Transient performance of power systems under power electronic control.
- Subsynchronous resonance and ferroresonance phenomena.

It must be emphasised, however, that the EMTP method was specifically devised to provide simple and efficient electromagnetic transient solutions and not to solve steady state problems. The EMTP method is therefore complementary to traditional power system load-flow, harmonic analysis and stability programs. However, it will be shown in later chapters that electromagnetic transient simulation can also play an important part in the areas of harmonic power flow and multimachine transient stability.

1.7 References

1 PETERSON, H. A.: 'An electric circuit transient analyser', *General Electric Review*, 1939, p. 394
2 BORGONOVO, G., CAZZANI, M., CLERICI, A., LUCCHINI, G. and VIDONI, G.: 'Five years of experience with the new C.E.S.I. TNA', *IEEE Canadian Communication and Power Conference, Montreal*, 1974
3 BICKFORD, J. P., MULLINEUX, N. and REED J. R.: '*Computation of power-systems transients*' (IEE Monograph Series 18, Peter Peregrinus Ltd., London, 1976)
4 DeRUSSO, P. M., ROY, R. J., CLOSE, C. M. and DESROCHERS, A. A.: '*State variables for engineers*' (John Wiley, New York, 2nd edition, 1998)
5 DOMMEL, H. W.: 'Digital computer solution of electromagnetic transients in single- and multi-phase networks', *IEEE Transactions on Power Apparatus and Systems*, 1969, **88** (2), pp. 734–71
6 BERGERON, L.: 'Du coup de Belier en hydraulique au coup de foudre en electricite' (Dunod, 1949). (English translation: 'Water Hammer in hydraulics and wave surges in electricity', ASME Committee, Wiley, New York, 1961.)
7 MOHAN, N., ROBBINS, W. P., UNDELAND, T. M., NILSSEN, R. and MO, O.: 'Simulation of power electronic and motion control systems – an overview', *Proceedings of the IEEE*, 1994, **82** (8), pp. 1287–1302

Chapter 2
Analysis of continuous and discrete systems

2.1 Introduction

Linear algebra and circuit theory concepts are used in this chapter to describe the formulation of the state equations of linear dynamic systems. The Laplace transform, commonly used in the solution of simple circuits, is impractical in the context of a large power system. Some practical alternatives discussed here are modal analysis, numerical integration of the differential equations and the use of difference equations.

An electrical power system is basically a continuous system, with the exceptions of a few auxiliary components, such as the digital controllers. Digital simulation, on the other hand, is by nature a discrete time process and can only provide solutions for the differential and algebraic equations at discrete points in time.

The discrete representation can always be expressed as a difference equation, where the output at a new time point is calculated from the output at previous time points and the inputs at the present and previous time points. Hence the digital representation can be synthesised, tuned, stabilised and analysed in a similar way as any discrete system.

Thus, as an introduction to the subject matter of the book, this chapter also discusses, briefly, the subjects of digital simulation of continuous functions and the formulation of discrete systems.

2.2 Continuous systems

An nth order linear dynamic system is described by an nth order linear differential equation which can be rewritten as n first-order linear differential equations, i.e.

$$\begin{aligned}
\dot{x}_1(t) &= a_{11}x_1(t) + a_{11}x_2(t) + \cdots + a_{1n}x_n(t) + b_{11}u_1(t) + b_{12}u_2(t) + \cdots + b_{1m}u_m(t) \\
\dot{x}_2(t) &= a_{21}x_1(t) + a_{22}x_2(t) + \cdots + a_{2n}x_n(t) + b_{21}u_1(t) + b_{22}u_2(t) + \cdots + b_{2m}u_m(t) \\
&\vdots \\
\dot{x}_n(t) &= a_{n1}x_1(t) + a_{n2}x_2(t) + \cdots + a_{nn}x_n(t) + b_{n1}u_1(t) + b_{n2}u_2(t) + \cdots + b_{nm}u_m(t)
\end{aligned} \tag{2.1}$$

Expressing equation 2.1 in matrix form, with parameter t removed for simplicity:

$$\begin{pmatrix} \dot{x}_1 \\ \dot{x}_2 \\ \vdots \\ \dot{x}_n \end{pmatrix} = \begin{bmatrix} a_{11} & a_{12} & \cdots & a_{1n} \\ a_{21} & a_{22} & \cdots & a_{2n} \\ \vdots & \vdots & \ddots & \vdots \\ a_{n1} & a_{n2} & \cdots & a_{nn} \end{bmatrix} \begin{pmatrix} x_1 \\ x_2 \\ \vdots \\ x_n \end{pmatrix} + \begin{bmatrix} b_{11} & b_{12} & \cdots & b_{1m} \\ b_{21} & b_{22} & \cdots & b_{2m} \\ \vdots & \vdots & \ddots & \vdots \\ b_{n1} & b_{n2} & \cdots & b_{nm} \end{bmatrix} \begin{pmatrix} u_1 \\ u_2 \\ \vdots \\ u_m \end{pmatrix} \quad (2.2)$$

or in compact matrix notation:

$$\dot{\mathbf{x}} = [A]\mathbf{x} + [B]\mathbf{u} \quad (2.3)$$

which is normally referred to as the state equation.

Also needed is a system of algebraic equations that relate the system output quantities to the state vector and input vector, i.e.

$$y_1(t) = c_{11}x_1(t) + c_{11}x_2(t) + \cdots + c_{1n}x_n(t) + d_{11}u_1(t) + d_{12}u_2(t) + \cdots + d_{1m}u_m(t)$$
$$y_2(t) = c_{21}x_1(t) + c_{22}x_2(t) + \cdots + c_{2n}x_n(t) + d_{21}u_1(t) + d_{22}u_2(t) + \cdots + d_{2m}u_m(t)$$
$$\vdots$$
$$y_o(t) = c_{01}x_1(t) + c_{02}x_2(t) + \cdots + c_{0n}x_n(t) + d_{01}u_1(t) + d_{02}u_2(t) + \cdots + d_{0m}u_m(t)$$
$$(2.4)$$

Writing equation 2.4 in matrix form (again with the parameter t removed):

$$\begin{pmatrix} y_1 \\ y_2 \\ \vdots \\ y_0 \end{pmatrix} = \begin{bmatrix} c_{11} & c_{12} & \cdots & c_{1n} \\ c_{21} & c_{22} & \cdots & c_{2n} \\ \vdots & \vdots & \ddots & \vdots \\ c_{01} & c_{02} & \cdots & c_{0n} \end{bmatrix} \begin{pmatrix} x_1 \\ x_2 \\ \vdots \\ x_n \end{pmatrix} + \begin{bmatrix} d_{11} & d_{12} & \cdots & d_{1m} \\ d_{21} & d_{22} & \cdots & d_{2m} \\ \vdots & \vdots & \ddots & \vdots \\ d_{01} & d_{02} & \cdots & d_{0m} \end{bmatrix} \begin{pmatrix} u_1 \\ u_2 \\ \vdots \\ u_m \end{pmatrix} \quad (2.5)$$

or in compact matrix notation:

$$\mathbf{y} = [C]\mathbf{x} + [D]\mathbf{u} \quad (2.6)$$

which is called the output equation.

Equations 2.3 and 2.6 constitute the standard form of the state variable formulation. If no direct connection exists between the input and output vectors then $[D]$ is zero.

Equations 2.3 and 2.6 can be solved by transformation methods, the convolution integral or numerically in an iterative procedure. These alternatives will be discussed in later sections. However, the form of the state variable equations is not unique and depends on the choice of state variables [1]. Some state variable models are more convenient than others for revealing system properties such as stability, controllability and observability.

2.2.1 State variable formulations

A transfer function is generally represented by the equation:

$$G(s) = \frac{a_0 + a_1 s + a_2 s^2 + a_3 s^3 + \cdots + a_N s^N}{b_0 + b_1 s + b_2 s^2 + b_3 s^3 + \cdots + b_n s^n} = \frac{Y(s)}{U(s)} \qquad (2.7)$$

where $n \geq N$.

Dividing numerator and denominator by b_n provides the standard form, such that the term s^n appears in the denominator with unit coefficient i.e.

$$G(s) = \frac{A_0 + A_1 s + A_2 s^2 + A_3 s^3 + \cdots + A_N s^N}{B_0 + B_1 s + B_2 s^2 + B_3 s^3 + \cdots + B_{n-1} s^{n-1} + s^n} = \frac{Y(s)}{U(s)} \qquad (2.8)$$

The following sections describe alternative state variable formulations based on equation 2.8.

2.2.1.1 Successive differentiation

Multiplying both sides of equation 2.8 by $D(s)$ (where $D(s)$ represents the polynomial in s that appears in the denominator, and similarly $N(s)$ is the numerator) to get the equation in the form $D(s)Y(s) = N(s)U(s)$ and replacing the s^k operator by its time domain equivalent d^k/dt^k yields [2]:

$$\frac{d^n y}{dt^n} + B_{n-1}\frac{d^{n-1} y}{dt^{n-1}} + \cdots + B_1 \frac{dy}{dt} + B_0 y = A_N \frac{d^N u}{dt^N} + A_{N-1}\frac{d^{N-1} u}{dt^{N-1}} + \cdots + A_1 \frac{du}{dt} + A_0 u \qquad (2.9)$$

To eliminate the derivatives of u the following n state variables are chosen [2]:

$$x_1 = y - C_0 u$$
$$x_2 = \dot{y} - C_0 \dot{u} - C_1 u = \dot{x}_1 - C_1 u$$
$$\vdots \qquad\qquad\qquad (2.10)$$
$$x_n = \frac{d^{n-1} y}{dt^{n-1}} - C_0 \frac{d^{n-1} u}{dt^{n-1}} - C_1 \frac{d^{n-2} u}{dt^{n-2}} - C_{n-2}\dot{u} - C_{n-1}$$
$$= \dot{x}_{n-1} - C_{n-1} u$$

where the relationship between the C's and A's is:

$$\begin{bmatrix} 1 & 0 & 0 & \cdots & 0 \\ B_{n-1} & 1 & 0 & \cdots & 0 \\ B_{n-2} & B_{n-1} & 1 & \cdots & 0 \\ \vdots & \vdots & \vdots & \ddots & 0 \\ B_0 & B_1 & \cdots & B_{n-1} & 1 \end{bmatrix} \begin{pmatrix} C_0 \\ C_1 \\ C_2 \\ \vdots \\ C_n \end{pmatrix} = \begin{pmatrix} A_n \\ A_{n-1} \\ A_{n-2} \\ \vdots \\ A_0 \end{pmatrix} \qquad (2.11)$$

The values $C_0, C_1, \ldots, C_n$ are determined from:

$$\begin{aligned}
C_0 &= A_n \\
C_1 &= A_{n-1} - B_{n-1}C_0 \\
C_2 &= A_{n-2} - B_{n-1}C_1 - B_{n-2}C_0 \\
C_3 &= A_{n-3} - B_{n-1}C_2 - B_{n-2}C_1 - B_{n-3}C_0 \\
&\vdots \\
C_n &= A_0 - B_{n-1}C_{n-1} - \cdots - B_1C_1 - B_0C_0
\end{aligned} \quad (2.12)$$

From this choice of state variables the state variable derivatives are:

$$\begin{aligned}
\dot{x}_1 &= x_2 + C_1 u \\
\dot{x}_2 &= x_3 + C_2 u \\
\dot{x}_3 &= x_4 + C_3 u \\
&\vdots \\
\dot{x}_{n-1} &= x_n + C_{n-1} u \\
\dot{x}_n &= -B_0 x_1 - B_1 x_2 - B_2 x_3 - \cdots - B_{n-1} x_n + C_n u
\end{aligned} \quad (2.13)$$

Hence the matrix form of the state variable equations is:

$$\begin{pmatrix} \dot{x}_1 \\ \dot{x}_2 \\ \vdots \\ \dot{x}_{n-1} \\ \dot{x}_n \end{pmatrix} = \begin{bmatrix} 0 & 1 & 0 & \cdots & 0 \\ 0 & 0 & 1 & \cdots & 0 \\ \vdots & \vdots & \vdots & \ddots & \vdots \\ 0 & 0 & 0 & \cdots & 1 \\ -B_0 & -B_1 & -B_2 & \cdots & -B_{n-1} \end{bmatrix} \begin{pmatrix} x_1 \\ x_2 \\ \vdots \\ x_{n-1} \\ x_n \end{pmatrix} + \begin{pmatrix} C_1 \\ C_2 \\ \vdots \\ C_{n-1} \\ C_n \end{pmatrix} u \quad (2.14)$$

$$y = \begin{pmatrix} 1 & 0 & \cdots & 0 & 0 \end{pmatrix} \begin{pmatrix} x_1 \\ x_2 \\ \vdots \\ x_{n-1} \\ x_n \end{pmatrix} + A_n u \quad (2.15)$$

This is the formulation used in PSCAD/EMTDC for control transfer functions.

2.2.1.2 Controller canonical form

This alternative, sometimes called the phase variable form [3], is derived from equation 2.8 by dividing the numerator by the denominator to get a constant (A_n) and a remainder, which is now a strictly proper rational function (i.e. the numerator order is less than the denominator's) [4]. This gives

$$G(s) = A_n$$
$$+ \frac{(A_0 - B_0 A_n) + (A_1 - B_1 A_n)s + (A_2 - B_2 A_n)s^2 + \cdots + (A_{n-1} - B_{n-1} A_n)s^{n-1}}{B_0 + B_1 s + B_2 s^2 + B_3 s^3 + \cdots + B_{n-1} s^{n-1} + s^n}$$
(2.16)

or

$$G(s) = A_n + \frac{Y_R(s)}{U(s)} \qquad (2.17)$$

where

$$Y_R(s) = U(s)$$
$$\times \frac{(A_0 - B_0 A_n) + (A_1 - B_1 A_n)s + (A_2 - B_2 A_n)s^2 + \cdots + (A_{n-1} - B_{n-1} A_n)s^{n-1}}{B_0 + B_1 s + B_2 s^2 + B_3 s^3 + \cdots + B_{n-1} s^{n-1} + s^n}$$

Equating 2.16 and 2.17 and rearranging gives:

$$Q(s) = \frac{U(s)}{B_0 + B_1 s + B_2 s^2 + B_3 s^3 + \cdots + B_{n-1} s^{n-1} + s^n}$$
$$= \frac{Y_R(s)}{(A_0 - B_0 A_n) + (A_1 - B_1 A_n)s + (A_2 - B_2 A_n)s^2 + \cdots + (A_{n-1} - B_{n-1} A_n)s^{n-1}}$$
(2.18)

From equation 2.18 the following two equations are obtained:

$$s^n Q(s) = U(s) - B_0 Q(s) - B_1 s Q(s) - B_2 s^2 Q(s) - B_3 s^3 Q(s)$$
$$- \cdots - B_{n-1} s^{n-1} Q(s) \qquad (2.19)$$
$$Y_R(s) = (A_0 - B_0 A_n) Q(s) + (A_1 - B_1 A_n) s Q(s) + (A_2 - B_2 A_n) s^2 Q(s)$$
$$+ \cdots + (A_{n-1} - B_{n-1} A_n) s^{n-1} Q(s) \qquad (2.20)$$

Taking as the state variables

$$X_1(s) = Q(s) \qquad (2.21)$$
$$X_2(s) = s Q(s) = s X_1(s) \qquad (2.22)$$
$$\vdots$$
$$X_n(s) = s^{n-1} Q(s) = s X_{n-1}(s) \qquad (2.23)$$

and replacing the operator s in the s-plane by the differential operator in the time domain:

$$\begin{aligned}\dot{x}_1 &= x_2 \\ \dot{x}_2 &= x_3 \\ &\vdots \\ \dot{x}_{n-1} &= x_n\end{aligned} \qquad (2.24)$$

The last equation for $\dot{x}_n$ is obtained from equation 2.19 by substituting in the state variables from equations 2.21–2.23 and expressing $sX_n(s) = s^n Q(S)$ as:

$$sX_n(s) = U(s) - B_0 X_1(s) - B_1 X_2(s) + B_3 s^3 X_3(s) - \cdots - B_{n-1} X_n(s) \quad (2.25)$$

The time domain equivalent is:

$$\dot{x}_n = u - B_0 x_1 - B_2 x_2 - B_3 x_3 + \cdots - B_{n-1} x_n \qquad (2.26)$$

Therefore the matrix form of the state equations is:

$$\begin{pmatrix} \dot{x}_1 \\ \dot{x}_2 \\ \vdots \\ \dot{x}_{n-1} \\ \dot{x}_n \end{pmatrix} = \begin{bmatrix} 0 & 1 & 0 & \cdots & 0 \\ 0 & 0 & 1 & \cdots & 0 \\ \vdots & \vdots & \vdots & \ddots & \vdots \\ 0 & 0 & 0 & \cdots & 1 \\ -B_0 & -B_1 & -B_2 & \cdots & -B_{n-1} \end{bmatrix} \begin{pmatrix} x_1 \\ x_2 \\ \vdots \\ x_{n-1} \\ x_n \end{pmatrix} + \begin{pmatrix} 0 \\ 0 \\ \vdots \\ 0 \\ 1 \end{pmatrix} u \quad (2.27)$$

Since $Y(s) = A_n U(s) + Y_R(s)$, equation 2.20 can be used to express $Y_R(s)$ in terms of the state variables, yielding the following matrix equation for Y:

$$y = \begin{pmatrix}(A_0 - B_0 A_n) & (A_1 - B_1 A_n) & \cdots & (A_{n-1} - B_{n-1} A_n)\end{pmatrix} \begin{pmatrix} x_1 \\ x_2 \\ \vdots \\ x_{n-1} \\ x_n \end{pmatrix} + A_0 u$$
$$(2.28)$$

2.2.1.3 Observer canonical form

This is sometimes referred to as the nested integration method [2]. This form is obtained by multiplying both sides of equation 2.8 by $D(s)$ and collecting like terms in s^k, to get the equation in the form $D(s)Y(s) - N(s)U(s) = 0$, i.e.

$$\begin{aligned}s^n(Y(s) - A_n U(s)) &+ s^{n-1}(B_{n-1} Y(s) - A_{n-1} U(s)) + \cdots \\ &+ s(B_1 Y(s) - A_1 U(s)) + (B_0 Y(s) - A_0 U(s)) = 0\end{aligned} \quad (2.29)$$

Dividing both sides of equation 2.29 by s^n and rearranging gives:

$$Y(s) = A_n U(s) + \frac{1}{s}(A_{n-1}U(s) - B_{n-1}Y(s)) + \cdots$$
$$+ \frac{1}{s^{n-1}}(A_1 U(s) - B_1 Y(s)) + \frac{1}{s^n}(A_0 U(s) - B_0 Y(s)) \qquad (2.30)$$

Choosing as state variables:

$$X_1(s) = \frac{1}{s}(A_0 U(s) - B_0 Y(s))$$
$$X_2(s) = \frac{1}{s}(A_1 U(s) - B_1 Y(s) + X_1(s)) \qquad (2.31)$$
$$\vdots$$
$$X_n(s) = \frac{1}{s}(A_{n-1}U(s) - B_{n-1}Y(s) + X_{n-1}(s))$$

the output equation is thus:

$$Y(s) = A_n U(s) + X_n(s) \qquad (2.32)$$

Equation 2.32 is substituted into equation 2.31 to remove the variable $Y(s)$ and both sides multiplied by s. The inverse Laplace transform of the resulting equation yields:

$$\begin{aligned}
\dot{x}_1 &= -B_0 x_n + (A_0 - B_0 A_n)u \\
\dot{x}_2 &= x_1 - B_1 x_n + (A_1 - B_1 A_n)u \\
&\vdots \\
\dot{x}_{n-1} &= x_{n-2} - B_{n-2} x_n + (A_{n-2} - B_{n-2} A_n)u \\
\dot{x}_n &= x_{n-1} - B_{n-1} x_n + (A_{n-1} - B_{n-1} A_n)u
\end{aligned} \qquad (2.33)$$

The matrix equations are:

$$\begin{pmatrix} \dot{x}_1 \\ \dot{x}_2 \\ \vdots \\ \dot{x}_{n-1} \\ \dot{x}_n \end{pmatrix} = \begin{bmatrix} 0 & 0 & \cdots & 0 & -B_0 \\ 1 & 0 & \cdots & 0 & -B_1 \\ 0 & 1 & \cdots & 0 & -B_2 \\ \vdots & \vdots & \ddots & \vdots & \vdots \\ 0 & 0 & \cdots & 1 & -B_{n-1} \end{bmatrix} \begin{pmatrix} x_1 \\ x_2 \\ \vdots \\ x_{n-1} \\ x_n \end{pmatrix} + \begin{pmatrix} A_0 - B_0 A_n \\ A_1 - B_1 A_n \\ A_2 - B_2 A_n \\ \vdots \\ A_{n-1} - B_{n-1} A_n \end{pmatrix} u \qquad (2.34)$$

$$y = \begin{pmatrix} 0 & 0 & \cdots & 0 & 1 \end{pmatrix} \begin{pmatrix} x_1 \\ x_2 \\ \vdots \\ x_{n-1} \\ x_n \end{pmatrix} + A_n u \qquad (2.35)$$

2.2.1.4 Diagonal canonical form

The diagonal canonical or Jordan form is derived by rewriting equation 2.7 as:

$$G(s) = \frac{A_0 + A_1 s + A_2 s^2 + A_3 s^3 + \cdots + A_N s^N}{(s - \lambda_1)(s - \lambda_2)(s - \lambda_3) \cdots (s - \lambda_n)} = \frac{Y(s)}{U(s)} \quad (2.36)$$

where λ_k are the poles of the transfer function. By partial fraction expansion:

$$G(s) = \frac{r_1}{(s - \lambda_1)} + \frac{r_2}{(s - \lambda_2)} + \frac{r_3}{(s - \lambda_3)} + \cdots + \frac{r_n}{(s - \lambda_n)} + D \quad (2.37)$$

or

$$G(s) = \frac{Y(s)}{U(s)} = r_1 \frac{p_1}{U(s)} + r_2 \frac{p_2}{U(s)} + r_3 \frac{p_3}{U(s)} + \cdots + r_n \frac{p_n}{U(s)} + D \quad (2.38)$$

where

$$p_i = \frac{U(s)}{(s - \lambda_i)} \qquad D = \begin{cases} A_n, & N = n \\ 0, & N < n \end{cases} \quad (2.39)$$

which gives

$$Y(s) = r_1 p_1 + r_2 p_2 + r_3 p_3 + \cdots + r_n p_n + DU(s) \quad (2.40)$$

In the time domain equation 2.39 becomes:

$$\dot{p}_i = \lambda_i p_i + u \quad (2.41)$$

and equation 2.40:

$$y = \sum_{1}^{n} r_i p_i + Du \quad (2.42)$$

for $i = 1, 2, \ldots, n$; or, in compact matrix notation,

$$\dot{\mathbf{p}} = [\lambda]\mathbf{p} + [\beta]\mathbf{u} \quad (2.43)$$
$$\mathbf{y} = [C]\mathbf{p} + Du \quad (2.44)$$

where

$$[\lambda] = \begin{bmatrix} \lambda_1 & 0 & \cdots & 0 \\ 0 & \lambda_2 & \cdots & 0 \\ \vdots & \vdots & \ddots & 0 \\ 0 & 0 & \cdots & \lambda_n \end{bmatrix}, \quad [\beta] = \begin{bmatrix} 1 \\ 1 \\ \vdots \\ 1 \end{bmatrix}, \quad [C] = \begin{bmatrix} r_1 \\ r_2 \\ \vdots \\ r_N \end{bmatrix}$$

and the λ terms in the Jordans' form are the eigenvalues of the matrix $[A]$.

2.2.1.5 Uniqueness of formulation

The state variable realisation is not unique; for example another possible state variable form for equation 2.36 is:

$$\begin{pmatrix} \dot{x}_1 \\ \dot{x}_2 \\ \vdots \\ \dot{x}_n \end{pmatrix} = \begin{bmatrix} -B_{n-1} & 1 & 0 & \cdots & 0 \\ -B_{n-2} & 0 & 1 & \cdots & 0 \\ \vdots & \vdots & \vdots & \ddots & \vdots \\ -B_1 & 0 & 0 & \cdots & 1 \\ -B_0 & 0 & 0 & \cdots & 0 \end{bmatrix} \begin{pmatrix} x_1 \\ x_2 \\ \vdots \\ x_D \end{pmatrix} + \begin{pmatrix} A_{n-1} - B_{n-1}A_n \\ A_{n-2} - B_{n-2}A_n \\ \vdots \\ A_1 - B_1 A_n \\ A_0 - B_0 A_n \end{pmatrix} u \quad (2.45)$$

However the transfer function is unique and is given by:

$$H(s) = [C](s[I] - [A])^{-1}[B] + [D] \quad (2.46)$$

For low order systems this can be evaluated using:

$$(s[I] - [A])^{-1} = \frac{\mathrm{adj}(s[I] - [A])}{|s[I] - [A]|} \quad (2.47)$$

where $[I]$ is the identity matrix.

In general a non-linear network will result in equations of the form:

$$\begin{aligned} \dot{\mathbf{x}} &= [A]\mathbf{x} + [B]\mathbf{u} + [B_1]\dot{\mathbf{u}} + ([B_2]\ddot{\mathbf{u}} + \cdots) \\ \mathbf{y} &= [C]\mathbf{x} + [D]\mathbf{u} + [D_1]\dot{\mathbf{u}} + ([D_2]\ddot{\mathbf{u}} + \cdots) \end{aligned} \quad (2.48)$$

For linear RLC networks the derivative of the input can be removed by a simple change of state variables, i.e.

$$\mathbf{x}' = \mathbf{x} - [B_1]\mathbf{u} \quad (2.49)$$

The state variable equations become:

$$\dot{\mathbf{x}}' = [A]\mathbf{x}' + [B]\mathbf{u} \quad (2.50)$$

$$\mathbf{y} = [C]\mathbf{x}' + [D]\mathbf{u} \quad (2.51)$$

However in general non-linear networks the time derivative of the forcing function appears in the state and output equations and cannot be readily eliminated.

Generally the differential equations for a circuit are of the form:

$$[M]\dot{\mathbf{x}} = [A_{(0)}]\mathbf{x} + [B_{(0)}]\mathbf{u} + ([B_{(0)1}]\dot{\mathbf{u}}) \quad (2.52)$$

To obtain the normal form, both sides are multiplied by the inverse of $[M]^{-1}$, i.e.

$$\begin{aligned} \dot{\mathbf{x}} &= [M]^{-1}[A_{(0)}]\mathbf{x} + [M]^{-1}[B_{(0)}]\mathbf{u} + \left([M]^{-1}[B_{(0)1}]\dot{\mathbf{u}}\right) \\ &= [A]\mathbf{x} + [B]\mathbf{u} + ([B_1]\dot{\mathbf{u}}) \end{aligned} \quad (2.53)$$

2.2.1.6 Example

Given the transfer function:

$$\frac{Y(s)}{U(s)} = \frac{s+3}{s^2+3s+2} = \frac{2}{(s+1)} + \frac{-1}{(s+2)}$$

derive the alternative state variable representations described in sections 2.2.1.1–2.2.1.4.

Successive differentiation:

$$\begin{pmatrix} \dot{x}_1 \\ \dot{x}_2 \end{pmatrix} = \begin{bmatrix} 0 & 1 \\ -2 & -3 \end{bmatrix} \begin{pmatrix} x_1 \\ x_2 \end{pmatrix} + \begin{pmatrix} 1 \\ 0 \end{pmatrix} u \quad (2.54)$$

$$y = \begin{bmatrix} 1 & 0 \end{bmatrix} \begin{pmatrix} x_1 \\ x_2 \end{pmatrix} \quad (2.55)$$

Controllable canonical form:

$$\begin{pmatrix} \dot{x}_1 \\ \dot{x}_2 \end{pmatrix} = \begin{bmatrix} 0 & 1 \\ -2 & -3 \end{bmatrix} \begin{pmatrix} x_1 \\ x_2 \end{pmatrix} + \begin{pmatrix} 0 \\ 1 \end{pmatrix} u \quad (2.56)$$

$$y = \begin{bmatrix} 3 & 1 \end{bmatrix} \begin{pmatrix} x_1 \\ x_2 \end{pmatrix} \quad (2.57)$$

Observable canonical form:

$$\begin{pmatrix} \dot{x}_1 \\ \dot{x}_2 \end{pmatrix} = \begin{bmatrix} 0 & -2 \\ 1 & -3 \end{bmatrix} \begin{pmatrix} x_1 \\ x_2 \end{pmatrix} + \begin{pmatrix} 3 \\ 1 \end{pmatrix} u \quad (2.58)$$

$$y = \begin{bmatrix} 0 & 1 \end{bmatrix} \begin{pmatrix} x_1 \\ x_2 \end{pmatrix} \quad (2.59)$$

Diagonal canonical form:

$$\begin{pmatrix} \dot{x}_1 \\ \dot{x}_2 \end{pmatrix} = \begin{bmatrix} -1 & 0 \\ 0 & -2 \end{bmatrix} \begin{pmatrix} x_1 \\ x_2 \end{pmatrix} + \begin{pmatrix} 1 \\ 1 \end{pmatrix} u \quad (2.60)$$

$$y = \begin{bmatrix} 2 & -1 \end{bmatrix} \begin{pmatrix} x_1 \\ x_2 \end{pmatrix} \quad (2.61)$$

Although all these formulations look different they represent the same dynamic system and their response is identical. It is left as an exercise to calculate $H(s) = [C](s[I] - [A])^{-1}[B] + [D]$ to show they all represent the same transfer function.

2.2.2 Time domain solution of state equations

The Laplace transform of the state equation is:

$$sX(s) - X(0_+) = [A]X(s) + [B]U(s) \quad (2.62)$$

Therefore

$$X(s) = (s[I] - [A])^{-1} X(0_+) + (s[I] - [A])^{-1}[B]U(s) \quad (2.63)$$

where $[I]$ is the identity (or unit) matrix.

Then taking the inverse Laplace transform will give the time response. However the use of the Laplace transform method is impractical to determine the transient response of large networks with arbitrary excitation.

The time domain solution of equation 2.63 can be expressed as:

$$x(t) = h(t)x(0_+) + \int_0^t h(t-T)[B]u(T)\,dT \qquad (2.64)$$

or, changing the lower limit from 0 to t_0:

$$x(t) = e^{[A](t-t_0)}x(t_0) + \int_{t_0}^t e^{[A](t-T)}[B]u(T)\,dT \qquad (2.65)$$

where $h(t)$, the impulse response, is the inverse Laplace transform of the transition matrix, i.e. $h(t) = L^{-1}((sI-[A])^{-1})$.

The first part of equation 2.64 is the homogeneous solution due to the initial conditions. It is also referred to as the natural response or the zero-input response, as calculated by setting the forcing function to zero (hence the homogeneous case). The second term of equation 2.64 is the forced solution or zero-state response, which can also be expressed as the convolution of the impulse response with the source. Thus equation 2.64 becomes:

$$x(t) = h(t)x(0_+) + h(t) \otimes [B]u(t) \qquad (2.66)$$

Only simple analytic solutions can be obtained by transform methods, as this requires taking the inverse Laplace transform of the impulse response transfer function matrix, which is difficult to perform. The same is true for the method of variation of parameters where integrating factors are applied.

The time convolution can be performed by numerical calculation. Thus by application of an integration rule a difference equation can be derived. The simplest approach is the use of an explicit integration method (such that the value at $t + \Delta t$ is only dependent on t values), however it suffers from the weaknesses of explicit methods. Applying the forward Euler method will give the following difference equation for the solution [5]:

$$x(t+\Delta t) = e^{[A]\Delta t}x(t) + [A]^{-1}(e^{[A]\Delta t} - I)[B]u(t) \qquad (2.67)$$

As can be seen the difference equation involves the transition matrix, which must be evaluated via its series expansion, i.e.

$$e^{[A]\Delta t} = I + [A]\Delta t + \frac{[A]^2 \Delta t^2}{2!} + \frac{[A]^3 \Delta t^3}{3!} + \cdots \qquad (2.68)$$

However this is not always straightforward and, even when convergence is possible, it may be very slow. Moreover, alternative terms of the series have opposite signs and these terms may have extremely high values.

The calculation of equation 2.68 may be aided by modal analysis. This is achieved by determining the eigenvalues and eigenvectors, hence the transformation matrix $[T]$, which will diagonalise the transition matrix i.e.

$$\dot{z} = [T]^{-1}[A][T]z + [T]^{-1}[B]u = [S]z + [T]^{-1}[B]u \qquad (2.69)$$

where

$$\mathbf{z} = [T]^{-1}\mathbf{x}$$

$$[S] = \begin{bmatrix} \lambda_1 & 0 & \cdots & 0 & 0 \\ 0 & \lambda_2 & \cdots & 0 & 0 \\ \vdots & \vdots & \ddots & \vdots & \vdots \\ 0 & 0 & \cdots & \lambda_{n-1} & 0 \\ 0 & 0 & \cdots & 0 & \lambda_n \end{bmatrix}$$

and $\lambda_1, \ldots, \lambda_n$ are the eigenvalues of the matrix.

The eigenvalues provide information on time constants, resonant frequencies and stability of a system. The time constants of the system $(1/\Re e(\lambda_{\min}))$ indicate the length of time needed to reach steady state and the maximum time step that can be used. The ratio of the largest to smallest eigenvalues $(\lambda_{\max}/\lambda_{\min})$ gives an indication of the stiffness of the system, a large ratio indicating that the system is mathematically stiff.

An alternative method of solving equation 2.65 is the use of numerical integration. In this case, state variable analysis uses an iterative procedure (predictor–corrector formulation) to solve for each time period. An implicit integration method, such as the trapezoidal rule, is used to calculate the state variables at time t, however this requires the value of the state variable derivatives at time t. The previous time step values can be used as an initial guess and once an estimate of the state variables has been obtained using the trapezoidal rule, the state equation is used to update the estimate of the state variable derivatives.

No matter how the differential equations are arranged and manipulated into different forms, the end result is only a function of whether a numerical integration formula is substituted in (discussed in section 2.2.3) or an iterative solution procedure adopted.

2.2.3 Digital simulation of continuous systems

As explained in the introduction, due to the discrete nature of the digital process, a difference equation must be developed to allow the digital simulation of a continuous system. Also the latter must be stable to be able to perform digital simulation, which implies that all the s-plane poles are in the left-hand half-plane, as illustrated in Figure 2.1.

However, the stability of the continuous system does not necessarily ensure that the simulation equations are stable. The equivalent of the s-plane for continuous signals is the z-plane for discrete signals. In the latter case, for stability the poles must lie inside the unit circle, as shown in Figure 2.4 on page 32. Thus the difference equations must be transformed to the z-plane to assess their stability. Time delay effects in the way data is manipulated must be incorporated and the resulting z-domain representation used to determine the stability of the simulation equations.

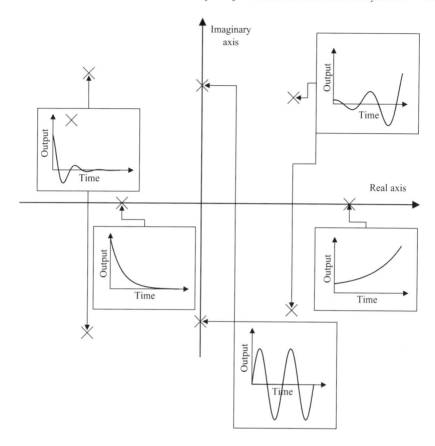

Figure 2.1 Impulse response associated with s-plane pole locations

A simple two-state variable system is used to illustrate the development of a difference equation suitable for digital simulation, i.e.

$$\begin{pmatrix}\dot{x}_1\\ \dot{x}_2\end{pmatrix} = \begin{bmatrix}a_{11} & a_{12}\\ a_{21} & a_{22}\end{bmatrix}\begin{pmatrix}x_1\\ x_2\end{pmatrix} + \begin{pmatrix}b_{11}\\ b_{21}\end{pmatrix}u \quad (2.70)$$

Applying the trapezoidal rule $(x_i(t) = x_i(t - \Delta t) + \Delta t/2(\dot{x}_i(t) + \dot{x}_i(t - \Delta t)))$ to the two rows of matrix equation 2.70 gives:

$$x_1(t) = x_1(t - \Delta t) + \frac{\Delta t}{2}[a_{11}x_1(t) + a_{12}x_2(t) + b_{11}u(t) + a_{11}x_1(t - \Delta t)$$
$$+ a_{12}x_2(t - \Delta t) + b_{11}u(t - \Delta t)] \quad (2.71)$$

$$x_2(t) = x_2(t - \Delta t) + \frac{\Delta t}{2}[a_{21}x_1(t) + a_{22}x_2(t) + b_{21}u(t) + a_{21}x_1(t - \Delta t)$$
$$+ a_{22}x_2(t - \Delta t) + b_{21}u(t - \Delta t)] \quad (2.72)$$

or in matrix form:

$$\begin{bmatrix} 1 - \frac{\Delta t}{2}a_{11} & -\frac{\Delta t}{2}a_{12} \\ -\frac{\Delta t}{2}a_{21} & 1 - \frac{\Delta t}{2}a_{22} \end{bmatrix} \begin{pmatrix} x_1(t) \\ x_2(t) \end{pmatrix} = \begin{bmatrix} 1 + \frac{\Delta t}{2}a_{11} & \frac{\Delta t}{2}a_{12} \\ \frac{\Delta t}{2}a_{21} & 1 + \frac{\Delta t}{2}a_{22} \end{bmatrix} \begin{pmatrix} x_1(t - \Delta t) \\ x_2(t - \Delta t) \end{pmatrix}$$
$$+ \begin{pmatrix} \frac{\Delta t}{2}b_{11} \\ \frac{\Delta t}{2}b_{21} \end{pmatrix} (u(t) + u(t - \Delta t)) \quad (2.73)$$

Hence the set of difference equations to be solved at each time point is:

$$\begin{pmatrix} x_1(t) \\ x_2(t) \end{pmatrix} = \begin{bmatrix} 1 - \frac{\Delta t}{2}a_{11} & -\frac{\Delta t}{2}a_{12} \\ -\frac{\Delta t}{2}a_{21} & 1 - \frac{\Delta t}{2}a_{22} \end{bmatrix}^{-1} \begin{bmatrix} 1 + \frac{\Delta t}{2}a_{11} & \frac{\Delta t}{2}a_{12} \\ \frac{\Delta t}{2}a_{21} & 1 + \frac{\Delta t}{2}a_{22} \end{bmatrix} \begin{pmatrix} x_1(t - \Delta t) \\ x_2(t - \Delta t) \end{pmatrix}$$
$$+ \begin{bmatrix} 1 - \frac{\Delta t}{2}a_{11} & -\frac{\Delta t}{2}a_{12} \\ -\frac{\Delta t}{2}a_{21} & 1 - \frac{\Delta t}{2}a_{22} \end{bmatrix}^{-1} \begin{pmatrix} \frac{\Delta t}{2}b_{11} \\ \frac{\Delta t}{2}b_{21} \end{pmatrix} (u(t) + u(t - \Delta t)) \quad (2.74)$$

This can be generalised for any state variable formulation by substituting the state equation ($\dot{\mathbf{x}} = [A]\mathbf{x} + [B]\mathbf{u}$) into the trapezoidal equation i.e.

$$\mathbf{x}(t) = \mathbf{x}(t - \Delta t) + \frac{\Delta t}{2} (\dot{\mathbf{x}}(t) + \dot{\mathbf{x}}(t - \Delta t))$$
$$= \mathbf{x}(t - \Delta t) + \frac{\Delta t}{2} ([A]\mathbf{x}(t) + [B]\mathbf{u}(t) + [A]\mathbf{x}(t - \Delta t) + [B]\mathbf{u}(t - \Delta t))$$
$$(2.75)$$

Collecting terms in $\mathbf{x}(t)$, $\mathbf{x}(t - \Delta t)$, $\mathbf{u}(t)$ and $\mathbf{u}(t - \Delta t)$ gives:

$$\left([I] - \frac{\Delta t}{2}[A]\right) \mathbf{x}(t) = \left([I] + \frac{\Delta t}{2}[A]\right) \mathbf{x}(t - \Delta t) + \frac{\Delta t}{2}[B] (\mathbf{u}(t) + \mathbf{u}(t - \Delta t))$$
$$(2.76)$$

Rearranging equation 2.76 to give $\mathbf{x}(t)$ in terms of previous time point values and present input yields:

$$\mathbf{x}(t) = \left[[I] - \frac{\Delta t}{2}[A]\right]^{-1} \left[[I] + \frac{\Delta t}{2}[A]\right] \mathbf{x}(t - \Delta t)$$
$$+ \left[[I] - \frac{\Delta t}{2}[A]\right]^{-1} \frac{\Delta t}{2}[B] (\mathbf{u}(t) + \mathbf{u}(t - \Delta t)) \quad (2.77)$$

The structure of $([I] - \Delta t/2[A])$ depends on the formulation, for example with the successive differentiation approach (used in PSCAD/EMTDC for transfer function representation) it becomes:

$$\begin{bmatrix} 1 & -\frac{\Delta t}{2} & 0 & \cdots & 0 & 0 \\ 0 & 1 & -\frac{\Delta t}{2} & \ddots & \vdots & \vdots \\ 0 & 0 & 1 & \ddots & 0 & 0 \\ \vdots & \vdots & \vdots & \ddots & -\frac{\Delta t}{2} & 0 \\ 0 & 0 & 0 & \cdots & 1 & -\frac{\Delta t}{2} \\ -B_0 & -B_1 & -B_2 & \cdots & -B_{n-2} & -B_{n-1} \end{bmatrix} \quad (2.78)$$

Similarly, the structure of $(I + \Delta t/2[A])$ is:

$$\begin{bmatrix} 1 & \frac{\Delta t}{2} & 0 & \cdots & 0 & 0 \\ 0 & 1 & \frac{\Delta t}{2} & \ddots & \vdots & \vdots \\ 0 & 0 & 1 & \ddots & 0 & 0 \\ \vdots & \vdots & \vdots & \ddots & \frac{\Delta t}{2} & 0 \\ 0 & 0 & 0 & \cdots & 1 & \frac{\Delta t}{2} \\ B_0 & B_1 & B_2 & \cdots & B_{n-2} & B_{n-1} \end{bmatrix} \quad (2.79)$$

The EMTP program uses the following internal variables for TACS:

$$x_1 = \frac{dy}{dt}, \quad x_2 = \frac{dx_1}{dt}, \quad \ldots, \quad x_n = \frac{dx_{n-1}}{dt} \quad (2.80)$$

$$u_1 = \frac{du}{dt}, \quad u_2 = \frac{du_1}{dt}, \quad \ldots, \quad u_N = \frac{du_{N-1}}{dt} \quad (2.81)$$

Expressing this in the s-domain gives:

$$x_1 = sy, \quad x_2 = sx_1, \quad \ldots, \quad x_n = sdx_{n-1} \quad (2.82)$$

$$u_1 = su, \quad u_2 = su_1, \quad \ldots, \quad u_N = su_{N-1} \quad (2.83)$$

Using these internal variables the transfer function (equation 2.8) becomes the algebraic equation:

$$b_0 y + b_1 x_1 + \cdots + b_n x_n = a_0 u + a_1 u_1 + \cdots + a_N u_N \quad (2.84)$$

Equations 2.80 and 2.81 are converted to difference equations by application of the trapezoidal rule, i.e.

$$x_i(t) = \frac{2}{\Delta t} x_{i-1}(t) - \underbrace{\left(x_i(t - \Delta t) + \frac{2}{\Delta t} x_{i-1}(t - \Delta t) \right)}_{\text{History term}} \quad (2.85)$$

for $i = 1, 2, \ldots, n$ and

$$u_k(t) = \frac{2}{\Delta t} u_{k-1}(t) - \underbrace{\left(u_k(t - \Delta t) + \frac{2}{\Delta t} u_{k-1}(t - \Delta t) \right)}_{\text{History term}} \quad (2.86)$$

for $k = 1, 2, \ldots, N$.

To eliminate these internal variables, x_n is expressed as a function of x_{n-1}, the latter as a function of $x_{n-2}, \ldots$ etc., until only y is left. The same procedure is used for u. This process yields a single output–input relationship of the form:

$$c \cdot x(t) = d \cdot u(t) + \text{History}(t - \Delta t) \quad (2.87)$$

After the solution at each time point is obtained, the n History terms must be updated to derive the single History term for the next time point (equation 2.87), i.e.

$$\text{hist}_1(t) = d_1 u(t) - c_1 x(t) - \text{hist}_1(t - \Delta t) - \text{hist}_2(t - \Delta t)$$
$$\vdots$$
$$\text{hist}_i(t) = d_i u(t) - c_i x(t) - \text{hist}_i(t - \Delta t) - \text{hist}_{i+1}(t - \Delta t) \quad (2.88)$$
$$\vdots$$
$$\text{hist}_{n-1}(t) = d_{n-1} u(t) - c_{n-1} x(t) - \text{hist}_{n-1}(t - \Delta t) - \text{hist}_n(t - \Delta t)$$
$$\text{hist}_n(t) = d_n u(t) - c_n x(t)$$

where History (equation 2.87) is equated to $\text{hist}_1(t)$ in equation 2.88.

The coefficients c_i and d_i are calculated once at the beginning, from the coefficients a_i and b_i. The recursive formula for c_i is:

$$c_i = c_{i-1} + (-2)^i \left(\binom{i}{i} \left(\frac{2}{\Delta t} \right)^i b_i + \binom{i+1}{i} \left(\frac{2}{\Delta t} \right)^{i+1} b_{i+1} \right.$$
$$\left. + \cdots + \binom{n}{i} \left(\frac{2}{\Delta t} \right)^n b_n \right) \quad (2.89)$$

where $\binom{n}{i}$ is the binomial coefficient.

The starting value is:

$$c_0 = \sum_{i=0}^{n} \left(\frac{2}{\Delta t} \right)^i b_i \quad (2.90)$$

Analysis of continuous and discrete systems   27

Similarly the recursive formula for d_i is:

$$d_i = d_{i-1} + (-2)^i \left(\binom{i}{i} \left(\frac{2}{\Delta t}\right)^i a_i + \binom{i+1}{i} \left(\frac{2}{\Delta t}\right)^{i+1} a_{i+1} \right.$$

$$\left. + \cdots + \binom{N}{i} \left(\frac{2}{\Delta t}\right)^N a_N \right) \quad (2.91)$$

2.2.3.1 Example

Use the trapezoidal rule to derive the difference equation that will simulate the lead–lag control block:

$$H(s) = \frac{100 + s}{500 + s} = \frac{1/5 + s/500}{1 + s/500} \quad (2.92)$$

The general form is

$$H(s) = \frac{a_0 + a_1 s}{1 + b_1 s} = \frac{A_0 + A_1 s}{B_0 + s}$$

where $a_0 = A_0/B_0 = 1/5$, $b_1 = 1/B_0 = 1/500$ and $a_1 = A_1/B_0 = 1/500$ for this case. Using the successive differentiation formulation (section 2.2.1.1) the equations are:

$$\dot{x}_1 = [-B_0]x_1 + [A_0 - B_0 A_1]u$$

$$y = [1]x_1 + [A_1]u$$

Using equation 2.77 gives the difference equation:

$$x_1(n\Delta t) = \frac{(1 - \Delta t B_0/2)}{(1 + \Delta t B_0/2)} x_1((n-1)\Delta t)$$

$$+ \frac{(\Delta t/2)(A_0 - B_0 A_1)}{(1 + \Delta t B_0/2)} (u(n\Delta t) + u((n-1)\Delta t))$$

Substituting the relationship $x_1 = y - A_1 u$ (equation 2.10) and rearranging yields:

$$y(n\Delta t) = \frac{(1 - \Delta t B_0/2)}{(1 + \Delta t B_0/2)} y((n-1)\Delta t)$$

$$+ \frac{(\Delta t/2)(A_0 - B_0 A_1)}{(1 + \Delta t B_0/2)} (u(n\Delta t) + u((n-1)\Delta t))$$

$$- \frac{A_1(1 - \Delta t B_0/2)}{(1 + \Delta t B_0/2)} u((n-1)\Delta t) + A_1 u(n\Delta t)$$

Expressing the latter equation in terms of a_0, a_1 and b_1, then collecting terms in $u(n\Delta t)$ and $u((n-1)\Delta t)$ gives:

$$y(n\Delta t) = \frac{(2b_1 - \Delta t)}{(2b_1 + \Delta t)} y((n-1)\Delta t)$$

$$+ \frac{(\Delta t a_0 + 2a_1)u(n\Delta t) + (\Delta t a_0 - 2a_1)u((n-1)\Delta t)}{(2b_1 + \Delta t)}$$

The equivalence between the trapezoidal rule and the bilinear transform (shown in section 5.2) provides another method for performing numerical integrator substitution (NIS) as follows.

Using the trapezoidal rule by making the substitution $s = (2/\Delta t)(1-z^{-1})/(1+z^{-1})$ in the transfer function (equation 2.92):

$$H(z) = \frac{Y(z)}{U(z)} = \frac{a_0 + a_1(2/\Delta t)(1-z^{-1})/(1+z^{-1})}{1 + b_1(2/\Delta t)(1-z^{-1})/(1+z^{-1})}$$

$$= \frac{a_0 \Delta t(1+z^{-1}) + 2a_1(1-z^{-1})}{\Delta t(1+z^{-1}) + 2b_1(1-z^{-1})}$$

$$= \frac{(a_0 \Delta t + 2a_1) + z^{-1}(a_0 \Delta t - 2a_1)}{(\Delta t + 2b_1) + z^{-1}(\Delta t - 2b_1)} \quad (2.93)$$

Multiplying both sides by the denominator:

$$Y(z)[(\Delta t + 2b_1) + z^{-1}(\Delta t - 2b_1)] = U(z)[(a_0 \Delta t + 2a_1) + z^{-1}(a_0 \Delta t - 2a_1)]$$

and rearranging gives the input–output relationship:

$$Y(z) = \frac{-(\Delta t - 2b_1)}{(\Delta t + 2b_1)} z^{-1} Y(z) + \frac{(a_0 \Delta t + 2a_1) + z^{-1}(a_0 \Delta t - 2a_1)}{(\Delta t + 2b_1)} U(z)$$

Converting from the z-domain to the time domain produces the following difference equation:

$$y(n\Delta t) = \frac{(2b_1 - \Delta t)}{(2b_1 + \Delta t)} y((n-1)\Delta t)$$
$$+ \frac{(a_0 \Delta t + 2a_1)u(n\Delta t) + (a_0 \Delta t - 2a_1)u((n-1)\Delta t)}{(\Delta t + 2b_1)}$$

and substituting in the values for a_0, a_1 and b_1:

$$y(n\Delta t) = \frac{(0.004 - \Delta t)}{(0.004 + \Delta t)} y((n-1)\Delta t)$$
$$+ \frac{(0.2\Delta t + 0.004)u(n\Delta t) + (0.2\Delta t - 0.004)u((n-1)\Delta t)}{(\Delta t + 0.004)}$$

This is a simple first order function and hence the same result would be obtained by substituting expressions for $\dot{y}(n\Delta t)$ and $\dot{y}((n-1)\Delta t)$, based on equation 2.92, into the trapezoidal rule (i.e. $y(n\Delta t) = y((n-1)\Delta t) + \Delta t/2(\dot{y}(n\Delta t) + \dot{y}((n-1)\Delta t)))$ i.e. from equation 2.92:

$$\dot{y}(n\Delta t) = \frac{-1}{b_1} x(n\Delta t) + \frac{a_0}{b_1} u(n\Delta t) + \frac{a_1}{b_1} \dot{u}(n\Delta t)$$

Figure 2.2 displays the step response of this lead–lag function for various lead time (a_1 values) constants, while Table 2.1 shows the numerical results for the first eight steps using a 50 μs time step.

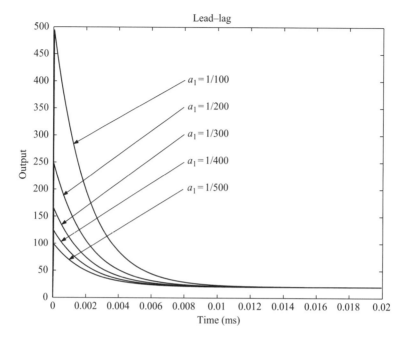

Figure 2.2   Step response of lead–lag function

Table 2.1   First eight steps for simulation of lead–lag function

Time (ms)	a_1				
	0.01	0.0050	0.0033	0.0025	0.0020
0.050	494.0741	247.1605	164.8560	123.7037	99.0123
0.100	482.3685	241.5516	161.2793	121.1431	97.0614
0.150	470.9520	236.0812	157.7909	118.6458	95.1587
0.200	459.8174	230.7458	154.3887	116.2101	93.3029
0.250	448.9577	225.5422	151.0704	113.8345	91.4930
0.300	438.3662	220.4671	147.8341	111.5176	89.7277
0.350	428.0361	215.5173	144.6777	109.2579	88.0060
0.400	417.9612	210.6897	141.5993	107.0540	86.3269
G_{eff}	4.94074	2.47160	1.64856	1.23704	0.99012

It should be noted that a first order lag function or an RL branch are special forms of lead–lag, where $a_1 = 0$, i.e.

$$H(s) = \frac{1/R}{1 + sL/R} = \frac{G}{1 + s\tau}$$

Thus in this case substitution of $b_1 = \tau = L/R$ and $a_0 = 1/R$ produces the well known difference equation of an RL branch:

$$y(n\Delta t) = \frac{(1 - \Delta t R/(2L))}{(1 + \Delta t R/(2L))} y((n-1)\Delta t)$$

$$+ \frac{\Delta t/(2L)}{(1 + \Delta t R/(2L))} (u(n\Delta t) + u((n-1)\Delta t))$$

or in terms of G and τ

$$y(n\Delta t) = \frac{(1 - \Delta t/(2\tau))}{(1 + \Delta t/(2\tau))} y((n-1)\Delta t) + \frac{(G\Delta t/(2\tau))}{(1 + \Delta t/(2\tau))} (u(n\Delta t) + u((n-1)\Delta t))$$

2.3 Discrete systems

A discrete system can be represented as a z-domain function, i.e.

$$H(z) = \frac{Y(z)}{U(z)} = \frac{a_0 + a_1 z^{-1} + a_2 z^{-2} + \cdots + a_N z^{-N}}{1 + b_1 z^{-1} + b_2 z^{-2} + \cdots + b_n z^{-n}} \quad (2.94)$$

When $H(z)$ is such that $a_i = 0$ for $i = 1, 2, \ldots, N$ but $a_0 \neq 0$ then equation 2.94 represents an all-pole model (i.e. no zeros), also called an autoregressive (AR) model, as the present output depends on the output at previous time points but not on the input at previous time points.

If $b_i = 0$ for $i = 1, 2, \ldots, n$ except $b_0 \neq 0$, equation 2.94 represents an all-zero model (no poles) or moving average (MA), as the current output is an average of the previous (and present) input but not of the previous output. In digital signal processing this corresponds to a finite impulse response (FIR) filter.

If both poles and zeros exist then equation 2.94 represents an ARMA model, which in digital signal processing corresponds to an infinite impulse response (IIR) filter [6], i.e.

$$U(z)\left(a_0 + a_1 z^{-1} + a_2 z^{-2} + \cdots + a_N z^{-N}\right)$$
$$= Y(z)\left(1 + b_1 z^{-1} + b_2 z^{-2} + \cdots + b_n z^{-n}\right) \quad (2.95)$$

$$Y(z) = -Y(z)\left(b_1 z^{-1} + b_2 z^{-2} + \cdots + b_n z^{-n}\right)$$
$$+ u(z)\left(a_0 + a_1 z^{-1} + a_2 z^{-2} + \cdots + a_N z^{-N}\right) \quad (2.96)$$

Transforming the last expression to the time domain, where $y(k)$ represents the kth time point value of y, gives:

$$y(k) = -(b_1 y(k-1) + b_2 y(k-2) + \cdots + b_n y(k-n))$$
$$+ (a_0 u_k + a_1 u(k-1) + a_2 u(k-2) + \cdots + a_N u(k-N)) \quad (2.97)$$

and rearranging to show the Instantaneous and History terms

$$y(k) = \overbrace{a_0 u(k)}^{\text{Instantaneous}} + \tag{2.98}$$

$$\overbrace{(a_1 u(k-1) + a_2 u(k-2) + \cdots + a_N u(k-N) - b_1 y(k-1) + a_2 y(k-2) + \cdots + a_n y(k-n))}^{\text{History term}}$$

This equation can then be represented as a Norton equivalent as depicted in Figure 2.3. The state variable equations for a discrete system are:

$$\mathbf{x}(k+1) = [A]\mathbf{x}(k) + [B]\mathbf{u}(k) \tag{2.99}$$

$$\mathbf{y}(k+1) = [C]\mathbf{x}(k) + [D]\mathbf{u}(k) \tag{2.100}$$

Taking the z-transform of the state equations and combining them shows the equivalence with the continuous time counterpart. i.e.

$$Y(z) = H(Z)U(z) \tag{2.101}$$

$$H(z) = [C](z[I] - [A])^{-1}[B] + [D] \tag{2.102}$$

where $[I]$ is the identity matrix.

The dynamic response of a discrete system is determined by the pole positions, which for stability must be inside the unit circle in the z-plane. Figure 2.4 displays the impulse response for various pole positions.

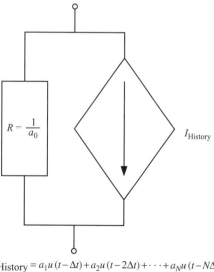

Figure 2.3 Norton of a rational function in z-domain

32 *Power systems electromagnetic transients simulation*

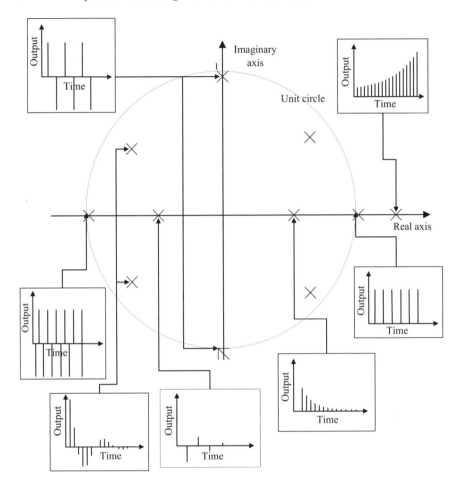

Figure 2.4 Data sequence associated with z-plane pole locations

2.4 Relationship of continuous and discrete domains

Figure 2.5 depicts the relationships between the continuous and discrete time processes as well as the s-domain and z-domain. Starting from the top left, in the time domain a continuous function can be expressed as a high order differential equation or a group of first order (state variable) equations. The equivalent of this exists in the discrete time case where the output can be related to the state at only the previous step and the input at the present and previous step. In this case the number of state variables, and hence equations, equals the order of the system. The alternative discrete time formulation is to express the output as a function of the output and input for a number of previous time steps (recursive formulation). In this case the number of previous time steps required equals the order of the system. To move from continuous time to discrete time requires a sampling process. The opposite process is a sample and hold.

Analysis of continuous and discrete systems 33

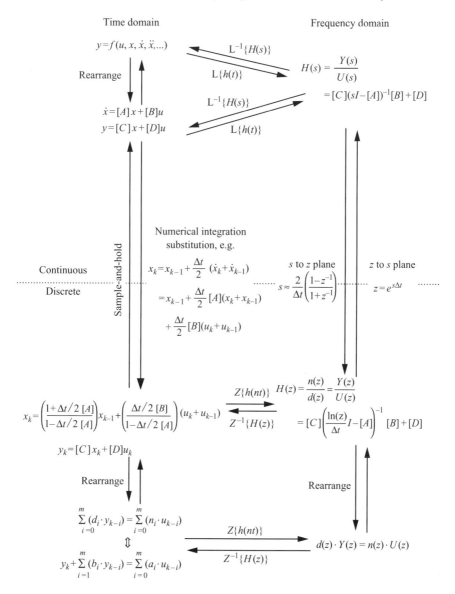

Figure 2.5 Relationship between the domains

Turning to the right-hand side of the figure, the Laplace transform of a continuous function is expressed in the s-plane. It can be converted to a z-domain function by using an equation that relates s to z. This equation is equivalent to numerical integrator substitution in the time domain and the equation will depend on the integration formula used. Note that when using an s-domain formulation (e.g. the state variable realisation $H(s) = [C](s[I] - [A])^{-1}[B] + [D]$), the solution requires a transition from the s to

z-domain. Often people make this transition without realising that they have done so. The z-domain is the discrete equivalent to the s-domain. Finally the z-transform and inverse z-transform are used to go between discrete time difference equations and a z-domain representation.

2.5 Summary

With the exceptions of a few auxiliary components, the electrical power system is a continuous system, which can be represented mathematically by a system of differential and algebraic equations.

A convenient form of these equations is the state variable formulation, in which a system of n first-order linear differential equations results from an nth order system. The state variable formulation is not unique and depends on the choice of state variables. The following state variable realisations have been described in this chapter: successive differentiation, controller canonical, observer canonical and diagonal canonical.

Digital simulation is by nature a discrete time process and can only provide solutions for the differential and algebraic equations at discrete points in time, hence this requires the formulation of discrete systems. The discrete representation can always be expressed as a difference equation, where the output at a new time point is calculated from the output at previous time points and the inputs at the present and previous time points.

2.6 References

1 KAILATH, T.: 'Linear systems' (Prentice Hall, Englewood Cliffs, 1980)
2 DeRUSSO, P. M., ROY, R. J., CLOSE, C. M. and DESROCHERS, A. A.: 'State variables for engineers' (John Wiley, New York, 2nd edition, 1998)
3 SMITH, J. M.: 'Mathematical modeling and digital simulation for engineers and scientists' (John Wiley, New York, 2nd edition, 1987)
4 OGATA, K.: 'Modern control engineering' (Prentice Hall International, Upper Saddle River, N. J., 3rd edition, 1997)
5 RAJAGOPALAN, V.: 'Computer-aided analysis of power electronic system' (Marcel Dekker, New York, 1987)
6 DORF, R. C. (Ed.): 'The electrical engineering handbook' (CRC Press, Boca Raton, FL, 2nd edition, 1997)

Chapter 3
State variable analysis

3.1 Introduction

State variables are the parameters of a system that completely define its energy storage state. State variable analysis was the dominant technique in transient simulation prior to the appearance of the numerical integration substitution method.

Early state variable programs used the 'central process' method [1] that breaks the switching operation down into similar consecutive topologies. This method requires many subroutines, each solving the set of differential equations arising from a particular network topology. It has very little versatility, as only coded topologies can be simulated, thus requiring a priori knowledge of all possible circuit configurations.

The application of Kron's tensor techniques [2] led to an elegant and efficient method for the solution of systems with periodically varying topology, such as an a.c.–d.c. converter. Its main advantages are more general applicability and a logical procedure for the automatic assembly and solution of the network equations. Thus the programmer no longer needs to be aware of all the sets of equations describing each particular topology.

The use of diakoptics, as proposed by Kron, considerably reduces the computational burden but is subject to some restrictions on the types of circuit topology that can be analysed. Those restrictions, the techniques used to overcome them and the computer implementation of the state variable method are considered in this chapter.

3.2 Choice of state variables

State variable (or state space) analysis represents the power system by a set of first order differential equations, which are then solved by numerical integration. Although the inductor current and capacitor voltage are the state variables normally chosen in textbooks, it is better to use the inductor's flux linkage (ϕ) and capacitor's charge (Q). Regardless of the type of numerical integration used, this variable selection reduces the propagation of local truncation errors [3]. Also any non-linearities present in the Q–V or ϕ–I characteristics can be modelled more easily.

The solution requires that the number of state variables must be equal to the number of independent energy-storage elements (i.e. independent inductors and capacitors). Therefore it is important to recognise when inductors and capacitors in a network are dependent or independent.

The use of capacitor charge or voltage as a state variable creates a problem when a set of capacitors and voltage sources forms a closed loop. In this case, the standard state variable formulation fails, as one of the chosen state variables is a linear combination of the others. This is a serious problem as many power system elements exhibit this property (e.g. the transmission line model). To overcome this problem the TCS (Transient Converter Simulation) program [4] uses the charge at a node rather than the capacitor's voltage as a state variable.

A dependent inductor is one with a current which is a linear combination of the current in k other inductors and current sources in the system. This is not always obvious due to the presence of the intervening network; an example of the difficulty is illustrated in Figure 3.1, where it is not immediately apparent that inductors 3, 4, 5, 6 and the current source form a cutset [5].

When only inductive branches and current sources are connected to a radial node, if the initialisation of state variables is such that the sum of the currents at this radial node was non-zero, then this error will remain throughout the simulation. The use of a phantom current source is one method developed to overcome the problem [6].

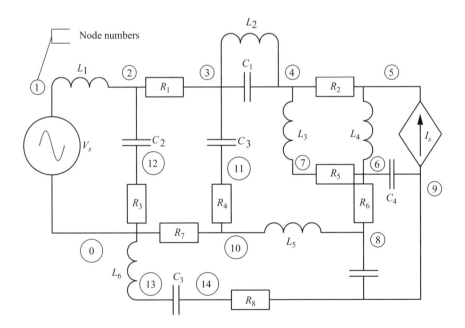

Figure 3.1   Non-trivial dependent state variables

Another approach is to choose an inductor at each node with only inductors connected to it, and make its flux a dependent rather than a state variable.

However, each method has some disadvantage. For instance the phantom current source can cause large voltage spikes when trying to compensate for the inaccurate initial condition. The partition of the inductor fluxes into state and dependent variables is complicated and time consuming. An inductor can still be dependent even if it is not directly connected to a radial node of inductive branches when there is an intervening resistor/capacitor network.

The identification of state variables can be achieved by developing a node–branch incidence matrix, where the branches are ordered in a particular pattern (e.g. current sources, inductors, voltage sources, capacitors, resistors) and Gaussian elimination performed. The resulting staircase columns represent state variables [3]. However, the computation required by this identification method has to be performed every time the system topology changes. It is therefore impractical when frequently switching power electronic components are present. One possible way to reduce the computation burden is to separate the system into constant and frequently switching parts, using voltage and current sources to interface the two [7].

Two state variable programs ATOSEC [8] and TCS (Transient Converter Simulator) [9], written in FORTRAN, have been used for system studies, the former for power electronic systems and the latter for power systems incorporating HVDC transmission. A toolkit for MATLAB using state variable techniques has also been developed.

3.3 Formation of the state equations

As already explained, the simplest method of formulating state equations is to accept all capacitor charges and inductor fluxes as state variables. Fictitious elements, such as the phantom current source and resistors are then added to overcome the dependency problem without affecting the final result significantly. However the elimination of the dependent variables is achieved more effectively with the transform and graph theory methods discussed in the sections that follow.

3.3.1 The transform method

A linear transformation can be used to reduce the number of state variables. The change from capacitor voltage to charge at the node, mentioned in section 3.2, falls within this category. Consider the simple loop of three capacitors shown in Figure 3.2, where the charge at the nodes will be defined, rather than the capacitor charge.

The use of a linear transformation changes the [C] matrix from a 3 × 3 matrix with only diagonal elements to a full 2 × 2 matrix. The branch–node incidence matrix, K_{bn}^t, is:

$$K_{bn}^t = \begin{bmatrix} 1 & 0 \\ 1 & -1 \\ 0 & 1 \end{bmatrix} \tag{3.1}$$

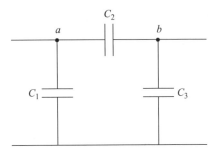

Figure 3.2 Capacitive loop

and the equation relating the three state variables to the capacitor voltages:

$$\begin{pmatrix} q_1 \\ q_2 \\ q_3 \end{pmatrix} = \begin{bmatrix} C_1 & 0 & 0 \\ 0 & C_2 & 0 \\ 0 & 0 & C_3 \end{bmatrix} \begin{pmatrix} v_1 \\ v_2 \\ v_3 \end{pmatrix} \quad (3.2)$$

Using the connection between node and capacitor charges (i.e. equation 3.1):

$$\begin{pmatrix} q_a \\ q_b \end{pmatrix} = \begin{bmatrix} 1 & 1 & 0 \\ 0 & -1 & 1 \end{bmatrix} \begin{pmatrix} q_1 \\ q_2 \\ q_3 \end{pmatrix} \quad (3.3)$$

and

$$\begin{pmatrix} v_1 \\ v_2 \\ v_3 \end{pmatrix} = \begin{bmatrix} 1 & 0 \\ 1 & -1 \\ 0 & 1 \end{bmatrix} \begin{pmatrix} v_a \\ v_b \end{pmatrix} \quad (3.4)$$

Substituting equations 3.2 and 3.4 in 3.3 yields:

$$\begin{pmatrix} q_a \\ q_b \end{pmatrix} = \begin{bmatrix} C_1 & C_2 & 0 \\ 0 & -C_2 & C_3 \end{bmatrix} \begin{bmatrix} 1 & 0 \\ 1 & -1 \\ 0 & 1 \end{bmatrix} \begin{pmatrix} v_a \\ v_b \end{pmatrix} = \begin{bmatrix} C_1+C_2 & -C_2 \\ -C_2 & C_2+C_3 \end{bmatrix} \begin{pmatrix} v_a \\ v_b \end{pmatrix}$$

(3.5)

Use of this transform produces a minimum set of state variables, and uses all the capacitor values at each iteration in the integration routine. However, there is a restriction on the system topology that can be analysed, namely all capacitor subnetworks must contain the reference node. For example, the circuit in Figure 3.3 (a) cannot be analysed, as this method defines two state variables and the [C] matrix is singular and cannot be inverted. i.e.

$$\begin{pmatrix} q_a \\ q_b \end{pmatrix} = \begin{bmatrix} C_1 & -C_1 \\ -C_1 & C_1 \end{bmatrix} \begin{pmatrix} v_a \\ v_b \end{pmatrix} \quad (3.6)$$

This problem can be corrected by adding a small capacitor, C_2, to the reference node (ground) as shown in Figure 3.3 (b). Thus the new matrix equation becomes:

$$\begin{pmatrix} q_a \\ q_b \end{pmatrix} = \begin{bmatrix} C_1+C_2 & -C_1 \\ -C_1 & C_1 \end{bmatrix} \begin{pmatrix} v_a \\ v_b \end{pmatrix} \quad (3.7)$$

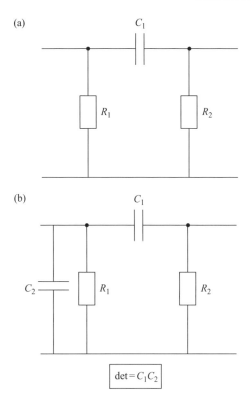

Figure 3.3 (a) *Capacitor with no connection to ground;* (b) *small capacitor added to give a connection to ground*

However this creates a new problem because C_2 needs to be very small so that it does not change the dynamics of the system, but this results in a small determinant for the $[C]$ matrix, which in turn requires a small time step for the integration routine to converge.

More generally, an initial state equation is of the form:

$$[M_{(0)}]\dot{\mathbf{x}}_{(0)} = [A_{(0)}]\mathbf{x}_{(0)} + [B_{(0)}]\mathbf{u} + [B_{1(0)}]\dot{\mathbf{u}} \quad (3.8)$$

where the vector $\mathbf{x}_{(0)}$ comprises all inductor fluxes and all capacitor charges. Equation 3.8 is then reduced to the normal form, i.e.

$$\dot{\mathbf{x}}_{(0)} = [A]\mathbf{x} + [B]\mathbf{u} + ([B_1]\dot{\mathbf{u}}\ldots) \quad (3.9)$$

by eliminating the dependent variables.

From equation 3.8 the augmented coefficient matrix becomes:

$$\begin{bmatrix} M_{(0)}, & A_{(0)}, & B_{(0)} \end{bmatrix} \quad (3.10)$$

Elementary row operations are performed on the augmented coefficient matrix to reduce it to echelon form [3]. If $M_{(0)}$ is non-singular the result will be an upper triangular matrix with non-zero diagonal elements. Further elementary row operations will reduce $M_{(0)}$ to the identity matrix. This is equivalent to pre-multiplying equation 3.10 by $M_{(0)}^{-1}$, i.e. reducing it to the form

$$[I, \quad A, \quad B] \tag{3.11}$$

If in the process of reducing to row echelon form the jth row in the first block becomes a row of all zeros then $M_{(0)}$ was singular. In this case three conditions can occur.

- The jth row in the other two submatrices are also zero, in which case the network has no unique solution as there are fewer constraint equations than unknowns.
- The jth row elements in the second submatrix (A) are zero, which gives an inconsistent network, as the derivatives of state variables relate only to input sources, which are supposed to be independent.
- The jth row elements in the second submatrix (originally $[A_{(0)}]$) are not zero (regardless of the third submatrix). Hence the condition is $[0, 0, \ldots, 0]\dot{\mathbf{x}} = [a_{j1}, a_{j2}, \ldots, a_{jn}]\mathbf{x} + [b_{j1}, b_{j2}, \ldots, b_{jm}]\mathbf{u}$. In this case there is at least one non-zero value a_{jk}, which allows state variable x_k to be eliminated. Rearranging the equation associated with the kth row of the augmented matrix 3.10 gives:

$$\begin{aligned} x_k = \frac{-1}{a_{jk}}(a_{j1}x_1 + a_{j2}x_2 + \cdots + a_{jk-1}x_{k-1} + a_{jk+1}x_{k+1} + \cdots + a_{jn}x_n \\ + b_{j1}u_1 + b_{j2}u_2 + \cdots + b_{jm}u_m) \end{aligned} \tag{3.12}$$

Substituting this for x_k in equation 3.8 and eliminating the equation associated with $\dot{x}_k$ yields:

$$[M_{(1)}]\dot{\mathbf{x}}_{(1)} = [A_{(1)}]\mathbf{x}_{(1)} + [B_{(1)}]\mathbf{u} + [B_{(1)}]\dot{\mathbf{u}} \tag{3.13}$$

This process is repeatedly applied until all variables are linearly independent and hence the normal form of state equation is achieved.

3.3.2 The graph method

This method solves the problem in two stages. In the first stage a tree, **T**, is found with a given preference to branch type and value for inclusion in the tree. The second stage forms the loop matrix associated with the chosen tree **T**.

The graph method determines the minimal and optimal state variables. This can be achieved either by:

(i) elementary row operations on the connection matrix, or
(ii) path search through a connection table.

The first approach consists of rearranging the rows of the incidence (or connection) matrix to correspond to the preference required, as shown in Figure 3.4. The dimension of the incidence matrix is $n \times b$, where n is the number of nodes (excluding the

State variable analysis   41

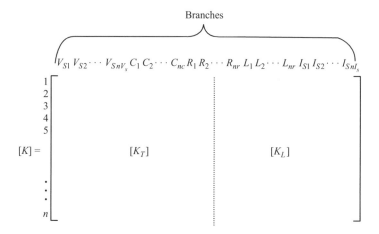

Figure 3.4 K matrix partition

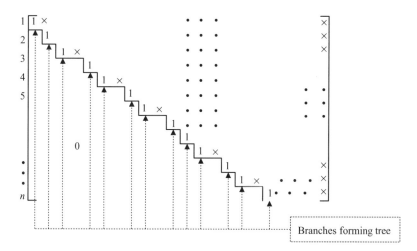

Figure 3.5 Row echelon form

reference) and b is the number of branches. The task is to choose n branches that correspond to linearly independent columns in $[K]$, to form the tree.

Since elementary row operations do not affect the linear dependence or independence of a set of columns, by reducing $[K]$ to echelon form through a series of elementary row operations the independent columns that are required to be part of the tree are easily found. The row echelon form is depicted in Figure 3.5. The branches above the step in the staircase (and immediately to the right of a vertical line) are linearly independent and form a tree. This method gives preference to branches to the left, therefore the closer to the left in the connection matrix the more likely a branch will be chosen as part of the tree. Since the ordering of the n branches in the connection matrix influences which branches become part of the tree, elements are grouped by type and within a type, by values, to obtain the best tree.

The net effect of identifying the dependent inductor fluxes and capacitor charges is to change the state variable equations to the form:

$$\dot{\mathbf{x}} = [A]\mathbf{x} + [B]\mathbf{u} + [E]\mathbf{z} \quad (3.14)$$

$$\mathbf{y} = [C]\mathbf{x} + [D]\mathbf{u} + [F]\mathbf{z} \quad (3.15)$$

$$\mathbf{z} = [G]\mathbf{x} + [H]\mathbf{u} \quad (3.16)$$

where

u is the vector of input voltages and currents
x is the vector of state variables
y is the vector of output voltages and currents
z is the vector of inductor fluxes (or currents) and capacitor charge (or voltages) that are not independent.

In equations 3.14–3.16 the matrices $[A]$, $[B]$, $[C]$, $[D]$, $[E]$, $[F]$, $[G]$ and $[H]$ are the appropriate coefficient matrices, which may be non-linear functions of x, y or z and/or time varying.

The attraction of the state variable approach is that non-linearities which are functions of time, voltage or current magnitude (i.e. most types of power system non-linearities) are easily handled. A non-linearity not easily simulated is frequency-dependence, as the time domain solution is effectively including all frequencies (up to the Nyquist frequency) every time a time step is taken. In graph terminology equation 3.14 can be restated as shown in Figure 3.6.

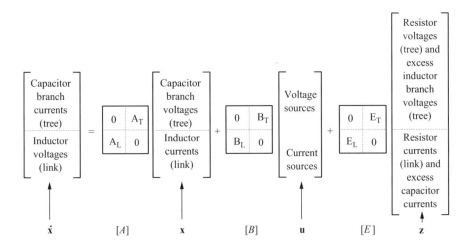

Figure 3.6 Modified state variable equations

3.4 Solution procedure

Figure 3.7 shows the structure of the state variable solution. Central to the solution procedure is the numerical integration technique. Among the possible alternatives,

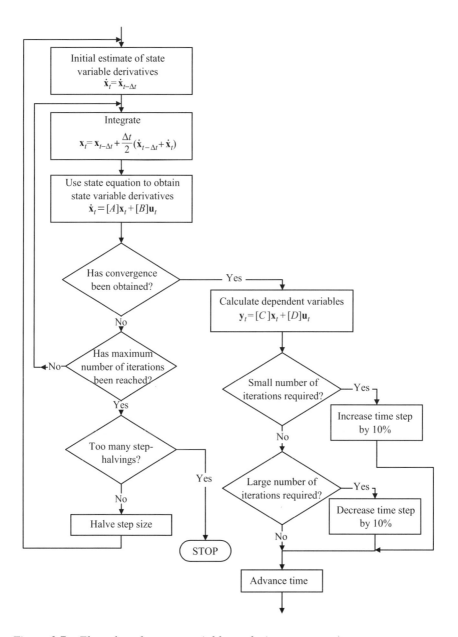

Figure 3.7 Flow chart for state variable analysis

the use of implicit trapezoidal integration has gained wide acceptance owing to its good stability, accuracy and simplicity [9], [10]. However, the calculation of the state variables at time t requires information on the state variable derivatives at that time.

As an initial guess the derivative at the previous time step is used, i.e.

$$\dot{x}_{n+1} = \dot{x}_n$$

An estimate of x_{n+1} based on the $\dot{x}_{n+1}$ estimate is then made, i.e.

$$x_{n+1} = x_n + \frac{\Delta t}{2}(\dot{x}_n + \dot{x}_{n+1})$$

Finally, the state variable derivative $\dot{x}_{n+1}$ is estimated from the state equation, i.e.

$$\dot{x}_{n+1} = f(t + \Delta t, x_{n+1}, u_{n+1})$$

$$\dot{\mathbf{x}}_{n+1} = [A]\mathbf{x} + [B]\mathbf{u}$$

The last two steps are performed iteratively until convergence has been reached. The convergence criterion will normally include the state variables and their derivatives.

Usually, three to four iterations will be sufficient, with a suitable step length. An optimisation technique can be included to modify the nominal step length. The number of iterations are examined and the step size increased or decreased by 10 per cent, based on whether that number is too small or too large. If convergence fails, the step length is halved and the iterative procedure is restarted. Once convergence is reached, the dependent variables are calculated.

The elements of matrices $[A]$, $[B]$, $[C]$ and $[D]$ in Figure 3.7 are dependent on the values of the network components R, L and C, but not on the step length. Therefore there is no overhead in altering the step. This is an important property for the modelling of power electronic equipment, as it allows the step length to be varied to coincide with the switching instants of the converter valves, thereby eliminating the problem of numerical oscillations due to switching errors.

3.5 Transient converter simulation (TCS)

A state space transient simulation algorithm, specifically designed for a.c.–d.c. systems, is TCS [4]. The a.c. system is represented by an equivalent circuit, the parameters of which can be time and frequency dependent. The time variation may be due to generator dynamics following disturbances or to component non-linear characteristics, such as transformer magnetisation saturation.

A simple a.c. system equivalent shown in Figure 3.8 was proposed for use with d.c. simulators [11]; it is based on the system short-circuit impedance, and the values of R and L selected to give the required impedance angle. A similar circuit is used as a default equivalent in the TCS program.

Of course this approach is only realistic for the fundamental frequency. Normally in HVDC simulation only the impedances at low frequencies (up to the fifth harmonic)

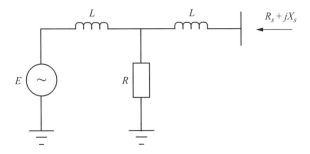

Figure 3.8 Tee equivalent circuit

are of importance, because the harmonic filters swamp the a.c. impedance at high frequencies. However, for greater accuracy, the frequency-dependent equivalents developed in Chapter 10 may be used.

3.5.1 Per unit system

In the analysis of power systems, per unit quantities, rather than actual values are normally used. This scales voltages, currents and impedances to the same relative order, thus treating each to the same degree of accuracy.

In dynamic analysis the instantaneous phase quantities and their derivatives are evaluated. When the variables change relatively rapidly large differences will occur between the order of a variable and its derivative. For example consider a sinusoidal function:

$$x = A \cdot \sin(\omega t + \phi) \tag{3.17}$$

and its derivative

$$\dot{x} = \omega A \cdot \cos(\omega t + \phi) \tag{3.18}$$

The relative difference in magnitude between x and $\dot{x}$ is ω, which may be high. Therefore a base frequency ω_0 is defined. All state variables are changed by this factor and this then necessitates the use of reactance and susceptance matrices rather than inductance and capacitance matrices,

$$\Psi_k = \omega_0 \psi_k = (\omega_0 l_k) \cdot I_k = L_k \cdot I_k \tag{3.19}$$

$$Q_k = \omega_0 q_k = (\omega_0 c_k) \cdot V_k = C_k \cdot I_k \tag{3.20}$$

where
  l_k is the inductance
  L_k the inductive reactance
  c_k is the capacitance
  C_k the capacitive susceptance
  ω_0 the base angular frequency.

The integration is now performed with respect to electrical angle rather than time.

3.5.2 Network equations

The nodes are partitioned into three possible groups depending on what type of branches are connected to them. The nodes types are:

α nodes: Nodes that have at least one capacitive branch connected
β nodes: Nodes that have at least one resistive branch connected but no capacitive branch
γ nodes: Nodes that have only inductive branches connected.

The resulting branch–node incidence (connection) matrices for the r, l and c branches are K^t_{rn}, K^t_{ln} and K^t_{cn} respectively. The elements in the branch–node incidence matrices are determined by:

$$K^t_{bn} = \begin{cases} 1 & \text{if node } n \text{ is the sending end of branch } b \\ -1 & \text{if node } n \text{ is the receiving end of branch } b \\ 0 & \text{if } b \text{ is not connected to node } n \end{cases}$$

Partitioning these branch–node incidence matrices on the basis of the above node types yields:

$$K^t_{ln} = \begin{bmatrix} K^t_{l\alpha} & K^t_{l\beta} & K^t_{l\gamma} \end{bmatrix} \quad (3.21)$$

$$K^t_{rn} = \begin{bmatrix} K^t_{r\alpha} & K^t_{r\beta} & 0 \end{bmatrix} \quad (3.22)$$

$$K^t_{cn} = \begin{bmatrix} K^t_{c\alpha} & 0 & 0 \end{bmatrix} \quad (3.23)$$

$$K^t_{sn} = \begin{bmatrix} K^t_{s\alpha} & K^t_{s\beta} & K^t_{s\gamma} \end{bmatrix} \quad (3.24)$$

The efficiency of the solution can be improved significantly by restricting the number of possible network configurations to those normally encountered in practice. The restrictions are:

(i) capacitive branches have no series voltage sources
(ii) resistive branches have no series voltage sources
(iii) capacitive branches are constant valued ($dC_c/dt = 0$)
(iv) every capacitive branch subnetwork has at least one connection to the system reference (ground node)
(v) resistive branch subnetworks have at least one connection to either the system reference or an α node.
(vi) inductive branch subnetworks have at least one connection to the system reference or an α or β node.

The fundamental branches that result from these restrictions are shown in Figure 3.9. Although the equations that follow are correct as they stand, with L and C being the inductive and capacitive matrices respectively, the TCS implementation uses instead the inductive reactance and capacitive susceptance matrices. As mentioned in the per unit section, this implies that the p operator (representing differentiation) relates to

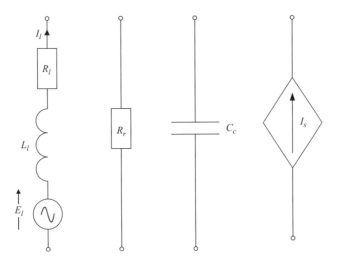

Figure 3.9 TCS branch types

electrical angle rather than time. Thus the following equations can be written:

Resistive branches
$$I_r = R_r^{-1}\left(K_{r\alpha}^t V_\alpha + K_{r\beta}^t V_\beta\right) \tag{3.25}$$

Inductive branches
$$E_l - p(L_l \cdot I_l) - R_l \cdot I_l + K_{l\alpha}^t V_\alpha + K_{l\beta}^t V_\beta + K_{l\gamma}^t V_\gamma = 0 \tag{3.26}$$

or
$$p\Psi_l = E_l - p(L_l) \cdot I_l - R_l \cdot I_l + K_{l\alpha}^t V_\alpha + K_{l\beta}^t V_\beta + K_{l\gamma}^t V_\gamma \tag{3.27}$$

where $\Psi_l = L_l \cdot I_l$.

Capacitive branches
$$I_c = \bar{C}_c p\left(K_{c\alpha}^t V_\alpha\right) \tag{3.28}$$

In deriving the nodal analysis technique Kirchhoff's current law is applied, the resulting nodal equation being:

$$K_{nc} I_c + K_{nr} I_r + K_{nl} I_l + K_{ns} I_s = 0 \tag{3.29}$$

where I are the branch current vectors and I_s the current sources. Applying the node type definitions gives rise to the following equations:

$$K_{\gamma l} I_l + K_{\gamma s} I_s = 0 \tag{3.30}$$

or taking the differential of each side:

$$K_{\gamma l} p(I_l) + K_{\gamma s} p(I_s) = 0 \tag{3.31}$$

$$K_{\beta r} I_r + K_{\beta l} I_l + K_{\beta s} I_s = 0 \tag{3.32}$$

$$K_{\alpha c}I_c + K_{\alpha r}I_r + K_{\alpha l}I_l + K_{\alpha s}I_s = 0 \qquad (3.33)$$

Pre-multiplying equation 3.28 by $K_{\alpha c}^t$ and substituting into equation 3.33 yields:

$$p(Q_\alpha) = -K_{\alpha l}I_l + K_{\alpha r}I_r - K_{\alpha r}I_r - K_{\alpha s}I_s \qquad (3.34)$$

where
$$Q_\alpha = C_\alpha V_\alpha$$
$$C_\alpha^{-1} = \left(K_{\alpha c}C_c K_{c\alpha}^t\right)^{-1}.$$

The dependent variables V_β, V_γ and I_r can be entirely eliminated from the solution so only I_l, V_α and the input variables are explicit in the equations to be integrated. This however is undesirable due to the resulting loss in computational efficiency even though it reduces the overall number of equations. The reasons for the increased computational burden are:

- loss of matrix sparsity
- incidence matrices no longer have values of -1, 0 or 1. This therefore requires actual multiplications rather than simple additions or subtractions when calculating a matrix product.
- Some quantities are not directly available, making it time-consuming to recalculate if it is needed at each time step.

Therefore V_β, V_γ and I_r are retained and extra equations derived to evaluate these dependent variables. To evaluate V_β equation 3.25 is pre-multiplied by $K_{\beta r}$ and then combined with equation 3.32 to give:

$$V_\beta = -R_\beta \left(K_{\beta s}I_s + K_{\beta l}I_l + K_{\beta r}R_r^{-1}K_{r\alpha}^t V_\alpha\right) \qquad (3.35)$$

where $R_\beta = (K_{\beta r}R_r^{-1}K_{r\beta}^t)^{-1}$.

Pre-multiplying equation 3.26 by $K_{\gamma l}$ and applying to equation 3.31 gives the following expression for V_γ:

$$V_\gamma = -L_\gamma K_{\gamma s}p(I_s) - L_\gamma K_{\gamma l}L_l^{-1}\left(E_l - p(L_l)\cdot I_l - R_l I_l + K_{l\alpha}^t V_\alpha + K_{l\beta}^t V_\beta\right) \qquad (3.36)$$

where $L_\gamma = (K_{\gamma l}L_l^{-1}K_{l\gamma}^t)^{-1}$ and I_r is evaluated by using equation 3.25.

Once the trapezoidal integration has converged the sequence of solutions for a time step is as follows: the state related variables are calculated followed by the dependent variables and lastly the state variable derivatives are obtained from the state equation. State related variables:

$$I_l = L_l^{-1}\Psi_l \qquad (3.37)$$

$$V_\alpha = C_\alpha^{-1}Q_\alpha \qquad (3.38)$$

Dependent variables:

$$V_\beta = -R_\beta \left(K_{\beta s} I_s + K_{\beta l} I_l + K_{\beta r} R_r^{-1} K_{r\alpha}^t V_\alpha \right) \qquad (3.39)$$

$$I_r = R_r^{-1} \left(K_{r\alpha}^t V_\alpha + K_{r\beta}^t V_\beta \right) \qquad (3.40)$$

$$V_\gamma = -L_\gamma K_{\gamma s} p(I_s) - L_\gamma K_{\gamma l} L_l^{-1} \left(E_l - p(L_l) \cdot I_l - R_l I_l + K_{l\alpha}^t V_\alpha + K_{l\beta}^t V_\beta \right) \qquad (3.41)$$

State equations:

$$p\Psi_l = E_l - p(L_l) \cdot I_l - R_l \cdot I_l + K_{l\alpha}^t V_\alpha + K_{l\beta}^t V_\beta + K_{l\gamma}^t V_\gamma \qquad (3.42)$$

$$p(Q_\alpha) = -K_{\alpha l} I_l + K_{\alpha r} I_r - K_{\alpha r} I_r - K_{\alpha s} I_s \qquad (3.43)$$

where
$$C_\alpha^{-1} = (K_{\alpha c} C_c K_{c\alpha}^t)^{-1}$$
$$L_\gamma = (K_{\gamma l} L_l^{-1} K_{l\gamma}^t)^{-1}$$
$$R_\beta = (K_{\beta r} R_r^{-1} K_{r\beta}^t)^{-1}.$$

3.5.3 Structure of TCS

To reduce the data input burden TCS suggests an automatic procedure, whereby the collation of the data into the full network is left to the computer. A set of control parameters provides all the information needed by the program to expand a given component data and to convert it to the required form. The component data set contains the initial current information and other parameters relevant to the particular component.

For example, for the converter bridges this includes the initial d.c. current, the delay and extinction angles, time constants for the firing control system, the smoothing reactor, converter transformer data, etc. Each component is then systematically expanded into its elementary *RLC* branches and assigned appropriate node numbers. Cross-referencing information is created relating the system busbars to those node numbers. The node voltages and branch currents are initialised to their specific instantaneous phase quantities of busbar voltages and line currents respectively. If the component is a converter, the bridge valves are set to their conducting states from knowledge of the a.c. busbar voltages, the type of converter transformer connection and the set initial delay angle.

The procedure described above, when repeated for all components, generates the system matrices in compact form with their indexing information, assigns node numbers for branch lists and initialises relevant variables in the system.

Once the system and controller data are assembled, the system is ready to begin execution. In the data file, the excitation sources and control constraints are entered followed by the fault specifications. The basic program flow chart is shown in Figure 3.10. For a simulation run, the input could be either from the data file or from a previous snapshot (stored at the end of a run).

50 *Power systems electromagnetic transients simulation*

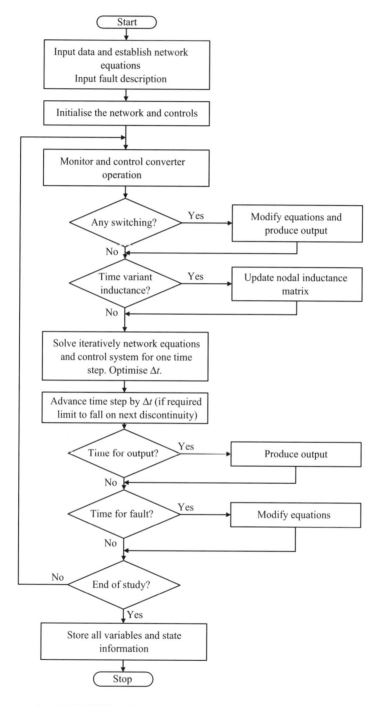

Figure 3.10   *TCS flow chart*

State variable analysis   51

Simple control systems can be modelled by sequentially assembling the modular building blocks available. Control block primitives are provided for basic arithmetic such as addition, multiplication and division, an integrator, a differentiator, pole–zero blocks, limiters, etc. The responsibility to build a useful continuous control system is obviously left to the user.

At each stage of the integration process, the converter bridge valves are tested for extinction, voltage crossover and conditions for firing. If indicated, changes in the valve states are made and the control system is activated to adjust the phase of firing. Moreover, when a valve switching occurs, the network equations and the connection matrix are modified to represent the new conditions.

During each conduction interval the circuit is solved by numerical integration of the state space model for the appropriate topology, as described in section 3.4.

3.5.4 Valve switchings

The step length is modified to fall exactly on the time required for turning ON switches. As some events, such as switching of diodes and thyristors, cannot be predicted the solution is interpolated back to the zero crossing. At each switching instance two solutions are obtained one immediately before and the other immediately after the switch changes state.

Hence, the procedure is to evaluate the system immediately prior to switching by restricting the time step or interpolating back. The connection matrices are modified to reflect the switch changing state, and the system resolved for the same time point using the output equation. The state variables are unchanged, as inductor flux (or current) and capacitor charge (or voltage) cannot change instantaneously. Inductor voltage and capacitor current can exhibit abrupt changes due to switching.

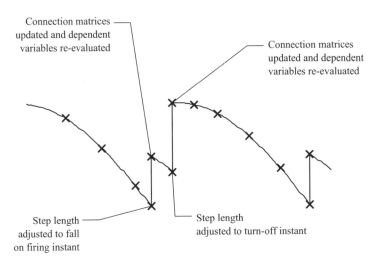

Figure 3.11 Switching in state variable program

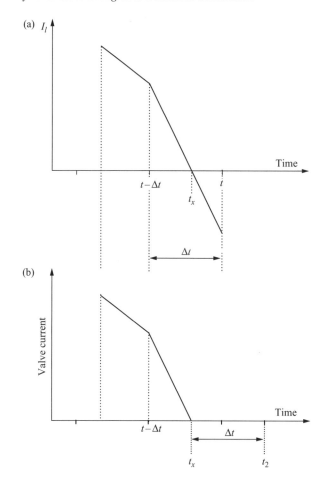

Figure 3.12 Interpolation of time upon valve current reversal

The time points produced are at irregular intervals with almost every consecutive time step being different. Furthermore, two solutions for the same time points do exist (as indicated in Figure 3.11). The irregular intervals complicate the post-processing of waveforms when an FFT is used to obtain a spectrum, and thus resampling and windowing is required. Actually, even with the regularly spaced time points produced by EMTP-type programs it is sometimes necessary to resample and use a windowed FFT. For example, simulating with a 50 μs time step a 60 Hz system causes errors because the period of the fundamental is not an integral multiple of the time step. (This effect produces a fictitious 2nd harmonic in the test system of ref [12].)

When a converter valve satisfies the conditions for conduction, i.e. the simultaneous presence of a sufficient forward voltage and a firing-gate pulse, it will be switched to the conduction state. If the valve forward voltage criterion is not satisfied the pulse is retained for a set period without upsetting the following valve.

Accurate prediction of valve extinctions is a difficult and time-consuming task which can degrade the solution efficiency. Sufficient accuracy is achieved by detecting extinctions after they have occurred, as indicated by valve current reversal; by linearly interpolating the step length to the instant of current zero, the actual turn-off instant is assessed as shown in Figure 3.12. Only one valve per bridge may be extinguished at any one time, and the earliest extinction over all the bridges is always chosen for the interpolation process. By defining the current (I) in the outgoing valve at the time of detection (t), when the step length of the previous integration step was Δt, the instant of extinction t_x will be given by:

$$t_x = t - \Delta t z \tag{3.44}$$

where

$$z = \frac{i_t}{i_t - i_{t-\Delta t}}$$

All the state variables are then interpolated back to t_x by

$$v_x = v_t = z(v_t - v_{t-\Delta t}) \tag{3.45}$$

The dependent state variables are then calculated at t_x from the state variables, and written to the output file. The next integration step will then begin at t_x with step length Δt as shown in Figure 3.12. This linear approximation is sufficiently accurate over periods which are generally less than one degree, and is computationally inexpensive. The effect of this interpolation process is clearly demonstrated in a case with an extended 1 ms time step in Figure 3.14 on page 55.

Upon switching any of the valves, a change in the topology has to be reflected back into the main system network. This is achieved by modifying the connection matrices. When the time to next firing is less than the integration step length, the integration time step is reduced to the next closest firing instant. Since it is not possible to integrate through discontinuities, the integration time must coincide with their occurrence. These discontinuities must be detected accurately since they cause abrupt changes in bridge-node voltages, and any errors in the instant of the topological changes will cause inexact solutions.

Immediately following the switching, after the system matrices have been reformed for the new topology, all variables are again written to the output file for time t_x. The output file therefore contains two sets of values for t_x, immediately preceding and after the switching instant. The double solution at the switching time assists in forming accurate waveshapes. This is specially the case for the d.c. side voltage, which almost contains vertical jump discontinuities at switching instants.

3.5.5 Effect of automatic time step adjustments

It is important that the switching instants be identified correctly, first for accurate simulations and, second, to avoid any numerical problems associated with such errors. This is a property of the algorithm rather than an inherent feature of the basic formulation. Accurate converter simulation requires the use of a very small time step, where the accuracy is only achieved by correctly reproducing the appropriate discontinuities. A smaller step length is not only needed for accurate switching but also for the simulation of other non-linearities, such as in the case of transformer saturation, around the knee point, to avoid introducing hysteresis due to overstepping. In the saturated region and the linear regions, a larger step is acceptable.

On the other hand, state variable programs, and TCS in particular, have the facility to adapt to a variable step length operation. The dynamic location of a discontinuity will force the step length to change between the maximum and minimum step sizes. The automatic step length adjustment built into the TCS program takes into account most of the influencing factors for correct performance. As well as reducing the step length upon the detection of a discontinuity, TCS also reduces the forthcoming step in anticipation of events such as an incoming switch as decided by the firing controller, the time for fault application, closing of a circuit breaker, etc.

To highlight the performance of the TCS program in this respect, a comparison is made with an example quoted as a feature of the NETOMAC program [13]. The example refers to a test system consisting of an ideal 60 Hz a.c. system (EMF sources) feeding a six-pulse bridge converter (including the converter transformer and smoothing reactor) terminated by a d.c. source; the firing angle is 25 degrees. Figure 3.13 shows the valve voltages and currents for 50 μs and 1 ms (i.e. 1 and 21 degrees) time steps respectively. The system has achieved steady state even with steps 20 times larger.

The progressive time steps are illustrated by the dots on the curves in Figure 3.13(b), where interpolation to the instant of a valve current reversal is made and from which a half time step integration is carried out. The next step reverts back to the standard trapezoidal integration until another discontinuity is encountered.

A similar case with an ideal a.c. system terminated with a d.c. source was simulated using TCS. A maximum time step of 1 ms was used also in this case. Steady state waveforms of valve voltage and current derived with a 1 ms time step, shown in Figure 3.14, illustrate the high accuracy of TCS, both in detection of the switching

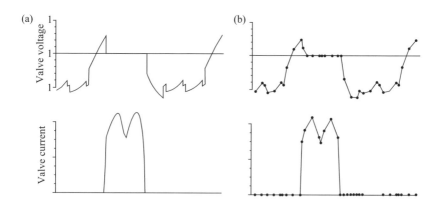

Figure 3.13   *NETOMAC simulation responses: (a) 50 μs time step; (b) 1 μs time step*

State variable analysis 55

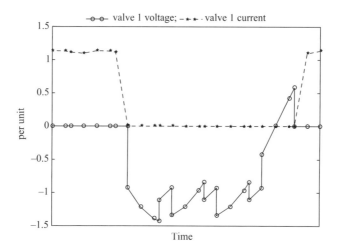

Figure 3.14   TCS simulation with 1 ms time step

discontinuities and the reproduction of the 50 μs results. The time step tracing points are indicated by dots on the waveforms.

Further TCS waveforms are shown in Figure 3.15 giving the d.c. voltage, valve voltage and valve current at 50 μs and 1 ms.

In the NETOMAC case, extra interpolation steps are included for the 12 switchings per cycle in the six pulse bridge. For the 60 Hz system simulated with a 1 ms time step, a total of 24 steps per cycle can be seen in the waveforms of Figure 3.13(b), where a minimum of 16 steps are required. The TCS cases shown in Figure 3.15 have been simulated with a 50 Hz system. The 50 μs case of Figure 3.15(a) has an average of 573 steps per cycle with the minimum requirement of 400 steps. On the other hand, the 1 ms time step needed only an average of 25 steps per cycle. The necessary sharp changes in waveshape are derived directly from the valve voltages upon topological changes.

When the TCS frequency was increased to 60 Hz, the 50 μs case used fewer steps per cycle, as would be expected, resulting in 418 steps compared to a minimum required of 333 steps per cycle. For the 1 ms case, an average of 24 steps were required, as for the NETOMAC case.

The same system was run with a constant current control of 1.225 p.u., and after 0.5 s a d.c. short-circuit was applied. The simulation results with 50 μs and 1 ms step lengths are shown in Figure 3.16. This indicates the ability of TCS to track the solution and treat waveforms accurately during transient operations (even with such an unusually large time step).

3.5.6   TCS converter control

A modular control system is used, based on Ainsworth's [14] phase-locked oscillator (PLO), which includes blocks of logic, arithmetic and transfer functions [15]. Valve firing and switchings are handled individually on each six-pulse unit. For twelve-pulse

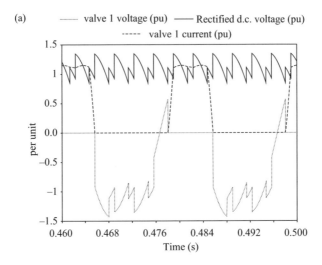

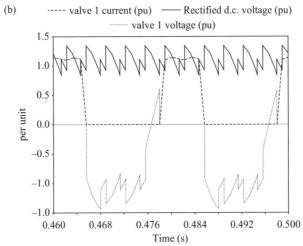

Figure 3.15   Steady state responses from TCS: (a) 50 μs time step; (b) 1 ms time step

units both bridges are synchronised and the firing controllers phase-locked loop is updated every 30 degrees instead of the 60 degrees used for the six-pulse converter.

The firing control mechanism is equally applicable to six or twelve-pulse valve groups; in both cases the reference voltages are obtained from the converter commutating bus voltages. When directly referencing to the commutating bus voltages any distortion in that voltage may result in a valve firing instability. To avoid this problem, a three-phase PLO is used instead, which attempts to synchronise the oscillator through a phase-locked loop with the commutating busbar voltages.

In the simplified diagram of the control system illustrated in Figure 3.17, the firing controller block (NPLO) consists of the following functional units:

State variable analysis 57

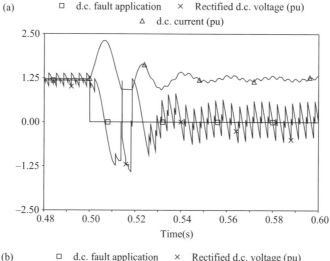

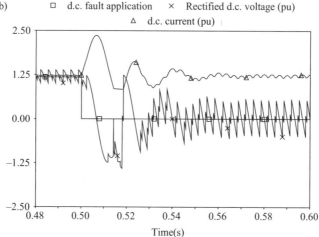

Figure 3.16 Transient simulation with TCS for a d.c. short-circuit at 0.5 s: (a) 1 ms time step; (b) 50 µs time step

(i) a zero-crossing detector
(ii) a.c. system frequency measurement
(iii) a phase-locked oscillator
(iv) firing pulse generator and synchronising mechanism
(v) firing angle (α) and extinction angle (γ) measurement unit.

Zero-crossover points are detected by the change of sign of the reference voltages and multiple crossings are avoided by allowing a space between the crossings. Distortion in the line voltage can create difficulties in zero-crossing detection, and therefore the voltages are smoothed before being passed to the zero-crossing detector.

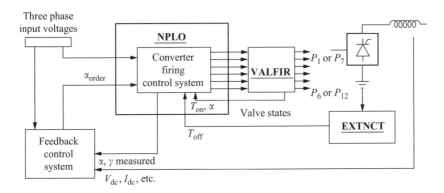

Figure 3.17   Firing control mechanism based on the phase-locked oscillator

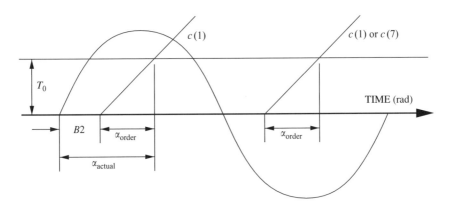

Figure 3.18   Synchronising error in firing pulse

The time between two consecutive zero crossings, of the positive to negative (or negative to positive) going waveforms of the same phase, is defined here as the half-period time, $T/2$. The measured periods are smoothed through a first order real-pole lag function with a user-specified time constant. From these half-period times the a.c. system frequency is estimated every 60 degrees (30 degrees) for a six (12) pulse bridge.

Normally the ramp for the firing of a particular valve ($c(1), \ldots, c(6)$) starts from the zero-crossing points of the voltage waveforms across the valve. After $T/6$ time ($T/12$ for twelve pulse), the next ramp starts for the firing of the following valve in sequence.

It is possible that during a fault or due to the presence of harmonics in the voltage waveform, the firing does not start from the zero-crossover point, resulting in a synchronisation error, B2, as shown in Figure 3.18. This error is used to update the phase-locked oscillator which, in turn, reduces the synchronising error, approaching zero at the steady state condition. The synchronisation error is recalculated every 60 deg for the six-pulse bridge.

The firing angle order (α_{order}) is converted to a level to detect the firing instant as a function of the measured a.c. frequency by

$$T_0 = \frac{\alpha_{\text{order}}(\text{rad.})}{f_{\text{ac}}(\text{p.u.})} \qquad (3.46)$$

As soon as the ramp $c(n)$ reaches the set level specified by T_0, as shown in Figure 3.18, valve n is fired and the firing pulse is maintained for 120 degrees. Upon having sufficient forward voltage with the firing-pulse enabled, the valve is switched on and the firing angle recorded as the time interval from the last voltage zero crossing detected for this valve.

At the beginning of each time-step, the valves are checked for possible extinctions. Upon detecting a current reversal, a valve is extinguished and its extinction angle counter is reset. Subsequently, from the corresponding zero-crossing instant, its extinction angle is measured, e.g. at valve 1 zero crossing, γ_2 is measured, and so on. (Usually, the lowest gamma angle measured for the converter is fed back to the extinction angle controller.) If the voltage zero-crossover points do not fall on the time step boundaries, a linear interpolation is used to derive them. As illustrated in Figure 3.17, the NPLO block coordinates the valve-firing mechanism, and VALFIR receives the firing pulses from NPLO and checks the conditions for firing the valves. If the conditions are met, VALFIR switches on the next incoming valve and measures the firing angle, otherwise it calculates the earliest time for next firing to adjust the step length. Valve currents are checked for extinction in EXTNCT and interpolation of all state variables is carried out. The valve's turn-on time is used to calculate the firing angle and the off time is used for the extinction angle.

By way of example, Figure 3.19 shows the response to a step change of d.c. current in the test system used earlier in this section.

3.6 Example

To illustrate the use of state variable analysis the simple *RLC* circuit of Figure 3.20 is used ($R = 20.0\,\Omega$, $L = 6.95\,\text{mH}$ and $C = 1.0\,\mu\text{F}$), where the switch is closed at 0.1 ms. Choosing $x_1 = v_C$ and $x_2 = i_L$ then the state variable equation is:

$$\begin{pmatrix} \dot{x}_1 \\ \dot{x}_2 \end{pmatrix} = \begin{bmatrix} 0 & \dfrac{1}{C} \\ \dfrac{-1}{L} & \dfrac{-R}{L} \end{bmatrix} \begin{pmatrix} x_1 \\ x_2 \end{pmatrix} + \begin{pmatrix} 0 \\ \dfrac{1}{L} \end{pmatrix} E_S \qquad (3.47)$$

The FORTRAN code for this example is given in Appendix G.1. Figure 3.21 displays the response from straight application of the state variable analysis using a 0.05 ms time step. The first plot compares the response with the analytic answer. The resonant frequency for this circuit is 1909.1 Hz (or a period of 0.5238 ms), hence having approximately 10 points per cycle. The second plot shows that the step length remained at

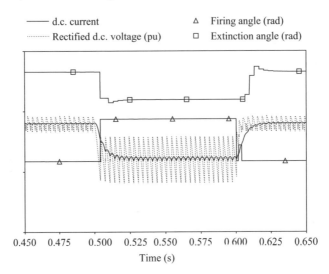

Figure 3.19   Constant $\alpha_{order}(15°)$ operation with a step change in the d.c. current

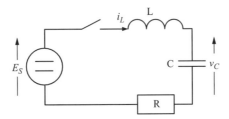

Figure 3.20   RLC test circuit

0.05 ms throughout the simulation and the third graph shows that 20–24 iterations were required to reach convergence. This is the worse case as increasing the nominal step length to 0.06 or 0.075 ms reduces the error as the algorithm is forced to step-halve (see Table 3.1). Figure 3.22 shows the resultant voltages and current in the circuit.

Adding a check on the state variable derivative substantially improves the agreement between the analytic and calculated responses so that there is no noticeable difference. Figure 3.23 also shows that the algorithm required the step length to be 0.025 in order to reach convergence of state variables and their derivatives.

Adding step length optimisation to the basic algorithm also improves the accuracy, as shown in Figure 3.24. Before the switch is closed the algorithm converges within one iteration and hence the optimisation routine increases the step length. As a result the first step after the switch closes requires more than 20 iterations and the optimisation routine starts reducing the step length until it reaches 0.0263 ms where it stays for the remainder of the simulation.

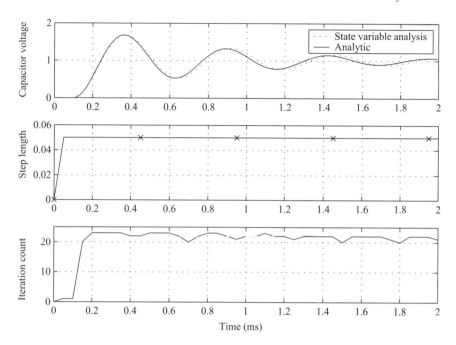

Figure 3.21 State variable analysis with 50 μs step length

Table 3.1 State variable analysis error

Condition	Maximum error (Volts)	Time (ms)
Base case	0.0911	0.750
$\dot{x}_{check}$	0.0229	0.750
Optimised Δt	0.0499	0.470
Both Opt. Δt and $\dot{x}_{check}$	0.0114	0.110
$\Delta t = 0.01$	0.0037	0.740
$\Delta t = 0.025$	0.0229	0.750
$\Delta t = 0.06$	0.0589	0.073
$\Delta t = 0.075$	0.0512	0.740
$\Delta t = 0.1$	0.0911	0.750

Combining both derivative of state variable checking and step length optimisation gives even better accuracy. Figure 3.25 shows that initially step-halving occurs when the switching occurs and then the optimisation routine takes over until the best step length is found.

A comparison of the error is displayed in Figure 3.26. Due to the uneven distribution of state variable time points, resampling was used to generate this comparison,

62 *Power systems electromagnetic transients simulation*

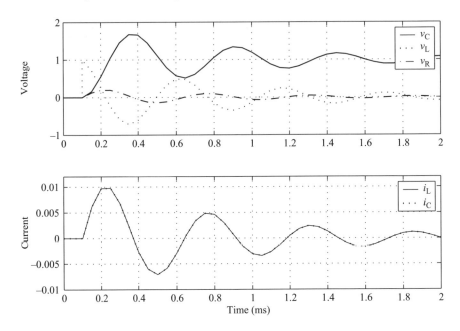

Figure 3.22 State variable analysis with 50 μs step length

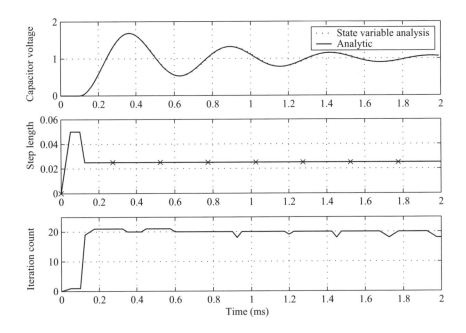

Figure 3.23 State variable analysis with 50 μs step length and $\dot{x}$ check

State variable analysis 63

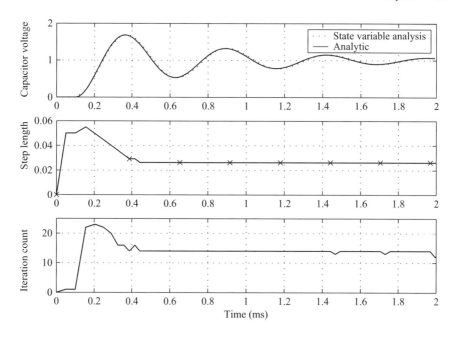

Figure 3.24 State variable with 50 µs step length and step length optimisation

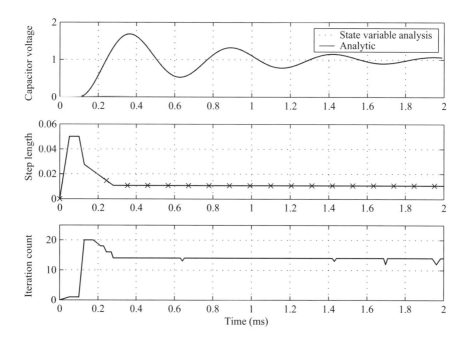

Figure 3.25 Both $\dot{x}$ check and step length optimisation

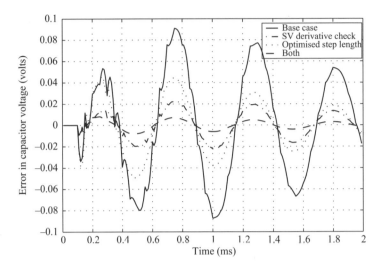

Figure 3.26 Error comparison

that is, the analytic solutions at 0.01 ms intervals were calculated and the state variable analysis results were interpolated on to this time grid, and the difference taken.

3.7 Summary

In the state variable solution it is the set of first order differential equations, rather than the system of individual elements, that is solved by numerical integration. The most popular numerical technique in current use is implicit trapezoidal integration, due to its simplicity, accuracy and stability. Solution accuracy is enhanced by the use of iterative methods to calculate the state variables.

State variable is an ideal method for the solution of system components with time-varying non-linearities, and particularly for power electronic devices involving frequent switching. This has been demonstrated with reference to the static a.c.–d.c. converter by an algorithm referred to as TCS (Transient Converter Simulation). Frequent switching, in the state variable approach, imposes no overhead on the solution. Moreover, the use of automatic step length adjustment permits optimising the integration step throughout the solution.

The main limitation is the need to recognise dependability between system variables. This process substantially reduces the effectiveness of the state variable algorithms, and makes them unsuited to very large systems. However, in a hybrid combination with the numerical integration substitution method, the state variable model can provide very accurate and efficient solutions. This subject is discussed in greater detail in Chapter 9.

3.8 References

1. HAY, J. L. and HINGORANI, N. G.: 'Dynamic simulation of multi-convertor HVdc systems by digital computer', *Proceedings of 6th PICA conference*, 1969, pp. 512–35
2. KRON, G.: 'Diakoptics – the piecewise solution' (MacDonald, London, 1963)
3. CHUA, L. O. and LIN, P. M.: 'Computer aided analysis of electronic circuits: algorithms and computational techniques' (Prentice Hall, Englewood Cliffs, CA, 1975)
4. ARRILLAGA, J., AL-KASHALI, H. J. and CAMPOS-BARROS, J. G.: 'General formulation for dynamic studies in power systems including static converters', *Proceedings of IEE*, 1977, **124** (11), pp. 1047–52
5. ROHRER, R. A.: 'Circuit theory, introduction to the state variable approach' (McGraw-Hill, Kogakusha, Tokyo, 1970)
6. JOOSTEN, A. P. B., ARRILLAGA, J., ARNOLD, C. P. and WATSON, N. R.: 'Simulation of HVdc system disturbances with reference to the magnetising history of the convertor transformer', *IEEE Transactions on Power Delivery*, 1990, **5** (1), pp. 330–6
7. KITCHEN, R. H.: 'New method for digital-computer evaluation of convertor harmonics in power systems using state-variable analysis', *Proceedings of IEE, Part C*, 1981, **128** (4), 196–207
8. RAJAGOPALAN, V.: 'Computer-aided analysis of power electronic system' (Marcel Dekker, New York, 1987)
9. ARRILLAGA, J., ARNOLD, C. P. and HARKER, B. J.: 'Computer modelling of electrical power systems' (John Wiley, Chicester, 1983)
10. GEAR, C. W.: 'Numerical initial value problems in ordinary differential equations' (Prentice Hall, Englewood Cliffs, 1971)
11. BOWLES, J. P.: 'AC system and transformer representation for HV-DC transmission studies', *IEEE Transactions on Power Apparatus and Systems*, 1970, **89** (7), pp. 1603–9
12. IEEE Task Force of Harmonics Modeling and Simulation: 'Test systems for harmonic modeling and simulation', *IEEE Transactions on Power Delivery*, 1999, **4** (2), pp. 579–87
13. KRUGER, K. H. and LASSETER, R. H.: 'HVDC simulation using NETOMAC', Proceedings, IEEE Montec '86 Conference on *HVDC Power Transmission*, Sept/Oct 1986, pp. 47–50
14. AINSWORTH, J. D.: 'The phase-locked oscillator – a new control system for controlled static convertors', *IEEE Transactions on Power Apparatus and Systems*, 1968, **87** (3), pp. 859–65
15. ARRILLAGA, J., SANKAR, S., ARNOLD, C. P. and WATSON, N. R.: 'Incorporation of HVdc controller dynamics in transient convertor simulation', *Trans. Inst. Prof. Eng. N.Z. Electrical/Mech/Chem. Engineering Section*, 1989, **16** (2), pp. 25–30

Chapter 4
Numerical integrator substitution

4.1 Introduction

A continuous function can be simulated by substituting a numerical integration formula into the differential equation and rearranging the function into an appropriate form. Among the factors to be taken into account in the selection of the numerical integrator are the error due to truncated terms, its properties as a differentiator, error propagation and frequency response.

Numerical integration substitution (NIS) constitutes the basis of Dommel's EMTP [1]–[3], which, as explained in the introductory chapter, is now the most generally accepted method for the solution of electromagnetic transients. The EMTP method is an integrated approach to the problems of:

- forming the network differential equations
- collecting the equations into a coherent system to be solved
- numerical solution of the equations.

The trapezoidal integrator (described in Appendix C) is used for the numerical integrator substitution, due to its simplicity, stability and reasonable accuracy in most circumstances. However, being based on a truncated Taylor's series, the trapezoidal rule can cause numerical oscillations under certain conditions due to the neglected terms [4]. This problem will be discussed further in Chapters 5 and 9.

The other basic characteristic of Dommel's method is the discretisation of the system components, given a predetermined time step, which are then combined in a solution for the nodal voltages. Branch elements are represented by the relationship which they maintain between branch current and nodal voltage.

This chapter describes the basic formulation and solution of the numerical integrator substitution method as implemented in the electromagnetic transient programs.

4.2 Discretisation of R, L, C elements

4.2.1 Resistance

The simplest circuit element is a resistor connected between nodes k and m, as shown in Figure 4.1, and is represented by the equation:

$$i_{km}(t) = \frac{1}{R}(v_k(t) - v_m(t)) \tag{4.1}$$

Resistors are accurately represented in the EMTP formulation provided R is not too small. If the value of R is too small its inverse in the system matrix will be large, resulting in poor conditioning of the solution at every step. This gives inaccurate results due to the finite precision of numerical calculations. On the other hand, very large values of R do not degrade the overall solution. In EMTDC version 3 if R is below a threshold (the default threshold value is 0.0005) then R is automatically set to zero and a modified solution method used.

4.2.2 Inductance

The differential equation for the inductor shown in Figure 4.2 is:

$$v_L = v_k - v_m = L\frac{di_{km}}{dt} \tag{4.2}$$

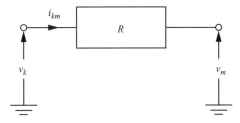

Figure 4.1 Resistor

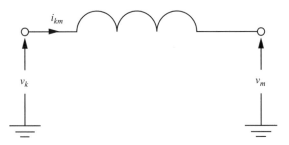

Figure 4.2 Inductor

Rearranging:

$$i_{km(t)} = i_{km(t-\Delta t)} + \int_{t-\Delta t}^{t} (v_k - v_m)\, dt \tag{4.3}$$

Applying the trapezoidal rule gives:

$$i_{km(t)} = i_{km(t-\Delta t)} + \frac{\Delta t}{2L}((v_k - v_m)_{(t)} + (v_k - v_m)_{(t-\Delta t)}) \tag{4.4}$$

$$= i_{km(t-\Delta t)} + \frac{\Delta t}{2L}(v_{k(t-\Delta t)} - v_{m(t-\Delta t)}) + \frac{\Delta t}{2L}(v_{k(t)} - v_{m(t)}) \tag{4.5}$$

$$i_{km}(t) = I_{\text{History}}(t - \Delta t) + \frac{1}{R_{\text{eff}}}(v_k(t) - v_m(t)) \tag{4.6}$$

This equation can be expressed in the form of a Norton equivalent (or companion circuit) as illustrated in Figure 4.3. The term relating the current contribution at the present time step to voltage at the present time step ($1/R_{\text{eff}}$) is a conductance (instantaneous term) and the contribution to current from the previous time step quantities is a current source (History term).

In equation 4.6 $I_{\text{History}}(t - \Delta t) = i_{km}(t - \Delta t) + (\Delta t/2L)(v_k(t - \Delta t) - v_m(t - \Delta t))$ and $R_{\text{eff}} = 2L/\Delta t$.

The term $2L/\Delta t$ is known as the instantaneous term as it relates the current to the voltage at the same time point, i.e. any change in one will instantly be reflected in the other. As an effective resistance, very small values of L or rather $2L/\Delta t$, can also result in poor conditioning of the conductance matrix.

Transforming equation 4.6 to the z-domain gives:

$$I_{km}(z) = z^{-1} I_{km}(z) + \frac{\Delta t}{2L}(1 + z^{-1})(V_k(z) - V_m(z))$$

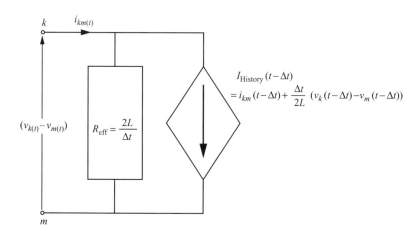

Figure 4.3 Norton equivalent of the inductor

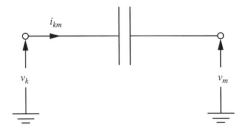

Figure 4.4 Capacitor

Rearranging gives the following transfer between current and voltage in the z-domain:

$$\frac{I_{km}(z)}{(V_k(z) - V_m(z))} = \frac{\Delta t}{2L} \frac{(1 + z^{-1})}{(1 - z^{-1})} \qquad (4.7)$$

4.2.3 Capacitance

With reference to Figure 4.4 the differential equation for the capacitor is:

$$i_{km}(t) = C \frac{d(v_k(t) - v_m(t))}{dt} \qquad (4.8)$$

Integrating and rearranging gives:

$$v_{km(t)} = (v_{k(t)} - v_{m(t)}) = (v_{k(t-\Delta t)} - v_{m(t-\Delta t)}) + \frac{1}{C} \int_{t-\Delta t}^{t} i_{km}\,dt \qquad (4.9)$$

and applying the trapezoidal rule:

$$v_{km}(t) = (v_k(t) - v_m(t)) = (v_k(t - \Delta t) - v_m(t - \Delta t)) + \frac{\Delta t}{2C}(i_{km}(t) + i_{km}(t - \Delta t)) \qquad (4.10)$$

Hence the current in the capacitor is given by:

$$i_{km}(t) = \frac{2C}{\Delta t}(v_k(t) - v_m(t)) - i_{km}(t - \Delta t) - \frac{2C}{\Delta t}(v_k(t - \Delta t) - v_m(t - \Delta t))$$

$$= \frac{1}{R_{\text{eff}}}[v_k(t) - v_m(t)] + I_{\text{History}}(t - \Delta t) \qquad (4.11)$$

which is again a Norton equivalent as depicted in Figure 4.5. The instantaneous term in equation 4.11 is:

$$R_{\text{eff}} = \frac{\Delta t}{2C} \qquad (4.12)$$

Thus very large values of C, although they are unlikely to be used, can cause ill conditioning of the conductance matrix.

Numerical integrator substitution 71

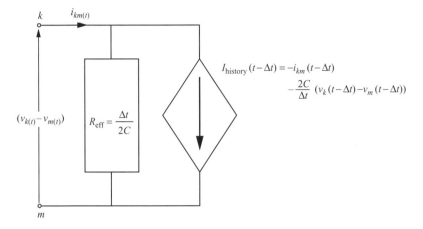

Figure 4.5   Norton equivalent of the capacitor

The History term represented by a current source is:

$$I_{\text{History}(t-\Delta t)} = -i_{km(t-\Delta t)} - \frac{2C}{\Delta t}(v_{k(t-\Delta t)} - v_{m(t-\Delta t)}) \quad (4.13)$$

Transforming to the z-domain gives:

$$I_{km} = -z^{-1}I_{km} - \frac{2C}{\Delta t}(V_k - V_m)z^{-1} + \frac{2C}{\Delta t}(V_k - V_m) \quad (4.14)$$

$$\frac{I_{km}}{(V_k - V_m)} = \frac{2C}{\Delta t}\frac{(1 - z^{-1})}{(1 + z^{-1})} \quad (4.15)$$

It should be noted that any implicit integration formula can be substituted into a differential equation to form a difference equation (and a corresponding Norton equivalent). Table 4.1 shows the Norton components that result from using three different integration methods.

4.2.4   Components reduction

Several components can be combined into a single Norton equivalent, thus reducing the number of nodes and hence the computation at each time point. Consider first the case of a simple RL branch.

The History term for the inductor is:

$$I_{\text{History}}(t - \Delta t) = i(t - \Delta t) + \frac{\Delta t}{2L}v_l(t - \Delta t) \quad (4.16)$$

where v_l is the voltage across the inductor. This is related to the branch voltage by:

$$v_l(t - \Delta t) = v(t - \Delta t) - i(t - \Delta t)R \quad (4.17)$$

Table 4.1 Norton components for different integration formulae

Integration method	R_{eq}	$I_{History}$
Inductor		
Backward Euler	$\dfrac{L}{\Delta t}$	i_{n-1}
Trapezoidal	$\dfrac{2L}{\Delta t}$	$i_{n-1} + \dfrac{\Delta t}{2L}v_{n-1}$
Gear 2nd order	$\dfrac{3L}{2\Delta t}$	$\tfrac{4}{3}i_{n-1} - \tfrac{1}{3}i_{n-2}$
Capacitor		
Backward Euler	$\dfrac{\Delta t}{C}$	$-\dfrac{C}{\Delta t}v_{n-1}$
Trapezoidal	$\dfrac{\Delta t}{2C}$	$-\dfrac{2C}{\Delta t}v_{n-1} - i_{n-1}$
Gear 2nd order	$\dfrac{2\Delta t}{3C}$	$-\dfrac{2C}{\Delta t}v_{n-1} - \dfrac{C}{2\Delta t}v_{n-2}$

Substituting equation 4.17 into equation 4.16 yields:

$$I_{History} = \frac{\Delta t}{2L}v(t - \Delta t) - \frac{\Delta t R}{2L}i(t - \Delta t) + i(t - \Delta t)$$

$$= \left(1 - \frac{\Delta t R}{2L}\right)i(t - \Delta t) + \frac{\Delta t}{2L}v(t - \Delta t) \quad (4.18)$$

The Norton equivalent circuit current source value for the complete RL branch is simply calculated from the short-circuit terminal current. The short-circuit circuit consists of a current source feeding into two parallel resistors (R and $2L/\Delta t$), with the current in R being the terminal current. This is given by:

$$I_{\text{short-circuit}} = \frac{(2L/\Delta t)I_{History}}{R + 2L/\Delta t}$$

$$= \frac{(2L/\Delta t)\left((1 - \Delta t R/(2L))\,i(t - \Delta t) + (\Delta t/(2L))v(t - \Delta t)\right)}{R + 2L/\Delta t}$$

$$= \frac{(2L/\Delta t - R)}{(R + 2L/\Delta t)}i(t - \Delta t) + \frac{1}{(R + 2L/\Delta t)}v(t - \Delta t)$$

$$= \frac{(1 - \Delta t R/(2L))}{(1 + \Delta t R/(2L))}i(t - \Delta t) + \frac{\Delta t/(2L)}{(1 + \Delta t R/(2L))}v(t - \Delta t) \quad (4.19)$$

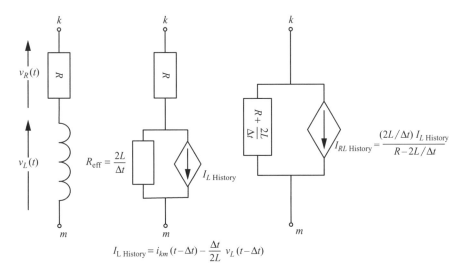

Figure 4.6   Reduction of RL branch

The instantaneous current term is obtained from the current that flows due to an applied voltage to the terminals (current source open circuited). This current is:

$$\frac{1}{(R+2L/\Delta t)}v(t) = \frac{\Delta t/(2L)}{(1+\Delta t R/(2L))}v(t) \qquad (4.20)$$

Hence the complete difference equation expressed in terms of branch voltage is obtained by adding equations 4.19 and 4.20, which gives:

$$i(t) = \frac{(1-\Delta t R/(2L))}{(1+\Delta t R/(2L))}i(t-\Delta t) + \frac{\Delta t/(2L)}{(1+\Delta t R/(2L))}(v(t-\Delta t)+v(t)) \qquad (4.21)$$

The corresponding Norton equivalent is shown in Figure 4.6.

The reduction of a tuned filter branch is illustrated in Figure 4.7, which shows the actual *RLC* components, their individual Norton equivalents and a single Norton representation of the complete filter branch. Parallel filter branches can be combined into one Norton by summing their current sources and conductance values. The reduction, however, hides the information on voltages across and current through each individual component. The mathematical implementation of the reduction process is carried out by first establishing the nodal admittance matrix of the circuit and then performing Gaussian elimination of the internal nodes.

4.3 Dual Norton model of the transmission line

A detailed description of transmission line modelling is deferred to Chapter 6. The single-phase lossless line [4] is used as an introduction at this stage, to illustrate the simplicity of Dommel's method.

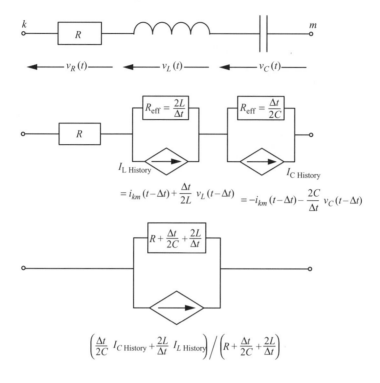

Figure 4.7 Reduction of RLC branch

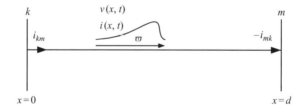

Figure 4.8 Propagation of a wave on a transmission line

Consider the lossless distributed parameter line depicted in Figure 4.8, where L' is the inductance and C' the capacitance per unit length. The wave propagation equations for this line are:

$$-\frac{\partial v(x,t)}{\partial x} = L'\frac{\partial i(x,t)}{\partial t} \tag{4.22}$$

$$-\frac{\partial i(x,t)}{\partial x} = C'\frac{\partial v(x,t)}{\partial t} \tag{4.23}$$

and the general solution:

$$i(x,t) = f_1(x - \varpi t) + f_2(x + \varpi t) \tag{4.24}$$

$$v(x,t) = Z \cdot f_1(x - \varpi t) - Z \cdot f_2(x + \varpi t) \tag{4.25}$$

with $f_1(x - \varpi t)$ and $f_2(x + \varpi t)$ being arbitrary functions of $(x - \varpi t)$ and $(x + \varpi t)$ respectively. $f_1(x - \varpi t)$ represents a wave travelling at velocity ϖ in a forward direction (depicted in Figure 4.8) and $f_2(x + \varpi t)$ a wave travelling in a backward direction. Z_C, the surge or characteristic impedance and ϖ, the phase velocity, are given by:

$$Z_C = \sqrt{\frac{L'}{C'}} \tag{4.26}$$

$$\varpi = \frac{1}{\sqrt{L'C'}} \tag{4.27}$$

Multiplying equation 4.24 by Z_C and adding it to, and subtracting it from, equation 4.25 leads to:

$$v(x,t) + Z_C \cdot i(x,t) = 2Z_C \cdot f_1(x - \varpi t) \tag{4.28}$$

$$v(x,t) - Z_C \cdot i(x,t) = -2Z_C \cdot f_2(x + \varpi t) \tag{4.29}$$

It should be noted that $v(x,t) + Z_C \cdot i(x,t)$ is constant when $(x - \varpi t)$ is constant. If d is the length of the line, the travelling time from one end (k) to the other end (m) of the line to observe a constant $v(x,t) + Z_C \cdot i(x,t)$ is:

$$\tau = d/\varpi = d\sqrt{L'C'} \tag{4.30}$$

Hence

$$v_k(t - \tau) + Z_C \cdot i_{km}(t - \tau) = v_m(t) + Z_C \cdot (-i_{mk}(t)) \tag{4.31}$$

Rearranging equation 4.31 gives the simple two-port equation for i_{mk}, i.e.

$$i_{mk}(t) = \frac{1}{Z_C} v_m(t) + I_m(t - \tau) \tag{4.32}$$

where the current source from past History terms is:

$$I_m(t - \tau) = -\frac{1}{Z_C} v_k(t - \tau) - i_{km}(t - \tau) \tag{4.33}$$

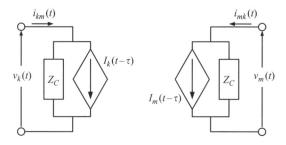

Figure 4.9 Equivalent two-port network for a lossless line

Similarly for the other end

$$i_{km}(t) = \frac{1}{Z_C} v_k(t) + I_k(t - \tau) \tag{4.34}$$

where

$$I_k(t - \tau) = -\frac{1}{Z_C} v_m(t - \tau) - i_{mk}(t - \tau)$$

The expressions $(x - \varpi t) = $ constant and $(x + \varpi t) = $ constant are called the characteristic equations of the differential equations.

Figure 4.9 depicts the resulting two-port model. There is no direct connection between the two terminals and the conditions at one end are seen indirectly and with time delays (travelling time) at the other through the current sources. The past History terms are stored in a ring buffer and hence the maximum travelling time that can be represented is the time step multiplied by the number of locations in the buffer. Since the time delay is not usually a multiple of the time-step, the past History terms on either side of the actual travelling time are extracted and interpolated to give the correct travelling time.

4.4 Network solution

With all the network components represented by Norton equivalents a nodal formulation is used to perform the system solution.

The nodal equation is:

$$[G]\mathbf{v}(t) = \mathbf{i}(t) + \mathbf{I}_{\text{History}} \tag{4.35}$$

where:
 [G] is the conductance matrix
 $\mathbf{v}(t)$ is the vector of nodal voltages
 $\mathbf{i}(t)$ is the vector of external current sources
 $\mathbf{I}_{\text{History}}$ is the vector current sources representing past history terms.

Numerical integrator substitution 77

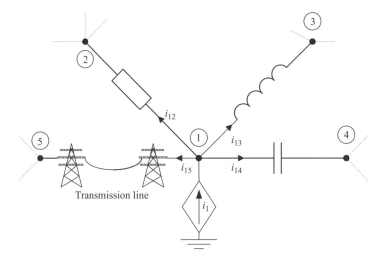

Figure 4.10   Node 1 of an interconnected circuit

The nodal formulation is illustrated with reference to the circuit in Figure 4.10 [5] where the use of Kirchhoff's current law at node 1 yields:

$$i_{12}(t) + i_{13}(t) + i_{14}(t) + i_{15}(t) = i_1(t) \tag{4.36}$$

Expressing each branch current in terms of node voltages gives:

$$i_{12}(t) = \frac{1}{R}(v_1(t) - v_2(t)) \tag{4.37}$$

$$i_{13}(t) = \frac{\Delta t}{2L}(v_1(t) - v_3(t)) + I_{13}(t - \Delta t) \tag{4.38}$$

$$i_{14}(t) = \frac{2C}{\Delta t}(v_1(t) - v_4(t)) + I_{14}(t - \Delta t) \tag{4.39}$$

$$i_{15}(t) = \frac{1}{Z}v_1(t) + I_{15}(t - \tau) \tag{4.40}$$

Substituting these gives the following equation for node 1:

$$\left(\frac{1}{R} + \frac{\Delta t}{2L} + \frac{2C}{\Delta t} + \frac{1}{Z}\right)v_1(t) - \frac{1}{R}v_2(t) - \frac{\Delta t}{2L}v_3(t) - \frac{2C}{\Delta t}v_4(t)$$
$$= I_1(t - \Delta t) - I_{13}(t - \Delta t) - I_{14}(t - \Delta t) - I_{15}(t - \Delta t) \tag{4.41}$$

Note that [G] is real and symmetric when incorporating network components. If control equations are incorporated into the same [G] matrix, the symmetry is lost; these are, however, solved separately in many programs. As the elements of [G] are dependent on the time step, by keeping the time step constant [G] is constant and triangular factorisation can be performed before entering the time step loop. Moreover, each node in a power system is connected to only a few other nodes and therefore

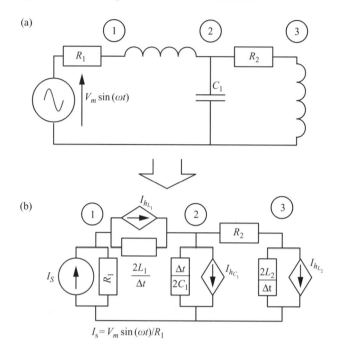

Figure 4.11 Example using conversion of voltage source to current source

the conductance matrix is sparse. This property is exploited by only storing non-zero elements and using optimal ordering elimination schemes.

Some of the node voltages will be known due to the presence of voltage sources in the system, but the majority are unknown. In the presence of series impedance with each voltage source the combination can be converted to a Norton equivalent and the algorithm remains unchanged.

Example: Conversion of voltage sources to current sources
To illustrate the incorporation of known voltages the simple network displayed in Figure 4.11(a) will be considered. The task is to write the matrix equation that must be solved at each time point.

Converting the components of Figure 4.11(a) to Norton equivalents (companion circuits) produces the circuit of Figure 4.11(b) and the corresponding nodal equation:

$$\begin{bmatrix} \dfrac{1}{R_1}+\dfrac{\Delta t}{2L_1} & -\dfrac{\Delta t}{2L_1} & 0 \\ -\dfrac{\Delta t}{2L_1} & \dfrac{\Delta t}{2L_1}+\dfrac{1}{R_2}+\dfrac{2C_1}{\Delta t} & -\dfrac{1}{R_2} \\ 0 & -\dfrac{1}{R_2} & \dfrac{1}{R_2}+\dfrac{\Delta t}{2L_2} \end{bmatrix} \begin{pmatrix} v_1 \\ v_2 \\ v_3 \end{pmatrix} = \begin{pmatrix} \dfrac{V_m \sin(\omega t)}{R_1} - I_{h_{L_1}} \\ I_{h_{L_1}} - I_{h_{C_1}} \\ -I_{h_{L_2}} \end{pmatrix}$$

(4.42)

Equation 4.42 is first solved for the node voltages and from these all the branch currents are calculated. Time is then advanced and the current sources representing History terms (previous time step information) are recalculated. The value of the voltage source is recalculated at the new time point and so is the matrix equation. The process of solving the matrix equation, calculating all currents in the system, advancing time and updating History terms is continued until the time range of the study is completed.

As indicated earlier, the conversion of voltage sources to Norton equivalents requires some series impedance, i.e. an ideal voltage source cannot be represented using this simple conductance method. A more general approach is to partition the nodal equation as follows:

$$\begin{bmatrix} [G_{UU}] & [G_{UK}] \\ [G_{KU}] & [G_{KK}] \end{bmatrix} \cdot \begin{pmatrix} \mathbf{v}_U(t) \\ \mathbf{v}_K(t) \end{pmatrix} = \begin{pmatrix} \mathbf{i}_U(t) \\ \mathbf{i}_K(t) \end{pmatrix} + \begin{pmatrix} \mathbf{I}_{U\,\text{History}} \\ \mathbf{I}_{K\,\text{History}} \end{pmatrix} = \begin{pmatrix} \mathbf{I}_U \\ \mathbf{I}_K \end{pmatrix} \quad (4.43)$$

where the subscripts U and K represent connections to nodes with unknown and known voltages, respectively. Using Kron's reduction the unknown voltage vector is obtained from:

$$[G_{UU}]\mathbf{v}_U(t) = \mathbf{i}_U(t) + \mathbf{I}_{U\,\text{History}} - [G_{UK}]\mathbf{v}_K(t) = \mathbf{I}'_U \quad (4.44)$$

The current in voltage sources can be calculated using:

$$[G_{KU}]\mathbf{v}_U(t) + [G_{KK}]\mathbf{v}_K(t) - \mathbf{I}_{K\,\text{History}} = \mathbf{i}_K(t) \quad (4.45)$$

The process for solving equation 4.44 is depicted in Figure 4.12. Only the right-hand side of this equation needs to be recalculated at each time step. Triangular factorisation is performed on the augmented matrix $[G_{UU}\ G_{UK}]$ before entering the time step loop. The same process is then extended to $\mathbf{i}_U(t) - \mathbf{I}_{\text{History}}$ at each time step (forward solution), followed by back substitution to get $\mathbf{V}_U(t)$. Once $\mathbf{V}_U(t)$ has been found, the History terms for the next time step are calculated.

4.4.1 Network solution with switches

To reflect switching operations or time varying parameters, matrices $[G_{UU}]$ and $[G_{UK}]$ need to be altered and retriangulated. By placing nodes with switches last, as illustrated in Figure 4.13, the initial triangular factorisation is only carried out for the nodes without switches [6]. This leaves a small reduced matrix which needs altering following a change. By placing the nodes with frequently switching elements in the lowest part the computational burden is further reduced.

Transmission lines using the travelling wave model do not introduce off-diagonal elements from the sending to the receiving end, and thus result in a block diagonal structure for $[G_{UU}]$, as shown in Figure 4.14. Each block represents a subsystem (a concept to be described in section 4.6), that can be solved independently of the rest of the system, as any influence from the rest of the system is represented by the History terms (i.e. there is no instantaneous term). This allows parallel computation of the

80 *Power systems electromagnetic transients simulation*

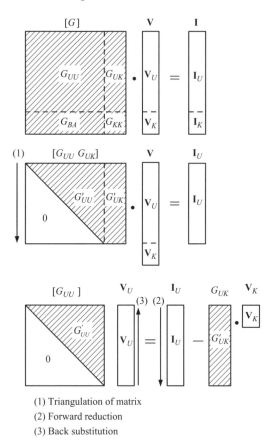

(1) Triangulation of matrix
(2) Forward reduction
(3) Back substitution

Figure 4.12 Network solution with voltage sources

solution, a technique that is used in the RTDS simulator. For non-linear systems, each non-linearity can be treated separately using the compensation approach provided that there is only one non-linearity per subsystem. Switching and interpolation are also performed on a subsystem basis.

In the PSCAD/EMTDC program, triangular factorisation is performed on a subsystem basis rather than on the entire matrix. Nodes connected to frequently switched branches (i.e. GTOs, thyristors, diodes and arrestors) are ordered last, but other switching branches (faults and breakers) are not. Each section is optimally ordered separately.

A flow chart of the overall solution technique is shown in Figure 4.15.

4.4.2 Example: voltage step applied to RL load

To illustrate the use of Kron reduction to eliminate known voltages the simple circuit shown in Figure 4.16 will be used.

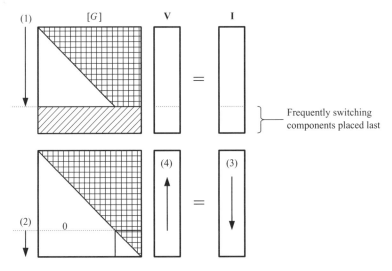

(1) Partial triangulation of matrix (prior to time step loop)
(2) Complete triangulation
(3) Forward reduction of current vector
(4) Back substitution for node voltages

Figure 4.13   Network solution with switches

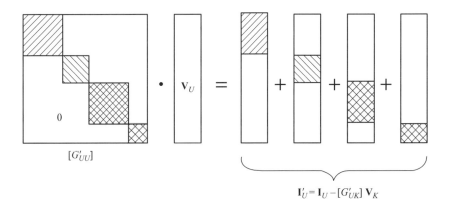

$$\mathbf{I}'_U = \mathbf{I}_U - [G'_{UK}]\mathbf{V}_K$$

Figure 4.14   Block diagonal structure

Figure 4.17 shows the circuit once the inductor is converted to its Norton equivalent. The nodal equation for this circuit is:

$$\begin{bmatrix} G_{\text{Switch}} & -G_{\text{Switch}} & 0 \\ -G_{\text{Switch}} & G_{\text{Switch}} + G_R & -G_R \\ 0 & -G_R & G_{L\_\text{eff}} + G_R \end{bmatrix} \begin{pmatrix} v_1 \\ v_2 \\ v_3 \end{pmatrix} = \begin{pmatrix} i_v \\ 0 \\ -I_{\text{History}} \end{pmatrix} \quad (4.46)$$

82 *Power systems electromagnetic transients simulation*

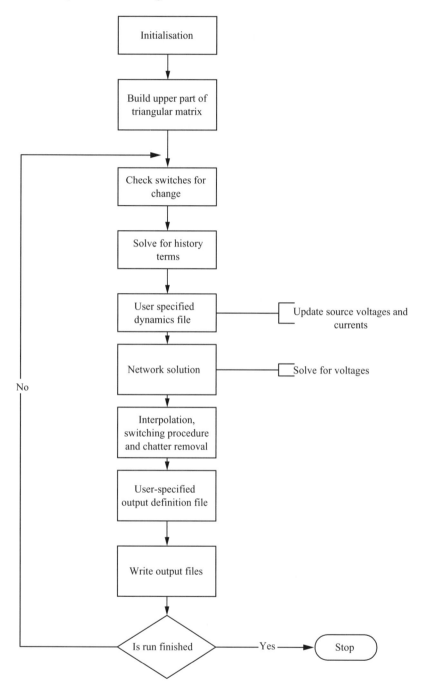

Figure 4.15 Flow chart of EMT algorithm

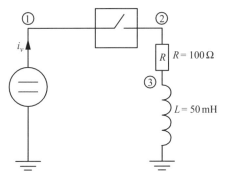

Figure 4.16 Simple switched RL load

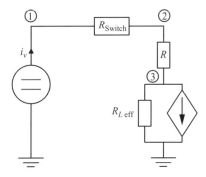

Figure 4.17 Equivalent circuit for simple switched RL load

As v_1 is a known voltage the conductance matrix is reordered by placing v_1 last in the column vector of nodal voltages and moving column 1 of $[G]$ to be column 3; then move row 1 (equation for current in voltage source) to become row 3. This then gives:

$$\left[\begin{array}{cc|c} G_{\text{Switch}} + G_R & -G_R & -G_{\text{Switch}} \\ -G_R & G_{L\_\text{eff}} + G_R & 0 \\ \hline -G_{\text{Switch}} & 0 & G_{\text{Switch}} \end{array}\right] \begin{pmatrix} v_2 \\ v_3 \\ v_1 \end{pmatrix} = \begin{pmatrix} 0 \\ -I_{\text{History}} \\ i_v \end{pmatrix} \quad (4.47)$$

which is of the form

$$\begin{bmatrix} [G_{UU}] & [G_{UK}] \\ [G_{KU}] & [G_{KK}] \end{bmatrix} \begin{pmatrix} \mathbf{v}_U(t) \\ \mathbf{v}_K(t) \end{pmatrix} = \begin{pmatrix} \mathbf{i}_U(t) \\ \mathbf{i}_K(t) \end{pmatrix} + \begin{pmatrix} \mathbf{I}_{U\_\text{History}} \\ \mathbf{I}_{K\_\text{History}} \end{pmatrix}$$

i.e.

$$\begin{bmatrix} G_{\text{Switch}} + G_R & -G_R & -G_{\text{Switch}} \\ -G_R & G_{L\_\text{eff}} + G_R & 0 \end{bmatrix} \begin{pmatrix} v_2 \\ v_3 \\ v_1 \end{pmatrix} = \begin{pmatrix} 0 \\ -I_{\text{History}} \end{pmatrix}$$

Note the negative I_{History} term as the current is leaving the node. Performing Gaussian elimination gives:

$$\begin{bmatrix} G_{\text{Switch}} + G_R & -G_R & -G_{\text{Switch}} \\ 0 & G_{L\_\text{eff}} + G_R - M(-G_R) & -M(-G_{\text{Switch}}) \end{bmatrix} \begin{pmatrix} v_2 \\ v_3 \\ v_1 \end{pmatrix} = \begin{pmatrix} 0 \\ -I_{\text{History}} \end{pmatrix}$$
(4.48)

where

$$M = \frac{-G_{L\_\text{eff}}}{G_{\text{Switch}} + G_R}$$

Moving the known voltage $v_1(t)$ to the right-hand side gives:

$$\begin{bmatrix} G_{\text{Switch}} + G_R & -G_R \\ 0 & G_{L\_\text{eff}} + G_R - M(-G_R) \end{bmatrix} \begin{pmatrix} v_2 \\ v_3 \end{pmatrix} = \begin{pmatrix} 0 \\ -I_{\text{History}} \end{pmatrix} - \begin{bmatrix} -G_{\text{Switch}} \\ -M(-G_{\text{Switch}}) \end{bmatrix} v_1$$
(4.49)

Alternatively, the known voltage could be moved to the right-hand side before performing the Gaussian elimination. i.e.

$$\begin{bmatrix} G_{\text{Switch}} + G_R & -G_R \\ -G_R & G_{L\_\text{eff}} + G_R \end{bmatrix} \begin{pmatrix} v_2 \\ v_3 \end{pmatrix} = \begin{pmatrix} 0 \\ -I_{\text{History}} \end{pmatrix} - \begin{bmatrix} -G_{\text{Switch}} \\ 0 \end{bmatrix} v_1 \quad (4.50)$$

Eliminating the element below the diagonal, and performing the same operation on the right-hand side will give equation 4.49 again. The implementation of these equations in FORTRAN is given in Appendix H.2 and MATLAB in Appendix F.3. The FORTRAN code in H.2 illustrates using a d.c. voltage source and switch, while the MATLAB version uses an a.c. voltage source and diode. Note that as Gaussian elimination is equivalent to performing a series of Norton–Thevenin conversion to produce one Norton, the RL branch can be modelled as one Norton. This is implemented in the FORTRAN code in Appendices H.1 and H.3 and MATLAB code in Appendices F.1 and F.2.

Table 4.2 compares the current calculated using various time steps with results from the analytic solution.

For a step response of an RL branch the analytic solution is given by:

$$i(t) = \frac{V}{R}\left(1 - e^{-tR/L}\right)$$

Note that the error becomes larger and a less damped response results as the time step increases. This information is graphically displayed in Figures 4.18(a)–4.19(b). As a rule of thumb the maximum time step must be one tenth of the smallest time constant in the system. However, the circuit time constants are not generally known a priori and therefore performing a second simulation with the time step halved will give a good indication if the time step is sufficiently small.

Table 4.2   Step response of RL circuit to various step lengths

Time (ms)	Current (amps)				
	Exact	$\Delta t = \tau/10$	$\Delta t = \tau$	$\Delta t = 5\tau$	$\Delta t = 10\tau$
1.0000	0	0	0	0	0
1.0500	63.2121	61.3082	33.3333	–	–
1.1000	86.4665	85.7779	77.7778	–	–
1.1500	95.0213	94.7724	92.5926	–	–
1.2000	98.1684	98.0785	97.5309	–	–
1.2500	99.3262	99.2937	99.1770	71.4286	–
1.3000	99.7521	99.7404	99.7257	–	–
1.3500	99.9088	99.9046	99.9086	–	–
1.4000	99.9665	99.9649	99.9695	–	–
1.4500	99.9877	99.9871	99.9898	–	–
1.5000	99.9955	99.9953	99.9966	112.2449	83.3333
1.5500	99.9983	99.9983	99.9989	–	–
1.6000	99.9994	99.9994	99.9996	–	–
1.6500	99.9998	99.9998	99.9999	–	–
1.7000	99.9999	99.9999	100.0000	–	–
1.7500	100.0000	100.0000	100.0000	94.7522	–
1.8000	100.0000	100.0000	100.0000	–	–
1.8500	100.0000	100.0000	100.0000	–	–
1.9000	100.0000	100.0000	100.0000	–	–
1.9500	100.0000	100.0000	100.0000	–	–
2.0000	100.0000	100.0000	100.0000	102.2491	111.1111
2.0500	100.0000	100.0000	100.0000	–	–
2.1000	100.0000	100.0000	100.0000	–	–
2.1500	100.0000	100.0000	100.0000	–	–
2.2000	100.0000	100.0000	100.0000	–	–
2.2500	100.0000	100.0000	100.0000	99.0361	–
2.3000	100.0000	100.0000	100.0000	–	–
2.3500	100.0000	100.0000	100.0000	–	–
2.4000	100.0000	100.0000	100.0000	–	–
2.4500	100.0000	100.0000	100.0000	–	–
2.5000	100.0000	100.0000	100.0000	100.4131	92.5926
2.5500	100.0000	100.0000	100.0000	–	–
2.6000	100.0000	100.0000	100.0000	–	–
2.6500	100.0000	100.0000	100.0000	–	–
2.7000	100.0000	100.0000	100.0000	–	–
2.7500	100.0000	100.0000	100.0000	99.8230	–
2.8000	100.0000	100.0000	100.0000	–	–
2.8500	100.0000	100.0000	100.0000	–	–
2.9000	100.0000	100.0000	100.0000	–	–
2.9500	100.0000	100.0000	100.0000	–	–
3.0000	100.0000	100.0000	100.0000	100.0759	104.9383

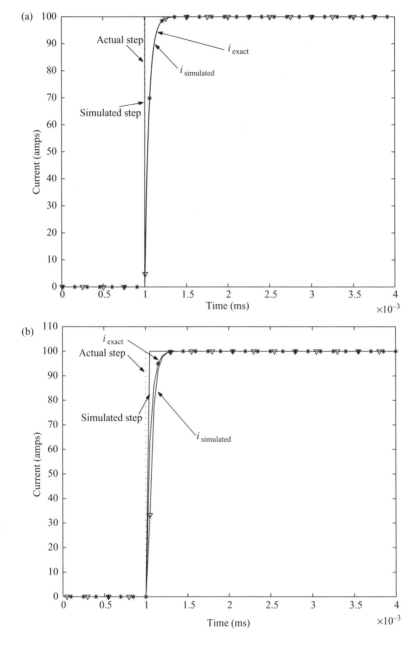

Figure 4.18 Step response of an RL branch for step lengths of: (a) $\Delta t = \tau/10$ and (b) $\Delta t = \tau$

Numerical integrator substitution 87

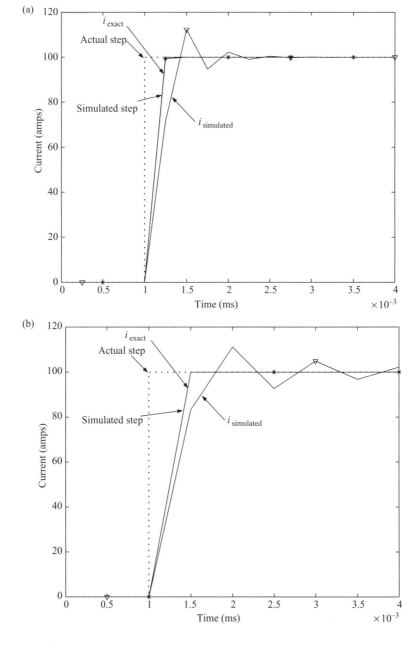

Figure 4.19 Step response of an RL branch for step lengths of: (a) $\Delta t = 5\tau$ and (b) $\Delta t = 10\tau$

The following data is used for this test system: $\Delta t = 50\,\mu\text{s}$, $R = 1.0\,\Omega$, $L = 0.05$ mH and $R_{\text{Switch}} = 10^{10}\,\Omega$ (OFF) $10^{-10}\,\Omega$ (ON) and $V_1 = 100$ V. Initially $I_{\text{History}} = 0$

$$\begin{bmatrix} 1.000000000 & -1.000000000 & -.1000000000E{-}09 \\ -1.000000000 & 1.500000000 & 0.000000000 \end{bmatrix} \begin{pmatrix} v_2 \\ v_3 \\ v_1 \end{pmatrix} = \begin{pmatrix} 0.000000000 \\ 0.000000000 \end{pmatrix}$$

The multiplier is -0.999999999900000. After forward reduction using this multiplier the G matrix becomes:

$$\begin{bmatrix} 1.000000000 & -1.000000000 & -.1000000000E{-}09 \\ 0.000000000 & 0.5000000001 & -.9999999999E{-}10 \end{bmatrix} \begin{pmatrix} v_2 \\ v_3 \\ v_1 \end{pmatrix} = \begin{pmatrix} 0.000000000 \\ 0.000000000 \end{pmatrix}$$

Moving the known voltage v_1 to the right-hand side gives

$$\begin{bmatrix} 1.000000000 & -1.000000000 \\ 0.000000000 & 0.5000000001 \end{bmatrix} \begin{pmatrix} v_2 \\ v_3 \end{pmatrix} = \begin{pmatrix} 0.000000000 \\ 0.000000000 \end{pmatrix} - \begin{bmatrix} -.1000000000E{-}09 \\ -.9999999999E{-}10 \end{bmatrix} v_1$$

Back substitution gives: $i = 9.9999999970E{-}009$ or essentially zero in the off state. When the switch is closed the G matrix is updated and the equation becomes:

$$\begin{bmatrix} 0.1000000000E{+}11 & -1.000000000 & -.1000000000E{+}11 \\ -1.000000000 & 1.500000000 & 0.000000000 \end{bmatrix} \begin{pmatrix} v_2 \\ v_3 \\ v_1 \end{pmatrix} = \begin{pmatrix} 0.000000000 \\ 0.000000000 \end{pmatrix}$$

After forward reduction:

$$\begin{bmatrix} 0.1000000000E{+}11 & -1.000000000 & -.1000000000E{+}11 \\ -1.000000000 & 1.500000000 & -.9999999999 \end{bmatrix} \begin{pmatrix} v_2 \\ v_3 \\ v_1 \end{pmatrix} = \begin{pmatrix} 0.000000000 \\ 0.000000000 \end{pmatrix}$$

Moving the known voltage v_1 to the right-hand side gives

$$\begin{bmatrix} 1.000000000 & -1.000000000 \\ 0.000000000 & 1.500000000 \end{bmatrix} \begin{pmatrix} v_2 \\ v_3 \end{pmatrix} = \begin{pmatrix} 0.000000000 \\ 0.000000000 \end{pmatrix} - \begin{bmatrix} -.1000000000E{+}11 \\ -.9999999999 \end{bmatrix} v_1$$

$$= \begin{pmatrix} 0.1000000000E{+}13 \\ 99.99999999 \end{pmatrix}$$

Hence back-substitution gives:

$$i_L = 33.333 \text{ A}$$
$$v_2 = 66.667 \text{ V}$$
$$v_3 = 33.333 \text{ V}$$

4.5 Non-linear or time varying parameters

The most common types of non-linear elements that need representing are inductances under magnetic saturation for transformers and reactors and resistances of surge arresters. Non-linear effects in synchronous machines are handled directly in the machine equations. As usually there are only a few non-linear elements, modification of the linear solution method is adopted rather than performing a less efficient non-linear solution method for the entire network. In the past, three approaches have been used, i.e.

- current source representation (with one time step delay)
- compensation methods
- piecewise linear (switch representation).

4.5.1 Current source representation

A current source can be used to model the total current drawn by a non-linear component, however by necessity this current has to be calculated from information at previous time steps. Therefore it does not have an instantaneous term and appears as an 'open circuit' to voltages at the present time step. This approach can result in instabilities and therefore is not recommended. To remove the instability a large fictitious Norton resistance would be needed, as well as the use of a correction source. Moreover there is a one time step delay in the correction source. Another option is to split the non-linear component into a linear component and non-linear source. For example a non-linear inductor is modelled as a linear inductor in parallel with a current source representing the saturation effect, as shown in Figure 4.20.

4.5.2 Compensation method

The compensation method can be applied provided there is only one non-linear element (it is, in general, an iterative procedure if more than one non-linear element is

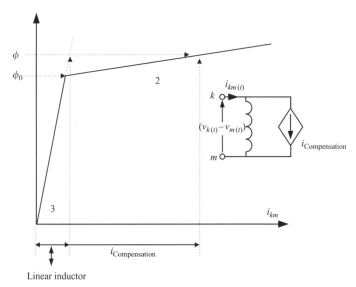

Figure 4.20 Piecewise linear inductor represented by current source

present). The compensation theorem states that a non-linear branch can be excluded from the network and be represented as a current source instead. Invoking the superposition theorem, the total network solution is equal to the value $\mathbf{v}_0(t)$ found with the non-linear branch omitted, plus the contribution produced by the non-linear branch.

$$\mathbf{v}(t) = \mathbf{v}_0(t) - \mathbf{R}_{\text{Thevenin}} \mathbf{i}_{km}(t) \qquad (4.51)$$

where

$\mathbf{R}_{\text{Thevenin}}$ is the Thevenin resistance of the network without a non-linear branch connected between nodes k and m.

$\mathbf{v}_0(t)$ is the open circuit voltage of the network, i.e. the voltage between nodes k and m without a non-linear branch connected.

The Thevenin resistance, $\mathbf{R}_{\text{Thevenin}}$, is a property of the linear network, and is calculated by taking the difference between the mth and k^{th} columns of $[G_{UU}]^{-1}$. This is achieved by solving $[G_{UU}]\mathbf{v}(t) = \mathbf{I}'_U$ with $\mathbf{I}'_U$ set to zero except -1.0 in the mth and 1.0 in the kth components. This can be interpreted as finding the terminal voltage when connecting a current source (of magnitude 1) between nodes k and m. The Thevenin resistance is pre-computed once, before entering the time step loop and only needs recomputing whenever switches open or close. Once the Thevenin resistance has been determined the procedure at each time step is thus:

(i) Compute the node voltages $\mathbf{v}_0(t)$ with the non-linear branch omitted. From this information extract the open circuit voltage between nodes k and m.
(ii) Solve the following two scalar equations simultaneously for i_{km}:

$$v_{km}(t) = v_{km0}(t) - \mathbf{R}_{\text{Thevenin}} i_{km} \qquad (4.52)$$

$$v_{km}(t) = f(i_{km}, di_{km}/dt, t, \ldots) \qquad (4.53)$$

This is depicted pictorially in Figure 4.21. If equation 4.53 is given as an analytic expression then a Newton–Raphson solution is used. When equation 4.53 is defined point-by-point as a piecewise linear curve then a search procedure is used to find the intersection of the two curves.
(iii) The final solution is obtained by superimposing the response to the current source i_{km} using equation 4.51. Superposition is permissible provided the rest of the network is linear.

The subsystem concept permits processing more than one non-linear branch, provided there is only one non-linear branch per subsystem.

If the non-linear branch is defined by $v_{km} = f(i_{km})$ or $v_{km} = R(t) \cdot i_{km}$ the solution is straightforward.

In the case of a non-linear inductor: $\lambda = f(i_{km})$, where the flux λ is the integral of the voltage with time, i.e.

$$\lambda(t) = \lambda(t - \Delta t) + \int_{t-\Delta t}^{t} v(u)\, du \qquad (4.54)$$

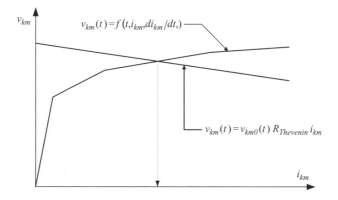

Figure 4.21   Pictorial view of simultaneous solution of two equations

The use of the trapezoidal rule gives:

$$\lambda(t) = \frac{\Delta t}{2} v(t) + \lambda_{\text{History}}(t - \Delta t) \quad (4.55)$$

where

$$\lambda_{\text{History}} = \lambda(t - \Delta t) + \frac{\Delta t}{2} v(t - \Delta)$$

Numerical problems can occur with non-linear elements if Δt is too large. The non-linear characteristics are effectively sampled and the characteristics between the sampled points do not enter the solution. This can result in artificial negative damping or hysteresis as depicted in Figure 4.22.

4.5.3   Piecewise linear method

The piecewise linear inductor characteristic, depicted in Figure 4.23, can be represented as a linear inductor in series with a voltage source. The inductance is changed (switched) when moving from one segment of the characteristic to the next. Although this model is easily implemented, numerical problems can occur as the need to change to the next segment is only recognised after the point exceeds the current segment (unless interpolation is used for this type of discontinuity). This is a switched model in that when the segment changes the branch conductance changes, hence the system conductance matrix must be modified.

A non-linear function can be modelled using a combination of piecewise linear representation and current source. The piecewise linear characteristics can be modelled with switched representation, and a current source used to correct for the difference between the piecewise linear characteristic and the actual.

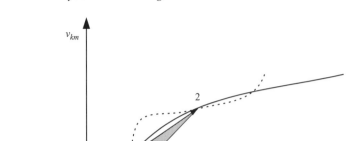

Figure 4.22 Artificial negative damping

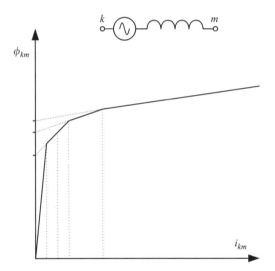

Figure 4.23 Piecewise linear inductor

4.6 Subsystems

Transmission lines and cables in the system being simulated introduce decoupling into the conductance matrix. This is because the transmission line model injects current at one terminal as a function of the voltage at the other at previous time steps. There is no instantaneous term (represented by a conductance in the equivalent models) that links one terminal to the other. Hence in the present time step, there is no dependency on the electrical conditions at the distant terminals of the line. This results in a block diagonal structure of the systems conductance matrix, i.e.

$$Y = \begin{bmatrix} [Y_1] & 0 & 0 \\ 0 & [Y_2] & 0 \\ 0 & 0 & [Y_3] \end{bmatrix}$$

Each decoupled block in this matrix is a subsystem, and can be solved at each time step independently of all other subsystems. The same effect can be approximated by introducing an interface into a coupled network. Care must be taken in choosing the interface point(s) to ensure that the interface variables must be sufficiently stable from one time point to the next, as one time step old values are fed across the interface. Capacitor voltages and inductor currents are the ideal choice for this purpose as neither can change instantaneously. Figure 4.24(a) illustrates coupled systems that are to be separated into subsystems. Each subsystem in Figure 4.24(b) is represented in the other by a linear equivalent. The Norton equivalent is constructed using information from the previous time step, looking into subsystem (2) from bus (A). The shunt connected at (A) is considered to be part of (1). The Norton admittance is:

$$Y_N = Y_A + \frac{(Y_B + Y_2)}{Z(1/Z + Y_B + Y_2)} \tag{4.56}$$

the Norton current:

$$I_N = I_A(t - \Delta t) + V_A(t - \Delta t)Y_A \tag{4.57}$$

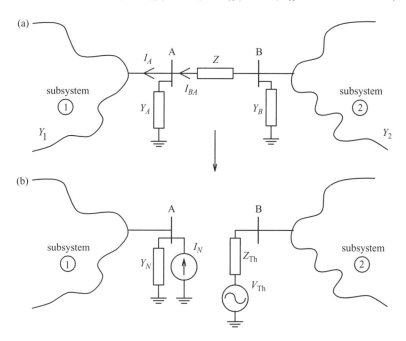

Figure 4.24 Separation of two coupled subsystems by means of linearised equivalent sources

94 Power systems electromagnetic transients simulation

the Thevenin impedance:

$$Z_{Th} = \frac{1}{Y_B} \left(\frac{Z + 1/(Y_1 + Y_A)}{Z + 1/(Y_1 + Y_A) + 1/Y_B} \right) \quad (4.58)$$

and the voltage source:

$$V_{Th} = V_B(t - \Delta t) + Z_{Th} I_{BA}(t - \Delta t) \quad (4.59)$$

The shunts (Y_N and Z_{Th}) represent the instantaneous (or impulse) response of each subsystem as seen from the interface busbar. If Y_A is a capacitor bank, Z is a series inductor, and Y_B is small, then

$Y_N \gg Y_A$ and $Y_N = I_{BA}(t - \Delta t)$ (the inductor current)
$Z_{Th} \gg Z$ and $V_{Th} = V_A(t - \Delta t)$ (the capacitor voltage)

When simulating HVDC systems, it can frequently be arranged that the subsystems containing each end of the link are small, so that only a small conductance matrix need be re-factored after every switching. Even if the link is not terminated at transmission lines or cables, a subsystem boundary can still be created by introducing a one time-step delay at the commutating bus. This technique was used in the EMTDC V2 B6P110 converter model, but not in version 3 because it can result in instabilities. A d.c. link subdivided into subsystems is illustrated in Figure 4.25.

Controlled sources can be used to interface subsystems with component models solved by another algorithm, e.g. components using numerical integration substitution on a state variable formulation. Synchronous machine and early non-switch-based

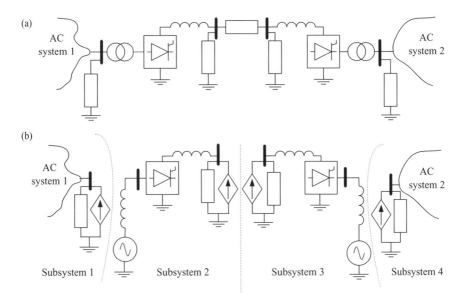

Figure 4.25 Interfacing for HVDC link

SVC models use a state variable formulation in PSCAD/EMTDC and appear to their parent subsystems as controlled sources. When interfacing subsystems, best results are obtained if the voltage and current at the point of connection are stabilised, and if each component/model is represented in the other as a linearised equivalent around the solution at the previous time step. In the case of synchronous machines, a suitable linearising equivalent is the subtransient reactance, which should be connected in shunt with the machine current injection. An RC circuit is applied to the machine interface as this adds damping to the high frequencies, which normally cause model instabilities, without affecting the low frequency characteristics and losses.

4.7 Sparsity and optimal ordering

The connectivity of power systems produces a conductance matrix $[G]$ which is large and sparse. By exploiting the sparsity, memory storage is reduced and significant solution speed improvement results. Storing only the non-zero elements reduces memory requirements and multiplying only by non-zero elements increases speed. It takes a computer just as long to multiply a number by zero as by any other number. Finding the solution of a system of simultaneous linear equations ($[G]V = I$) using the inverse is very inefficient as, although the conductance matrix is sparse, the inverse is full. A better approach is the triangular decomposition of a matrix, which allows repeated direct solutions without repeating the triangulation (provided the $[G]$ matrix does not change). The amount of fill-in that occurs during the triangulation is a function of the node ordering and can be minimised using optimal ordering [7].

To illustrate the effect of node ordering consider the simple circuit shown in Figure 4.26. Without optimal ordering the $[G]$ matrix has the structure:

$$\begin{bmatrix} X & X & X & X & X \\ X & X & 0 & 0 & 0 \\ X & 0 & X & 0 & 0 \\ X & 0 & 0 & X & 0 \\ X & 0 & 0 & 0 & X \end{bmatrix}$$

After processing the first row the structure is:

$$\begin{bmatrix} 1 & X & X & X & X \\ 0 & X & X & X & X \\ 0 & X & X & X & X \\ 0 & X & X & X & X \\ 0 & X & X & X & X \end{bmatrix}$$

When completely triangular the upper triangular structure is full

$$\begin{bmatrix} 1 & X & X & X & X \\ 0 & 1 & X & X & X \\ 0 & 0 & 1 & X & X \\ 0 & 0 & 0 & 1 & X \\ 0 & 0 & 0 & 0 & 1 \end{bmatrix}$$

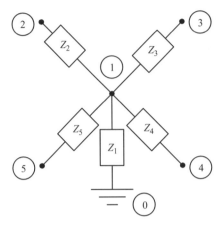

Figure 4.26   Example of sparse network

If instead node 1 is ordered last then the [G] matrix has the structure:

$$\begin{bmatrix} X & 0 & 0 & 0 & X \\ 0 & X & 0 & 0 & X \\ 0 & 0 & X & 0 & X \\ 0 & 0 & 0 & X & X \\ X & X & X & X & X \end{bmatrix}$$

After processing the first row the structure is:

$$\begin{bmatrix} 1 & 0 & 0 & 0 & X \\ 0 & X & 0 & 0 & X \\ 0 & 0 & X & 0 & X \\ 0 & 0 & 0 & X & X \\ 0 & X & X & X & X \end{bmatrix}$$

When triangulation is complete the upper triangular matrix now has less fill-in.

$$\begin{bmatrix} 1 & 0 & 0 & 0 & X \\ 0 & 1 & 0 & 0 & X \\ 0 & 0 & 1 & 0 & X \\ 0 & 0 & 0 & 1 & X \\ 0 & 0 & 0 & 0 & 1 \end{bmatrix}$$

This illustration uses the standard textbook approach of eliminating elements below the diagonal on a column basis; instead, a mathematically equivalent row-by-row elimination is normally performed that has programming advantages [5]. Moreover symmetry in the [G] matrix allows only half of it to be stored. Three ordering schemes have been published [8] and are now commonly used in transient programs. There is a tradeoff between the programming complexity, computation effort and level of optimality achieved by these methods, and the best scheme depends on the network topology, size and number of direct solutions required.

4.8 Numerical errors and instabilities

The trapezoidal rule contains a truncation error which normally manifests itself as chatter or simply as an error in the waveforms when the time step is large. This is particularly true if cutsets of inductors and current sources, or loops of capacitors and voltage sources exist. Whenever discontinuities occur (switching of devices, or modification of non-linear component parameters, ...) care is needed as these can initiate chatter problems or instabilities. Two separate problems are associated with discontinuities. The first is the error in making changes at the next time point after the discontinuity, for example current chopping in inductive circuits due to turning OFF a device at the next time point after the current has gone to zero, or proceeding on a segment of a piecewise linear characteristic one step beyond the knee point. Even if the discontinuity is not stepped over, chatter can occur due to error in the trapezoidal rule. These issues, as they apply to power electronic circuits, are dealt with further in Chapter 9.

Other instabilities can occur because of time step delays inherent in the model. For example this could be due to an interface between a synchronous machine model and the main algorithm, or from feedback paths in control systems (Chapter 8). Instabilities can also occur in modelling non-linear devices due to the sampled nature of the simulation as outlined in section 4.5. Finally 'bang–bang' instability can occur due to the interaction of power electronic device non-linearity and non-linear devices such as surge arresters. In this case the state of one influences the other and finding the appropriate state can be difficult.

4.9 Summary

The main features making numerical integration substitution a popular method for the solution of electromagnetic transients are: simplicity, general applicability and computing efficiency.

Its simplicity derives from the conversion of the individual power system elements (i.e. resistance, inductance and capacitance) and the transmission lines into Norton equivalents easily solvable by nodal analysis. The Norton current source represents the component past History terms and the Norton impedance consists of a pure conductance dependent on the step length.

By selecting the appropriate integration step, numerical integration substitution is applicable to all transient phenomena and to systems of any size. In some cases, however, the inherent truncation error of the trapezoidal method may lead to oscillations; improved numerical techniques to overcome this problem will be discussed in Chapters 5 and 9.

Efficient solutions are possible by the use of a constant integration step length throughout the study, which permits performing a single conductance matrix triangular factorisation before entering the time step loop. Further efficiency is achieved by exploiting the large sparsity of the conductance matrix.

An important concept is the use of subsystems, each of which, at a given time step, can be solved independently of the others. The main advantage of subsystems is the performance improvement when multiple time-steps/interpolation algorithms are used. Interpolating back to discontinuities is performed only on one subsystem. Subsystems also allow parallel processing hence real-time applications as well as interfacing different solution algorithms. If sparsity techniques are not used (early EMTDC versions) then subsystems also greatly improve the performance.

4.10 References

1 DOMMEL, H. W.: 'Digital computer solution of electromagnetic transients in single- and multi-phase networks', *IEEE Transactions on Power Apparatus and Systems*, 1969, **88** (2), pp. 734–41
2 DOMMEL, H. W.: 'Nonlinear and time-varying elements in digital simulation of electromagnetic transients', *IEEE Transactions on Power Apparatus and Systems*, 1971, **90** (6), pp. 2561–7
3 DOMMEL, H. W.: 'Techniques for analyzing electromagnetic transients', *IEEE Computer Applications in Power*, 1997, **10** (3), pp. 18–21
4 BRANIN, F. H.: 'Computer methods of network analysis', *Proceedings of IEEE*, 1967, **55**, pp. 1787–1801
5 DOMMEL, H. W.: 'Electromagnetic transients program reference manual: EMTP theory book' (Bonneville Power Administration, Portland, OR, August 1986).
6 DOMMEL, H. W.: 'A method for solving transient phenomena in multiphase systems', *Proc. 2nd Power System Computation Conference*, Stockholm, 1966, Rept. 5.8
7 SATO, N. and TINNEY, W. F.: 'Techniques for exploiting the sparsity of the network admittance matrix', *Transactions on Power Apparatus and Systems*, 1963, **82**, pp. 944–50
8 TINNEY, W. F. and WALKER, J. W.: 'Direct solutions of sparse network equations by optimally ordered triangular factorization', *Proceedings of IEEE*, 1967, **55**, pp. 1801–9

Chapter 5
The root-matching method

5.1 Introduction

The integration methods based on a truncated Taylor's series are prone to numerical oscillations when simulating step responses.

An interesting alternative to numerical integration substitution that has already proved its effectiveness in the control area, is the exponential form of the difference equation. The implementation of this method requires the use of root-matching techniques and is better known by that name.

The purpose of the root-matching method is to transfer correctly the poles and zeros from the s-plane to the z-plane, an important requirement for reliable digital simulation, to ensure that the difference equation is suitable to simulate the continuous process correctly.

This chapter describes the use of root-matching techniques in electromagnetic transient simulation and compares its performance with that of the conventional numerical integrator substitution method described in Chapter 4.

5.2 Exponential form of the difference equation

The application of the numerical integrator substitution method, and the trapezoidal rule, to a series RL branch produces the following difference equation for the branch:

$$i_k = \frac{(1 - \Delta t R/(2L))}{(1 + \Delta t R/(2L))} i_{k-1} + \frac{\Delta t/(2L)}{(1 + \Delta t R/(2L))} (v_k + v_{k-1}) \qquad (5.1)$$

Careful inspection of equation 5.1 shows that the first term is a first order approximation of e^{-x}, where $x = \Delta t R/L$ and the second term is a first order approximation of $(1 - e^{-x})/2$ [1]. This suggests that the use of the exponential expressions in the difference equation should eliminate the truncation error and thus provide accurate and stable solutions regardless of the time step.

Equation 5.1 can be expressed as:

$$i_k = e^{-\Delta t R/L} i_{k-1} + \left(1 - e^{-\Delta t R/L}\right) v_k \qquad (5.2)$$

Although the exponential form of the difference equation can be deduced from the difference equation developed by the numerical integrator substitution method, this approach is unsuitable for most transfer functions or electrical circuits, due to the difficulty in identifying the form of the exponential that has been truncated. The root-matching technique provides a rigorous method.

Numerical integrator substitution provides a mapping from continuous to discrete time, or equivalently from the s to the z-domain. The integration rule used will influence the mapping and hence the error. Table 5.1 shows the characteristics of forward rectangular, backward rectangular (implicit or backward Euler) and trapezoidal integrators, including the mapping of poles in the left-hand half s-plane into the z-plane. If the continuous system is stable (has all its poles in the left-hand half s-plane) then under forward Euler the poles in the z-plane can lie outside the unit circle and hence an unstable discrete system can result. Both backward Euler and the trapezoidal rule give stable discrete systems, however stability gives no indication of the accuracy of the representation.

The use of the trapezoidal integrator is equivalent to the bilinear transform (or Tustin method) for transforming from a continuous to a discrete system, the former being the time representation of the latter. To illustrate this point the bilinear transform will be next derived from the trapezoidal rule.

In the s-plane the expression for integration is:

$$\frac{Y(s)}{X(s)} = \frac{1}{s} \qquad (5.3)$$

In discrete time the trapezoidal rule is expressed as:

$$y_n = y_{n-1} + \frac{\Delta t}{2}(x_n + x_{n-1}) \qquad (5.4)$$

Transforming equation 5.4 to the z-plane gives:

$$Y(z) = z^{-1} Y(z) + \frac{\Delta t}{2}(X(z) + X(z)z^{-1}) \qquad (5.5)$$

Rearranging gives for integration in the z-domain:

$$\frac{Y(z)}{X(z)} = \frac{\Delta t}{2} \frac{(1 + z^{-1})}{(1 - z^{-1})} \qquad (5.6)$$

Equating the two integration expressions (i.e. equations 5.3 and 5.6) gives the well known bilinear transform equation:

$$s \approx \frac{2}{\Delta t} \frac{(1 - z^{-1})}{(1 + z^{-1})} \qquad (5.7)$$

Table 5.1 *Integrator characteristics*

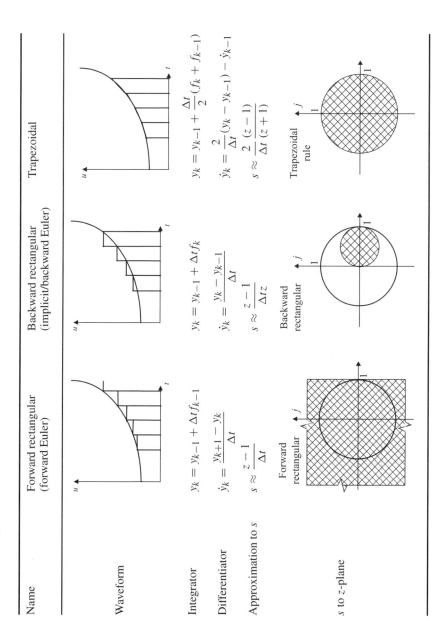

Hence the trapezoidal rule and the bilinear transform give the same mapping between the s and z-planes and are therefore identical.

Equation 5.7 can also be derived from an approximation of an exponential. The actual relationship between s and z is:

$$z = e^{s\Delta t} \tag{5.8}$$

Hence

$$z^{-1} = e^{-s\Delta t} \tag{5.9}$$

Expressing $e^{-s\Delta t}$ as two exponential functions and then using the series approximation gives:

$$z^{-1} = e^{-s\Delta t} = \frac{e^{-s\Delta t/2}}{e^{s\Delta t/2}} \approx \frac{(1 - s\Delta t/2)}{(1 + s\Delta t/2)} \tag{5.10}$$

Rearranging for s gives:

$$s \approx \frac{2}{\Delta t} \cdot \frac{(1 - z^{-1})}{(1 + z^{-1})} \tag{5.11}$$

which is identical to equation 5.7. Hence the trapezoidal rule (and many other integration rules for that matter) can be considered as a truncated series approximation of the exact relationship between s and z.

5.3 z-domain representation of difference equations

Digital simulation requires the use of the z-domain, either in the form of a transfer function or as an equivalent difference equation.

In the transfer function approach:

$$H(z) = \frac{a_0 + a_1 \cdot z^{-1} + a_2 \cdot z^{-2} + \cdots + a_m \cdot z^{-m}}{1 + b_1 \cdot z^{-1} + b_2 \cdot z^{-2} + \cdots + b_m \cdot z^{-m}} = \frac{Y(z)}{U(z)} \tag{5.12}$$

or expressed as a two-sided recursion [2]

$$\left(a_0 + a_1 \cdot z^{-1} + a_2 \cdot z^{-2} + \cdots + a_m \cdot z^{-m}\right) U(z)$$
$$= \left(1 + b_1 \cdot z^{-1} + b_2 \cdot z^{-2} + \cdots + b_m \cdot z^{-m}\right) Y(z) \tag{5.13}$$

Equation 5.13 can be implemented directly and without any approximation as a Norton equivalent.

Rearranging equation 5.13 gives:

$$Y(z) = \left(a_0 + a_1 \cdot z^{-1} + a_2 \cdot z^{-2} + \cdots + a_m \cdot z^{-m}\right) U(z)$$
$$- \left(b_1 \cdot z^{-1} + b_2 \cdot z^{-2} + \cdots + b_m \cdot z^{-m}\right) Y(z) \tag{5.14}$$

The corresponding difference equation is:

$$y(k\Delta t) = (a_0 \cdot u + a_1 \cdot u_{-1} + a_2 \cdot u_{-2} + \cdots + a_m \cdot u_{-m})$$
$$- (b_1 \cdot y_{-1} + b_2 \cdot y_{-2} + \cdots + b_m \cdot y_{-m}) \tag{5.15}$$

The first term on the right side of equation 5.15 is the instantaneous term between input and output, while the other terms are history terms. Hence the conductance is a_0 and the history term is:

$$a_1 u_{-1} + a_2 u_{-2} + \cdots + a_m u_{-m} - b_1 y_{-1} + b_2 y_{-2} + \cdots + b_m y_{-m} \tag{5.16}$$

Whereas in the s-domain stability is ensured if poles are in the left-hand half-plane, the equivalent criterion in the z-plane is that the poles must reside inside the unit circle.

In the transformation from the s to z-plane, as required by digital simulation, the poles and zeros must be transformed correctly and this is the purpose of the root-matching technique. In other words, to ensure that a difference equation is suitable to simulate a continuous process the poles, zeros and final value of the difference equation should match those of the actual system. If these conditions are met the difference equations are intrinsically stable, provided the actual system is stable, regardless of the step size. The difference equations generated by this method involve exponential functions, as the transform equation $z^{-1} = e^{-s\Delta t}$ is used rather than some approximation to it.

When integrator substitution is used to derive a difference equation, the poles and zeros usually are not inspected, and these can therefore be poorly positioned or there can even be extra poles and zeros. Because the poles and zeros of the difference equation do not match well those of the continuous system, there are situations when the difference equation is a poor representation of the continuous system.

The steps followed in the application of the root-matching technique are:

1. Determine the transfer function in the s-plane, $H(s)$ and the position of its poles and zeros.
2. Write the transfer function $H(s)$ in the z-plane using the mapping $z = e^{s\Delta t}$, thus ensuring the poles and zeros are in the correct place. Also add a constant to allow adjustment of the final value.
3. Use the final value theorem to compute the final value of $H(s)$ for a unit step input.
4. Determine the final value of $H(z)$ for unit step input and adjust the constant to be the correct value.
5. Add extra zeros depending on the assumed input variation between solution points.
6. Write the resulting z-domain equation in the form of a difference equation.

The final value of $H(s)$ must not be zero to allow the final value matching constant in $H(z)$ to be determined. When that happens the final value is matched for a different input. For example some systems respond to the derivative of the input and in such cases the final value for a unit ramp input is used.

Appendix E (sections E.1 and E.2) illustrate the use of the above procedure with a single order lag function and a first order differential pole, respectively. Table 5.2

Table 5.2 *Exponential form of difference equation*

Transfer function	Expression for Norton
$H(s) = \dfrac{G}{1+s\tau}$	$R = 1/k$ $I_{\text{History}} = e^{-\Delta t/\tau} \cdot I_{t-\Delta t}$ $k = G \cdot (1 - e^{-\Delta t/\tau})$
$H(s) = G \cdot (1+s\tau)$	$R = 1/k$ $I_{\text{History}} = -k \cdot e^{-\Delta t/\tau} \cdot V_{t-\Delta t}$ $k = \dfrac{G}{(1 - e^{-\Delta t/\tau})}$
$H(s) = \dfrac{G \cdot s}{1+s\tau}$	$R = 1/k$ $I_{\text{History}} = e^{-\Delta t/\tau} \cdot I_{t-\Delta t} - k \cdot V_{t-\Delta t}$ $k = \dfrac{G \cdot (1 - e^{-\Delta t/\tau})}{\Delta t}$
$H(s) = \dfrac{G \cdot (1+s\tau_1)}{(1+s\tau_2)}$	$R = 1/k$ $I_{\text{History}} = e^{-\Delta t/\tau_2} \cdot I_{t-\Delta t} - k \cdot V_{t-\Delta t} \cdot e^{-\Delta t/\tau_1}$ $k = \dfrac{G \cdot (1 - e^{-\Delta t/\tau_1})}{(1 - e^{-\Delta t/\tau_2})}$
$H(s) = \dfrac{G \cdot \omega_n^2}{s^2 + 2\zeta\omega_n s + \omega_n^2}$	$R = 1/k$ $I_{\text{History}} = A \cdot I_{t-\Delta t} - B \cdot I_{t-2\Delta t}$ $k = G \cdot (1 - e^{\Delta t \cdot p1}) \cdot (1 - e^{\Delta t \cdot p2})$ $ = G \cdot (1 - A + B)$
$H(s) = \dfrac{G \cdot s\omega_n^2}{s^2 + 2\zeta\omega_n s + \omega_n^2}$	$R = 1/k$ $I_{\text{History}} = -k \cdot V_{t-\Delta t} + A \cdot I_{t-\Delta t} - B \cdot I_{t-2\Delta t}$ $k = \dfrac{G \cdot (1 - e^{\Delta t \cdot p1}) \cdot (1 - e^{\Delta t \cdot p2})}{\Delta t}$ $ = \dfrac{G \cdot (1 - A + B)}{\Delta t}$
$H(s) = \dfrac{G \cdot (s^2 + 2\zeta\omega_n + \omega_n^2)}{s\omega_n}$	$R = k$ $I_{\text{History}} = I_{t-\Delta t} - \dfrac{A}{k} \cdot V_{t-\Delta t} + \dfrac{B}{k} \cdot I_{t-2\Delta t}$ $k = \dfrac{G \cdot (1 - e^{\Delta t \cdot p1}) \cdot (1 - e^{\Delta t \cdot p2})}{\Delta t}$ $ = \dfrac{G \cdot (1 - A + B)}{\Delta t}$

gives expressions of the exponential form of difference equation for various s-domain transfer functions.

In Table 5.2, A and B are as follows:
If two real roots ($\zeta > 1$):

$$A = 2e^{-\zeta \omega_n \Delta t} \left(e^{\Delta t \omega_n \sqrt{\zeta^2 - 1}} + e^{-\Delta t \omega_n \sqrt{\zeta^2 - 1}} \right)$$

$$B = e^{-2\zeta \omega_n \Delta t}$$

If two repeated roots ($\zeta = 1$):

$$A = 2e^{-\omega_n \Delta t}$$

$$B = e^{-2\omega_n \Delta t}$$

If complex roots ($\zeta < 1$):

$$A = 2e^{-\zeta \omega_n \Delta t} \cos\left(\omega_n \Delta t \sqrt{1 - \zeta^2} \right)$$

$$B = e^{-2\zeta \omega_n \Delta t}$$

By using the input form shown in Figure 5.13(a) on page 113, the homogeneous solution of the difference equation matches the homogeneous solution of the differential equation exactly. It also generates a solution of the differential equation's response that is exact for the step function and a good approximation for an arbitrary forcing function.

5.4 Implementation in EMTP algorithm

The exponential form of the difference equation can be viewed as a Norton equivalent in just the same way as the difference equation developed by Dommel's method, the only difference being the formula used for the derivation of the terms. Figure 5.1 illustrates this by showing the Norton equivalents of a series RL branch developed using Dommel's method and the exponential form respectively. Until recently it has not been appreciated that the exponential form of the difference equation can be applied to the main electrical components as well as control equations, in time domain simulation. Both can be formed into Norton equivalents, entered in the conductance matrix and solved simultaneously with no time step delay in the implementation.

To remove all the numerical oscillations when the time step is large compared to the time constant, the difference equations developed by root-matching techniques must be implemented for all series and parallel RL, RC, LC and RLC combinations.

The network solution of Dommel's method is:

$$[G]\mathbf{v}(t) = \mathbf{i}(t) + \mathbf{I}_{\text{History}} \tag{5.17}$$

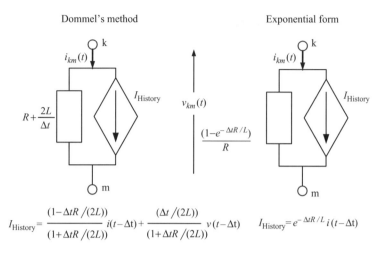

Figure 5.1   Norton equivalent for RL branch

Structurally the root-matching algorithm is the same as Dommel's, the only difference being in the formula used for the derivation of the conductance and past history terms. Moreover, although the root-matching technique can also be applied to single L or C elements, there is no need for that, as in such cases the response is no longer of an exponential form. Hence Dommel's algorithm is still used for converting individual L and C elements to a Norton equivalent. This allows difference equations, hence Norton equivalents, based on root-matching methods to be used in existing electromagnetic transient programs easily, yet giving unparalleled improvement in accuracy, particularly for large time steps.

In the new algorithm, I_{History} includes the history terms of both Dommel's and the root-matching method. Similarly the conductance matrix, which contains the conductance terms of the Norton equivalents, includes some terms from Dommel's technique and others of the exponential form developed from the root-matching technique.

The main characteristics of the exponential form that permit an efficient implementation are:

- The exponential term is calculated and stored prior to entering the time step loop.
- During the time step loop only two multiplications and one addition are required to calculate the I_{History} term. It is thus more efficient than NIS using the trapezoidal rule.
- Fewer previous time step variables are required. Only the previous time step current is needed for an RL circuit, while Dommel's method requires both current and voltage at the previous time-step.

Three simple test cases are used to illustrate the algorithm's capability [3]. The first case shown in Figure 5.2 relates to the switching of a series RL branch. Using a $\Delta t = \tau$ time step (τ being the time constant of the circuit), Figure 5.3 shows the current response derived from Dommel's method, the exponential method and

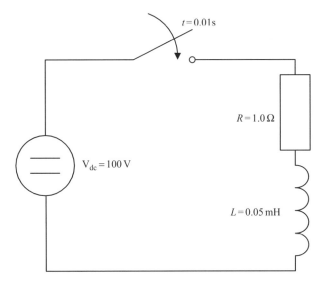

Figure 5.2   Switching test system

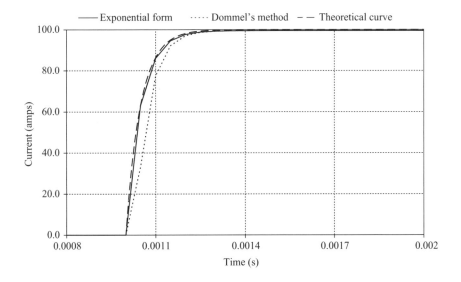

Figure 5.3   Step response of switching test system for $\Delta t = \tau$

continuous analysis (theoretical result). At this time step, Dommel's method does not show numerical oscillations, but introduces considerable error. The results shown in Figure 5.4 correspond to a time step of $\Delta t = 5\tau$ ($\tau = 50\,\mu s$). Dommel's method now exhibits numerical oscillations due to truncation errors, whereas the exponential form gives the correct answer at each solution point. Increasing the time step to

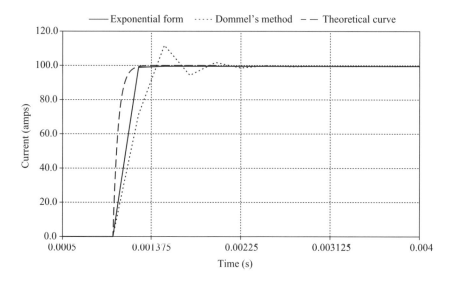

Figure 5.4 Step response of switching test system for $\Delta t = 5\tau$

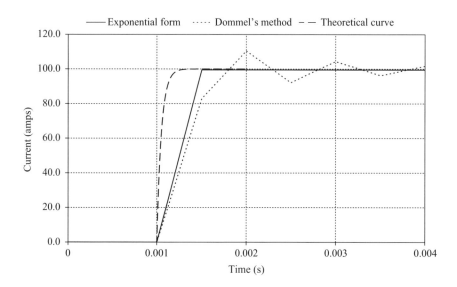

Figure 5.5 Step response of switching test system for $\Delta t = 10\tau$

$\Delta t = 10\tau$ results in much greater numerical oscillation for Dommel's method, while the exponential form continues to give the exact answer (Figure 5.5).

The second test circuit, shown in Figure 5.6, consists of a *RLC* circuit with a resonant frequency of 10 kHz, excited by a 5 kHz current source. Figures 5.7 and 5.8

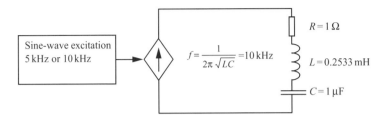

Figure 5.6   Resonance test system

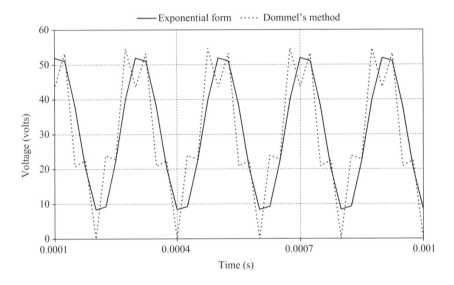

Figure 5.7   Comparison between exponential form and Dommel's method to a 5 kHz excitation for resonance test system. $\Delta t = 25\,\mu s$

show the voltage response using 25 μs and 10 μs time steps, respectively. Considerable deviation from the expected sinusoidal waveform is evident for Dommel's method. Figure 5.9 shows the comparison when the excitation contains a 10 kHz component of 1 A peak for a time-step of 10 μs. At that frequency the inductance and capacitance cancel out and the exponential form gives the correct response, i.e. a 2 V peak-to-peak 10 kHz sinusoid on top of the d.c. component (shown in Figure 5.10), whereas Dommel's method oscillates. The inductor current leads the capacitor voltage by 90 degrees. Therefore, when initialising the current to zero the capacitor voltage should be at its maximum negative value. If the capacitor voltage is also initialised to zero a d.c. component of voltage ($|V| = I/\omega C$) is effectively added, which is equivalent to an additional charge on the capacitor to change its voltage from maximum negative to zero.

A third test circuit is used to demonstrate the numerical problem of current chopping in inductive circuits. A common example is the modelling of power electronic

110 *Power systems electromagnetic transients simulation*

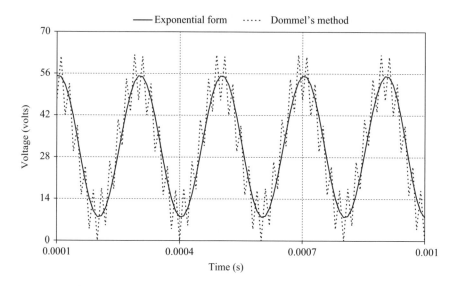

Figure 5.8 *Comparison between exponential form and Dommel's method to a 5 kHz excitation for resonance test system.* $\Delta t = 10\,\mu s$

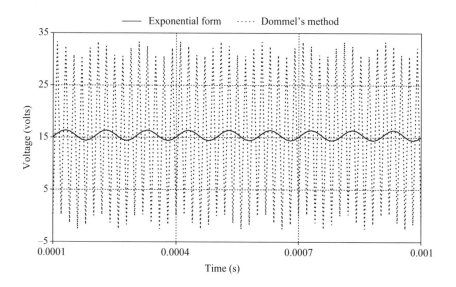

Figure 5.9 *Comparison between exponential form and Dommel's method to 10 kHz excitation for resonance test system*

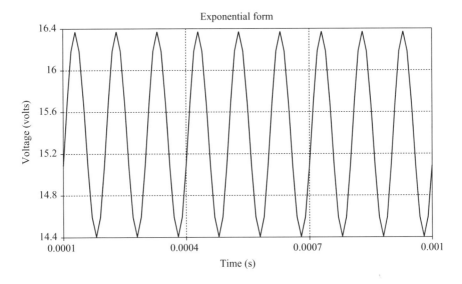

Figure 5.10 Response of resonance test system to 10 kHz excitation, blow-up of exponential form's response

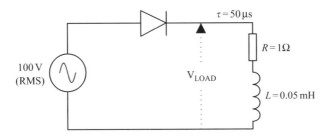

Figure 5.11 Diode test system

devices such as diodes and thyristors. Although the changes of state are constrained to occur at multiples of the step length, the current falls to zero between these points [4]; thus the change occurs at the time point immediately after and hence effectively turning the device off with a slight negative current. To demonstrate this effect Figure 5.11 uses a simple system where an a.c. voltage source supplies power to an RL load via a diode. Figure 5.12(a) shows the load voltage for the exponential form and Dommel's method using a time-step of 500 μs. This clearly shows the superiority of the exponential form of difference equation. The numerical oscillation at switch-off depends on how close to a time point the current drops to zero, and hence the size of negative current at the switching point. The negative current at switching is clearly evident in the load current waveform shown in Figure 5.12(b).

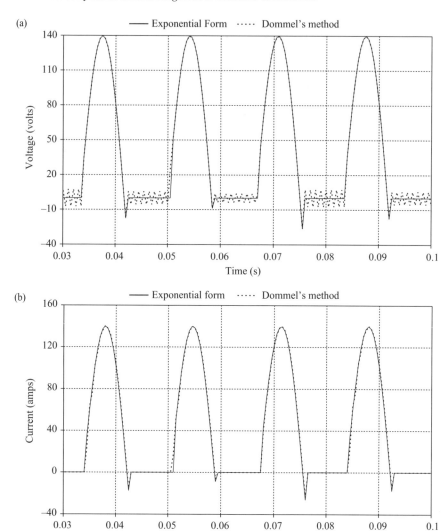

Figure 5.12 Response to diode test system (a) Voltage (b) Current

These three test circuits clearly demonstrate the accuracy and stability of the exponential form of the difference equation regardless of the time step.

5.5 Family of exponential forms of the difference equation

In the root-matching technique used to derive the exponential form of a difference equation the poles and zeros of the s-domain function are matched in the z-domain

The root-matching method 113

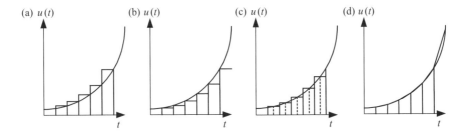

Figure 5.13   Input as function of time

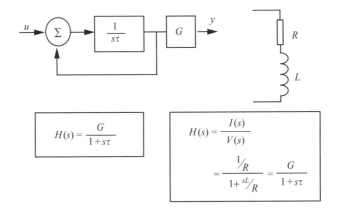

Figure 5.14   Control or electrical system as first order lag

function. Extra zeros are added based on the assumed discretisation on the input, which is continuous [5]. Figure 5.13 shows some of the possible discretisations and these result in a family of exponential forms of the difference equation. The root-matching technique is equally applicable to equations representing control or electrical systems [6]. For each of the discretisation types, with reference to the first order lag function shown in Figure 5.14, the use of the root-matching technique expressed as a rational function in z^{-1} produces the following exponential form difference equations.

Input type (a):
$$\frac{y(z)}{u(z)} = \frac{b/a(1 - e^{-a\Delta t})}{(1 - z^{-1}e^{-a\Delta t})}$$

Input type (b):
$$\frac{y(z)}{u(z)} = \frac{b/a(1 - e^{-a\Delta t})z^{-1}}{(1 - z^{-1}e^{-a\Delta t})}$$

Input type (c):
$$\frac{y(z)}{u(z)} = \frac{b/(2a)(1 - e^{-a\Delta t})(1 + z^{-1})}{(1 - z^{-1}e^{-a\Delta t})}$$

Input type (d):

$$\frac{y(z)}{u(z)} = \frac{b/a(-e^{-a\Delta t} + 1/(a\Delta t)(1 - e^{-a\Delta t}))z^{-1} + b/a(1 - 1/(a\Delta t)(1 - e^{-a\Delta t}))}{(1 - z^{-1}e^{-a\Delta t})}$$

Table E.3 (Appendix E) summarises the resulting difference equation for the family of exponential forms developed using root-matching techniques. The table also contains the difference equations derived from trapezoidal integrator substitution. The difference equations are then converted to the form:

$$(a_0 + a_1 z^{-1})/(b_0 + b_1 z^{-1}) \text{ or } (a'_0 + a'_1 z^{-1})/(1 + b_1 z^{-1}) \quad \text{if } b_0 \text{ is non-zero}$$

Tables E.1 and E.2 give the coefficients of a rational function in z^{-1} that represents each difference equation for the family of exponential forms, for admittance and impedance respectively. It can be shown that the difference equation obtained assuming type input (d) is identical to that obtained from the recursive convolution technique developed by Semlyen and Dabuleanu [7].

5.5.1 Step response

A comparison of step responses is made here with reference to the simple switching of a series RL branch, shown in Figure 5.2. Figure 5.15 shows the current magnitude using the difference equations generated by Dommel's method, root-matching for input types (a), (b), (c) and (d) and the theoretical result for $\Delta t = \tau$ ($\tau = 50\,\mu s$). Figures 5.16 and 5.17 show the same comparison for $\Delta t = 5\tau$ and $\Delta t = 10\tau$, respectively. Note that in the latter cases Dommel's method exhibits numerical oscillation. Root-matching type (a) gives the exact answer at each time point as its discretisation of the input is exact. Root-matching type (b) gives the exact values but one time step late as its discretisation of the input is a step occurring one time step later. Root-matching type (c) is an average between the previous two root-matching techniques.

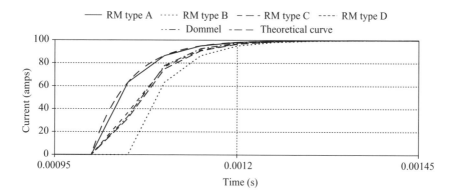

Figure 5.15 Comparison of step response of switching test system for $\Delta t = \tau$

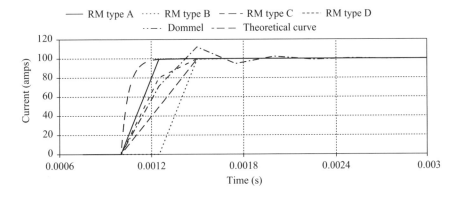

Figure 5.16 Comparison of step response of switching test system for $\Delta t = 5\tau$

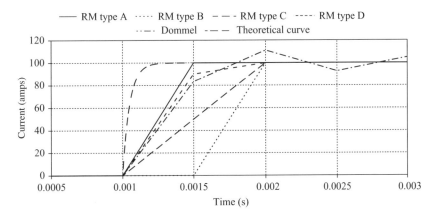

Figure 5.17 Comparison of step response of switching test system for $\Delta t = 10\tau$

Although from Figure 5.13 it would seem that root-matching type (d) should provide the best approximation to an arbitrary waveform, this input resulted in significant inaccuracies. The reason is that this discretisation is unable to model a pure step, i.e. there will always be a slope, which is a function of Δt, as depicted in Figure 5.18. However if Δt is sufficiently small then this method will provide a good approximation to a step response.

Root-matching type (b) results in terms from the previous time step only, that is only a current source but no parallel conductance. This can cause simulation problems if a non-ideal switch model is used. If a switch is modelled by swapping between high and low resistance states then even when it is OFF, a very small current flow is calculated. This current is then multiplied by $e^{-\Delta t/\tau}$ and injected into the high impedance switch and source, which results in a voltage appearing at the terminals. If an ideal switch cannot be modelled, judicious selection of the switch parameters can

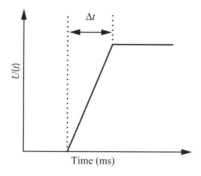

Figure 5.18 Root-matching type (d) approximation to a step

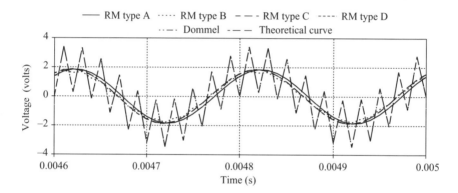

Figure 5.19 Comparison with a.c. excitation (5 kHz) ($\Delta t = \tau$)

remove the problem, however a better solution is to use a controlled voltage source when applying the step in voltage.

5.5.2 Steady-state response

The second test system, shown in Figure 5.6, consists of an RL branch, excited by a 5 kHz current source. Figure 5.19 shows the voltage response using a 10 μs step length for each of the difference equations. The theoretical answer is $1.86 \sin(\omega t - \phi)$, where $\phi = -57.52°$. Root-matching types (a), (b) and (d) give good answers; however, root-matching type (c) gives results indistinguishable from Dommel's method.

It should be noted that as the excitation is a current source and root-matching type (b) is also a pure current source, there are two current sources connected to one node. Hence, in order to get answers for this system a parallel conductance must be added to enable Kirchhoff's current law to be satisfied. The conductance value must be large enough so as not to influence the solution significantly but not too large otherwise instability will occur. However, from a stability viewpoint the poles in the z-plane for the complete solution fall outside the unit circle when parallel

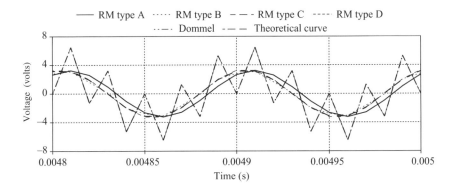

Figure 5.20 Comparison with a.c. excitation (10 kHz) ($\Delta t = \tau$)

resistance is increased. Using a voltage source rather than current source excitation would eliminate the need for a parallel resistor in the root-matching type (b).

The same conclusions are found from a simulation using 10 kHz as the excitation frequency and a step length of 10 μs. The theoretical answer is $3.30 \sin(\Delta t - \phi)$, where $\phi = -72.43°$. In this case root-matching types (a), (b) and type (d) give good answers, and again, root-matching type (c) gives results indistinguishable from Dommel's method (this is shown in Figure 5.20).

5.5.3 Frequency response

The frequency response of each difference equation can be reconstructed from the rational function by using the following equation:

$$Z(f) = \frac{\sum_{i=0}^{n} a_i e^{-j\omega i \Delta t}}{\sum_{i=0}^{n} b_i e^{-j\omega i \Delta t}} \quad (5.18)$$

which for root-matching, simplifies to:

$$\frac{1}{G_{equ}} - \frac{e^{-\Delta t/\tau} e^{-j\omega \Delta t}}{G_{equ}} = ((1 - \cos(\omega \Delta t)e^{-\Delta t/\tau}) + j \sin(\omega \Delta t))/G_{equ} \quad (5.19)$$

The magnitude and phase components are:

$$|Z(f)| = \sqrt{(1 - \cos(\omega \Delta t)e^{-\Delta t/\tau})^2 + \sin(\omega \Delta t)^2}/G_{equ}$$

and

$$\angle Z(f) = \tan^{-1}\left(\frac{\sin(\omega \Delta t)}{1 - \cos(\omega \Delta t)e^{-\Delta t/\tau}}\right)$$

The corresponding equation for an s-domain function is:

$$h(f) = \frac{\sum_{k=0}^{n} a_k (j\omega)^k}{1 + \sum_{k=1}^{n} b_k (j\omega)^k}$$

118 *Power systems electromagnetic transients simulation*

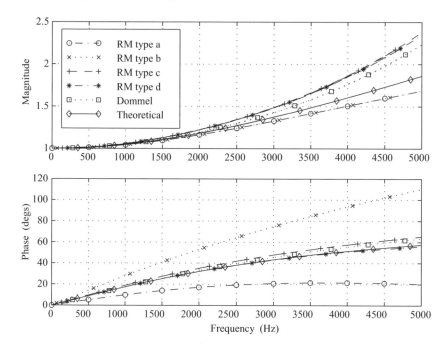

Figure 5.21 Frequency response for various simulation methods

The results of calculations performed using MATLAB (the code is given in Appendix F.4), are displayed in Figure 5.21. These results were verified by performing injection tests into the appropriate difference equation using PSCAD/EMTDC simulation software. As expected, the root-matching methods (a) and (b) provide the closest match to the theoretical magnitude response, while root-matching methods (c) and (d) are similar to the trapezoidal rule. The phase response clearly shows the phase advance and phase lag inherent in the various discretisations used in the various root-matching methods. Root-matching method (a) shows the phase advance and method (b) the phase lag expected. Root-matching methods (c) and (d) and trapezoidal integration show a considerably better phase response.

The trapezoidal rule assumes a linear variation between time points. An exponential form of the difference equation can also be derived assuming constant input between solution points. Hence the exponential form of a circuit or transfer function is not unique but depends on the assumed variation in input between time points.

5.6 Example

For the test system shown in Figure 5.2, if the switch is closed at $t = 1.0$ s the exact solution is:

$$i(t) = \frac{V_{DC}}{R}(1 - e^{-(t-1.0)/\tau})$$

for $t \geq 1.0$ where $\tau = L/R$.

For this example the parameters of the circuit are: $R = 100\,\Omega$, $L = 0.05\,\text{mH}$, $V_{DC} = 100\,\text{V}$. If $\Delta t = \tau = 50\,\mu\text{s}$ the difference equation obtained using the trapezoidal rule is:

$$i(t + \Delta t) = \tfrac{1}{3}i(t + \Delta t) + \tfrac{1}{3}(v(t + \Delta t) + v(t))$$

For root-matching the difference equation is:

$$i(t + \Delta t) = i(t)e^{-1} + v(t + \Delta t)(1 - e^{-1})$$

and the results are summarised in Table 5.3.

For $\Delta t = 5\tau = 250\,\mu\text{s}$ the difference equations are:

$i(t + \Delta t) = \tfrac{-3}{7}i(t + \Delta t) + \tfrac{5}{7}(v(t + \Delta t) + v(t))$ – for the trapezoidal rule
$i(t + \Delta t) = i(t)e^{-5} + v(t + \Delta t)(1 - e^{-5})$ – for the root-matching method

and the corresponding results are summarised in Table 5.4.

Finally for $\Delta t = 10\tau = 500\,\mu\text{s}$ the difference equations are:

$i(t + \Delta t) = \tfrac{-2}{3}i(t + \Delta t) + \tfrac{5}{6}(v(t + \Delta t) + v(t))$ – for the trapezoidal rule
$i(t + \Delta t) = i(t)e^{-10} + v(t + \Delta t)(1 - e^{-10})$ – for the root-matching method

and the results are summarised in Table 5.5.

Table 5.3 Response for $\Delta t = \tau = 50\,\mu\text{s}$

	Exact solution	Trapezoidal rule	Root-matching
1.0	0.0	0.0	0.0
$1.0 + \Delta t$	63.212056	33.333333	63.212056
$1.0 + 2\Delta t$	86.466472	77.777777	86.466472
$1.0 + 3\Delta t$	95.021293	92.2530864	95.021293
$1.0 + 4\Delta t$	98.168436	97.530864	98.168436
$1.0 + 5\Delta t$	99.326205	99.176955	99.326205

Table 5.4 Response for $\Delta t = 5\tau = 250\,\mu\text{s}$

	Exact solution	Trapezoidal rule	Root-matching
1.0	0.0	0.0	0.0
$1.0 + \Delta t$	99.326205	71.428571	99.326205
$1.0 + 2\Delta t$	99.995460	112.244898	99.995460
$1.0 + 3\Delta t$	99.999969	94.752187	99.999969
$1.0 + 4\Delta t$	100.000000	102.249063	100.000000

120 Power systems electromagnetic transients simulation

Table 5.5 Response for $\Delta t = 10\tau = 500\,\mu s$

	Exact solution	Trapezoidal rule	Root-matching
1.0	0.0	0.0	0.0
$1.0 + \Delta t$	99.995460	83.333333	99.995460
$1.0 + 2\Delta t$	100.000000	111.111111	100.000000
$1.0 + 3\Delta t$	100.000000	92.592593	100.000000

To demonstrate why root-matching is so good let us consider the exact response at a discrete time t_k, i.e.

$$i(t_k) = \frac{V_{dc}}{R}(1 - e^{-(t_k - 1.0)/\tau}) \tag{5.20}$$

which, expressed as a function of a previous time point at $t_k - \Delta t$, becomes:

$$i(t_k) = \frac{V_{dc}}{R}(1 - e^{-(t_k - 1.0)/\tau}) = \frac{V_{dc}}{R}(1 - e^{-\Delta t/\tau} e^{-(t_k - \Delta t - 1.0)/\tau}) \tag{5.21}$$

Now the same must be true for the previous time point, hence from equation 5.20:

$$i(t_k - \Delta t) = \frac{V_{dc}}{R}(1 - e^{-(t_k - \Delta t - 1.0)/\tau}) \tag{5.22}$$

Hence

$$e^{-(t_k - \Delta t - 1.0)/\tau} = 1 - \frac{R}{V_{dc}} i(t_k - \Delta t) \tag{5.23}$$

Substituting equation 5.23 in equation 5.21 gives:

$$i(t_k) = \frac{V_{dc}}{R}\left(1 - e^{-\Delta t/\tau}\left(1 - \frac{R}{V_{dc}} i(t_k - \Delta t)\right)\right)$$

$$= e^{-\Delta t/\tau} i(t_k - \Delta t) + \frac{V_{dc}}{R}(1 - e^{-\Delta t/\tau}) \tag{5.24}$$

which is exactly the difference equation for the root-matching method.

5.7 Summary

An alternative to the difference equation using the trapezoidal integration developed in Chapter 4 for the solution of the differential equations has been described in this chapter. It involves the exponential form of the difference equation and has been developed using the root-matching technique. The exponential form offers the following advantages:

- Eliminates truncation errors, and hence numerical oscillations, regardless of the step length used.
- Can be applied to both electrical networks and control blocks.
- Can be viewed as a Norton equivalent in exactly the same way as the difference equation developed by the numerical integration substitution (NIS) method.
- It is perfectly compatible with NIS and the matrix solution technique remains unchanged.
- Provides highly efficient and accurate time domain simulation.

The exponential form can be implemented for all series and parallel RL, RC, LC and RLC combinations, but not arbitrary components and hence is not a replacement for NIS but a supplement.

5.8 References

1 WATSON, N. R. and IRWIN, G. D.: 'Electromagnetic transient simulation of power systems using root-matching techniques', *Proceedings IEE, Part C*, 1998, **145** (5), pp. 481–6
2 ANGELIDIS, G. and SEMLYEN, A.: 'Direct phase-domain calculation of transmission line transients using two-sided recursions', *IEEE Transactions on Power Delivery*, 1995, **10** (2), pp. 941–7
3 WATSON, N. R. and IRWIN, G. D.: 'Accurate and stable electromagnetic transient simulation using root-matching techniques', *International Journal of Electrical Power & Energy Systems*, Elsevier Science Ltd, 1999, **21** (3), pp. 225–34
4 CAMPOS-BARROS, J. G. and RANGEL, R. D.: 'Computer simulation of modern power systems: the elimination of numerical oscillation caused by valve action', Proceedings of 4th International Conference on *AC and DC Power Transmission*, London, 1985, Vol. IEE Conf. Publ., **255**, pp. 254–9
5 WATSON, N. R. and IRWIN, G. D.: 'Comparison of root-matching techniques for electromagnetic transient simulation', *IEEE Transactions on Power Delivery*, 2000, **15** (2), pp. 629–34
6 WATSON, N. R., IRWIN, G. D. and NAYAK, O.: 'Control modelling in electromagnetic transient simulations', Proceedings of International Conference on *Power System Transients (IPST'99)*, June 1999, pp. 544–8
7 SEMLYEN, A. and DABULEANU, A.: 'Fast and accurate switching transient calculations on transmission lines with ground return using recursive convolutions', *IEEE Transactions on Power Apparatus and Systems*, 1975, **94** (2), pp. 561–71

Chapter 6

Transmission lines and cables

6.1 Introduction

Approximate nominal PI section models are often used for short transmission lines (of the order of 15 km), where the travel time is less than the solution time-step, but such models are unsuitable for transmission distances. Instead, travelling wave theory is used in the development of more realistic models.

A simple and elegant travelling wave model of the lossless transmission line has already been described in Chapter 4 in the form of a dual Norton equivalent. The model is equally applicable to overhead lines and cables; the main differences arise from the procedures used in the calculation of the electrical parameters from their respective physical geometries. Carson's solution [1] forms the basis of the overhead line parameter calculation, either as a numerical integration of Carson's equation, the use of a series approximation or in the form of a complex depth of penetration. Underground cable parameters, on the other hand, are calculated using Pollack's equations [2], [3].

Multiconductor lines have been traditionally accommodated in the EMTP by a transformation to natural modes to diagonalise the matrices involved. Original stability problems were thought to be caused by inaccuracies in the modal domain representation, and thus much of the effort went into the development of more accurate fitting techniques. More recently, Gustavsen and Semlyen [4] have shown that, although the phase domain is inherently stable, its associated modal domain may be inherently unstable regardless of the fitting. This revelation has encouraged a return to the direct modelling of lines in the phase domain.

Figure 6.1 displays a decision tree for the selection of the appropriate transmission line model. The minimum limit for travel time is Length/c where the c is the speed of light, and this can be compared to the time step to see if a PI section or travelling wave model is appropriate. Various PI section models exist, however the nominal (or coupled) PI, displayed in Figure 6.2, is the preferred option for transient solutions. The exact equivalent PI is only adequate for steady-state solution where only one frequency is considered.

124 *Power systems electromagnetic transients simulation*

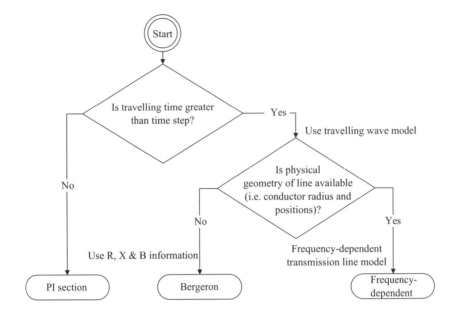

Figure 6.1 Decision tree for transmission line model selection

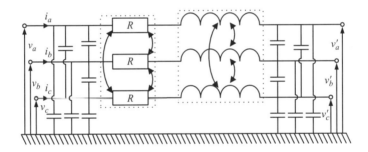

Figure 6.2 Nominal PI section

6.2 Bergeron's model

Bergeron's model [5] is a simple, constant frequency method based on travelling wave theory. It is basically the model described in Chapter 4. Here, the line is still treated as lossless but its distributed series resistance is added in lump form. Although the lumped resistances can be inserted throughout the line by dividing its total length into several sections, it makes little difference to do so and the use of just two sections at the ends is perfectly adequate. This lumped resistance model, shown in Figure 6.3, gives reasonable answers provided that $R/4 \ll Z_C$, where Z_C is the characteristic (or surge) impedance. However, for high frequency studies (e.g. power line carrier) this lumped resistance model may not be adequate.

Transmission lines and cables 125

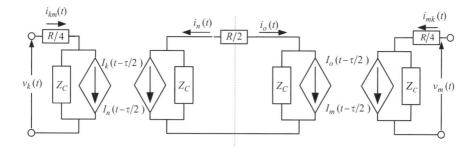

Figure 6.3 Equivalent two-port network for line with lumped losses

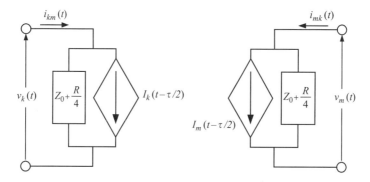

Figure 6.4 Equivalent two-port network for half-line section

By assigning half of the mid-point resistance to each line section, a model of half the line is depicted in Figure 6.4, where:

$$i_{km}(t) = \frac{1}{Z_C + R/4} v_k(t) + I_k(t - \tau/2) \tag{6.1}$$

and

$$I_k(t - \tau/2) = \frac{-1}{Z_C + R/4} v_m(t - \tau/2) - \left(\frac{Z_C - R/4}{Z_C + R/4}\right) i_m(t - \tau/2) \tag{6.2}$$

Finally, by cascading two half-line sections and eliminating the mid-point variables, as only the terminals are of interest, the model depicted in Figure 6.5 is obtained. It has the same form as the previous models but the current source representing the history terms is more complicated as it contains conditions from both ends on the line at time $(t - \tau/2)$. For example the expression for the current source at end k is:

$$I'_k(t - \tau) = \frac{-Z_C}{(Z_C + R/4)^2} (v_m(t - \tau) + (Z_C - R/4) i_{mk}(t - \tau))$$
$$+ \frac{-R/4}{(Z_C + R/4)^2} (v_k(t - \tau) + (Z_C - R/4) i_{km}(t - \tau)) \tag{6.3}$$

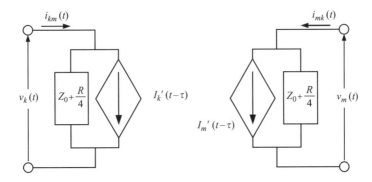

Figure 6.5 Bergeron transmission line model

In the EMTDC program the line model separates the propagation into low and high frequency paths, so that the line can have a higher attenuation to higher frequencies. This was an early attempt to provide frequency dependence, but newer models (in the phase domain) are now preferred.

6.2.1 Multiconductor transmission lines

Equations 4.22 and 4.23 are also applicable to multiconductor lines by replacing the scalar voltages and currents by vectors and using inductance and capacitance matrices. The wave propagation equations in the frequency domain are:

$$-\left[\frac{dV_{\text{phase}}}{dx}\right] = \left[Z'_{\text{phase}}\right][I_{\text{phase}}] \tag{6.4}$$

$$-\left[\frac{dI_{\text{phase}}}{dx}\right] = \left[Y'_{\text{phase}}\right][V_{\text{phase}}] \tag{6.5}$$

By differentiating a second time, one vector, either the voltage or current, may be eliminated giving:

$$-\left[\frac{d^2V_{\text{phase}}}{dx^2}\right] = \left[Z'_{\text{phase}}\right]\left[\frac{dI_{\text{phase}}}{dx}\right] = -\left[Z'_{\text{phase}}\right][Y'_{\text{phase}}][V_{\text{phase}}] \tag{6.6}$$

$$-\left[\frac{d^2I_{\text{phase}}}{dx^2}\right] = \left[Y'_{\text{phase}}\right]\left[\frac{dV_{\text{phase}}}{dx}\right] = -\left[Y'_{\text{phase}}\right][Z'_{\text{phase}}][I_{\text{phase}}] \tag{6.7}$$

Traditionally the complication of having off-diagonal elements in the matrices of equations 6.6 and 6.7 is overcome by transforming into natural modes. Eigenvalue analysis is applied to produce diagonal matrices, thereby transforming from coupled equations in the phase domain to decoupled equations in the modal domain. Each equation in the modal domain is solved as for a single phase line by using modal travelling time and modal surge impedance.

Transmission lines and cables 127

The transformation matrices between phase and modal quantities are different for voltage and current, i.e.

$$[V_{\text{phase}}] = [T_v][V_{\text{mode}}] \quad (6.8)$$

$$[I_{\text{phase}}] = [T_i][I_{\text{mode}}] \quad (6.9)$$

Substituting equation 6.8 in 6.6 gives:

$$\left[\frac{d^2[T_v]V_{\text{mode}}}{dx^2}\right] = \left[Z'_{\text{phase}}\right]\left[Y'_{\text{phase}}\right][T_v][V_{\text{mode}}] \quad (6.10)$$

Hence

$$\left[\frac{d^2 V_{\text{mode}}}{dx^2}\right] = [T_v]^{-1}\left[Z'_{\text{phase}}\right]\left[Y'_{\text{phase}}\right][T_v][V_{\text{mode}}] = [\Lambda][V_{\text{mode}}] \quad (6.11)$$

To find the matrix $[T_v]$ that diagonalises $[Z'_{\text{phase}}][Y'_{\text{phase}}]$ its eigenvalues and eigenvectors must be found. However the eigenvectors are not unique as when multiplied by a non-zero complex constant they are still valid eigenvectors, therefore some normalisation is desirable to allow the output from different programs to be compared. PSCAD/EMTDC uses the root squaring technique developed by Wedepohl for eigenvalue analysis [6]. To enable us to generate frequency-dependent line models the eigenvectors must be consistent from one frequency to the next, such that the eigenvectors form a continuous function of frequency so that curve fitting can be applied. A Newton–Raphson algorithm has been developed for this purpose [6].

Once the eigenvalue analysis has been completed then:

$$[Z_{\text{mode}}] = [T_v]^{-1}[Z_{\text{phase}}][T_i] \quad (6.12)$$

$$[Y_{\text{mode}}] = [T_i]^{-1}[Y_{\text{phase}}][T_v] \quad (6.13)$$

$$[Z_{\text{surge } i}] = \sqrt{\frac{Z_{\text{mode}}(i,i)}{Y_{\text{mode}}(i,i)}} \quad (6.14)$$

where $[Z_{\text{mode}}]$ and $[Y_{\text{mode}}]$ are diagonal matrices.

As the products $[Z'_{\text{phase}}][Y'_{\text{phase}}]$ and $[Y'_{\text{phase}}][Z'_{\text{phase}}]$ are different so are their eigenvectors, even though their eigenvalues are identical. They are, however, related, such that $[T_i] = ([T_v]^T)^{-1}$ (assuming a normalised Euclidean norm, i.e. $\sum_{j=1}^{n} T_{ij}^2 = 1$) and therefore only one of them needs to be calculated. Looking at mode i, i.e. taking the ith equation from 6.11, gives:

$$\left[\frac{d^2 V_{\text{mode } i}}{dx^2}\right] = \Lambda_{ii} V_{\text{mode } i} \quad (6.15)$$

and the general solution at point x in the line is:

$$V_{\text{mode } i}(x) = e^{-\gamma_i x} V^F_{\text{mode } i}(k) + e^{\gamma_i x} V^B_{\text{mode } i}(m) \quad (6.16)$$

where

$\gamma_i = \sqrt{\Lambda_{ii}}$
V^F is the forward travelling wave
V^B is the backward travelling wave.

Equation 6.16 contains two arbitrary constants of integration and therefore n such equations (n being the number of conductors) require $2n$ arbitrary constants. This is consistent with there being $2n$ boundary conditions, one for each end of each conductor. The corresponding matrix equation is:

$$\mathbf{V}_{\text{mode}}(x) = [e^{-\gamma x}]\mathbf{V}^F_{\text{mode}}(k) + [e^{\gamma x}]\mathbf{V}^B_{\text{mode}}(m) \tag{6.17}$$

An n-conductor line has n natural modes. If the transmission line is perfectly balanced the transformation matrices are not frequency dependent and the three-phase line voltage transformation becomes:

$$[T_v] = \frac{1}{k}\begin{bmatrix} 1 & 1 & -1 \\ 1 & 0 & 2 \\ 1 & -1 & -1 \end{bmatrix}$$

Normalising and rearranging the rows will enable this matrix to be seen to correspond to Clarke's components ($\alpha, \beta, 0$) [7], i.e.

$$\begin{pmatrix} V_a \\ V_b \\ V_c \end{pmatrix} = \begin{bmatrix} 1 & 0 & 1 \\ -\frac{1}{2} & \frac{\sqrt{3}}{2} & 1 \\ -\frac{1}{2} & -\frac{\sqrt{3}}{2} & 1 \end{bmatrix} \begin{pmatrix} V_\alpha \\ V_\beta \\ V_0 \end{pmatrix}$$

$$\begin{pmatrix} V_\alpha \\ V_\beta \\ V_0 \end{pmatrix} = \begin{bmatrix} \frac{2}{3} & -\frac{1}{3} & -\frac{1}{3} \\ 0 & \frac{1}{\sqrt{3}} & -\frac{1}{\sqrt{3}} \\ \frac{1}{3} & \frac{1}{3} & \frac{1}{3} \end{bmatrix} \begin{pmatrix} V_a \\ V_b \\ V_c \end{pmatrix} = \frac{1}{3}\begin{bmatrix} 2 & -1 & -1 \\ 0 & \sqrt{3} & -\sqrt{3} \\ 1 & 1 & 1 \end{bmatrix} \begin{pmatrix} V_a \\ V_b \\ V_c \end{pmatrix}$$

Reintroducing phase quantities with the use of equation 6.8 gives:

$$\mathbf{V}_x(\omega) = [e^{-\Gamma x}]\mathbf{V}^F + [e^{\Gamma x}]\mathbf{V}^B \tag{6.18}$$

where $[e^{-\Gamma x}] = [T_v][e^{-\gamma x}][T_v]^{-1}$ and $[e^{\Gamma x}] = [T_v][e^{\gamma x}][T_v]^{-1}$.

The matrix $A(\omega) = [e^{-\Gamma x}]$ is the wave propagation (comprising of attenuation and phase shift) matrix.

The corresponding equation for current is:

$$\mathbf{I}_x(\omega) = [e^{-\Gamma x}] \cdot \mathbf{I}^F - [e^{\Gamma x}] \cdot \mathbf{I}^B = Y_C \left([e^{-\Gamma x}] \cdot \mathbf{V}^F - [e^{\Gamma x}] \cdot \mathbf{V}^B\right) \tag{6.19}$$

where

I^F is the forward travelling wave
I^B is the backward travelling wave.

Transmission lines and cables 129

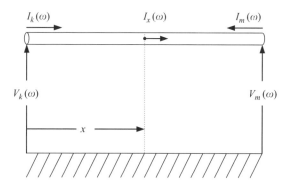

Figure 6.6 Schematic of frequency-dependent line

The voltage and current vectors at end k of the line are:

$$\mathbf{V}_k(\omega) = (\mathbf{V}^F + \mathbf{V}^B)$$
$$\mathbf{I}_k(\omega) = (\mathbf{I}^F + \mathbf{I}^B) = Y_C(\mathbf{V}^F - \mathbf{V}^B)$$

and at end m:

$$\mathbf{V}_m(\omega) = [e^{-\Gamma l}] \cdot \mathbf{V}^F + [e^{\Gamma x}] \cdot \mathbf{V}^B \tag{6.20}$$
$$\mathbf{I}_m(\omega) = -Y_C([e^{-\Gamma l}] \cdot \mathbf{V}^F - [e^{\Gamma l}] \cdot \mathbf{V}^B) \tag{6.21}$$

Note the negative sign due to the reference direction for current at the receiving end (see Figure 6.6).

Hence the expression for the forward and backward travelling waves at k are:

$$\mathbf{V}^F = (\mathbf{V}_k(\omega) + Z_C \mathbf{I}_k(\omega))/2 \tag{6.22}$$
$$\mathbf{V}^B = (\mathbf{V}_k(\omega) - Z_C \mathbf{I}_k(\omega))/2 \tag{6.23}$$

Also, since

$$[Y_C] \cdot \mathbf{V}_k(\omega) + \mathbf{I}_k(\omega) = 2\mathbf{I}^F = 2[e^{-\Gamma l}] \cdot \mathbf{I}^B \tag{6.24}$$

and

$$[Y_C] \cdot \mathbf{V}_m(\omega) + \mathbf{I}_m(\omega) = 2\mathbf{I}^B = 2[e^{-\Gamma l}] \cdot \mathbf{I}^F = [e^{-\Gamma l}]([Y_C] \cdot \mathbf{V}_k(\omega) + \mathbf{I}_k(\omega)) \tag{6.25}$$

the forward and backward travelling current waves at k are:

$$\mathbf{I}^F = ([Y_C] \cdot \mathbf{V}_k(\omega) + \mathbf{I}_k(\omega))/2 \tag{6.26}$$
$$\mathbf{I}^B = [e^{-\Gamma l}]([Y_C] \cdot \mathbf{V}_k(\omega) - \mathbf{I}_k(\omega))/2 \tag{6.27}$$

6.3 Frequency-dependent transmission lines

The line frequency-dependent surge impedance (or admittance) and line propagation matrix are first calculated from the physical line geometry. To obtain the time domain response, a convolution must be performed as this is equivalent to a multiplication in the frequency domain. It can be achieved efficiently using recursive convolutions (which can be shown to be a form of root-matching, even though this is not generally recognised). This is performed by fitting a rational function in the frequency domain to both the frequency-dependent surge impedance and propagation constant.

As the line parameters are functions of frequency, the relevant equations should first be viewed in the frequency domain, making extensive use of curve fitting to incorporate the frequency-dependent parameters into the model. Two important frequency-dependent parameters influencing wave propagation are the characteristic impedance Z_C and propagation constant γ. Rather than looking at Z_C and γ in the frequency domain and considering each frequency independently, they are expressed by continuous functions of frequency that need to be approximated by a fitted rational function.

The characteristic impedance is given by:

$$Z_C(\omega) = \sqrt{\frac{R'(\omega) + j\omega L'(\omega)}{G'(\omega) + j\omega C'(\omega)}} = \sqrt{\frac{Z'(\omega)}{Y'(\omega)}} \qquad (6.28)$$

while the propagation constant is:

$$\gamma(\omega) = \sqrt{(R'(\omega) + j\omega L'(\omega))(G'(\omega) + j\omega C'(\omega))} = \alpha(\omega) + j\beta(\omega) \qquad (6.29)$$

The frequency dependence of the series impedance is most pronounced in the zero sequence mode, thus making frequency-dependent line models more important for transients where appreciable zero sequence voltages and zero sequence currents exist, such as in single line-to-ground faults.

Making use of the following relationships

$$\cosh(\Gamma l) = (e^{-\Gamma l} + e^{\Gamma l})/2$$
$$\sinh(\Gamma l) = (e^{\Gamma l} - e^{-\Gamma l})/2$$
$$\operatorname{cosech}(\Gamma l) = 1/\sinh(\Gamma l)$$
$$\coth(\Gamma l) = 1/\tanh(\Gamma l) = \cosh(\Gamma l)/\sinh(\Gamma l)$$

allows the following input–output matrix equation to be written:

$$\begin{pmatrix} V_k \\ I_{km} \end{pmatrix} = \begin{bmatrix} A & B \\ C & D \end{bmatrix} \cdot \begin{pmatrix} V_m \\ -I_{mk} \end{pmatrix} = \begin{bmatrix} \cosh(\Gamma l) & Z_C \cdot \sinh(\Gamma l) \\ Y_C \cdot \sinh(\Gamma l) & \cosh(\Gamma l) \end{bmatrix} \begin{pmatrix} V_m \\ -I_{mk} \end{pmatrix} \qquad (6.30)$$

Rearranging equation 6.30 leads to the following two-port representation:

$$\begin{pmatrix} I_{km} \\ I_{mk} \end{pmatrix} = \begin{bmatrix} D \cdot B^{-1} & C - D \cdot B^{-1} A \\ -B & B^{-1} A \end{bmatrix} \begin{pmatrix} V_k \\ V_m \end{pmatrix}$$

$$= \begin{bmatrix} Y_C \cdot \coth(\Gamma l) & -Y_C \cdot \operatorname{cosech}(\Gamma l) \\ -Y_C \cdot \operatorname{cosech}(\Gamma l) & Y_C \cdot \coth(\Gamma l) \end{bmatrix} \begin{pmatrix} V_k \\ V_m \end{pmatrix} \quad (6.31)$$

and using the conversion between the modal and phase domains, i.e.

$$[\coth(\Gamma l)] = [T_v] \cdot [\coth(\gamma(\omega)l)] \cdot [T_v]^{-1} \quad (6.32)$$

$$[\operatorname{cosech}(\Gamma l)] = [T_v] \cdot [\operatorname{cosech}(\gamma(\omega)l)] \cdot [T_v]^{-1} \quad (6.33)$$

the exact a.c. steady-state input–output relationship of the line at any frequency is:

$$\begin{pmatrix} V_k(\omega) \\ I_{km}(\omega) \end{pmatrix} = \begin{bmatrix} \cosh(\gamma(\omega)l) & Z_C \sinh(\gamma(\omega)l) \\ \dfrac{1}{Z_C} \sinh(\gamma(\omega)l) & \cosh(\gamma(\omega)l) \end{bmatrix} \begin{pmatrix} V_m(\omega) \\ -I_{mk}(\omega) \end{pmatrix} \quad (6.34)$$

This clearly shows the Ferranti effect in an open circuit line, because the ratio

$$V_m(\omega)/V_k(\omega) = 1/\cosh(\gamma(\omega)l)$$

increases with line length and frequency.

The forward and backward travelling waves at end k are:

$$F_k(\omega) = V_k(\omega) + Z_C(\omega)I_k(\omega) \quad (6.35)$$

$$B_k(\omega) = V_k(\omega) - Z_C(\omega)I_k(\omega) \quad (6.36)$$

and similarly for end m:

$$F_m(\omega) = V_m(\omega) + Z_C(\omega)I_m(\omega) \quad (6.37)$$

$$B_m(\omega) = V_m(\omega) - Z_C(\omega)I_m(\omega) \quad (6.38)$$

Equation 6.36 can be viewed as a Thevenin circuit (shown in Figure 6.7) where $V_k(\omega)$ is the terminal voltage, $B_k(\omega)$ the voltage source and characteristic or surge impedance, $Z_C(\omega)$, the series impedance.

The backward travelling wave at k is the forward travelling wave at m multiplied by the wave propagation matrix, i.e.

$$B_k(\omega) = A(\omega)F_m(\omega) \quad (6.39)$$

Rearranging equation 6.35 to give $V_k(\omega)$, and substituting in equation 6.39, then using equation 6.37 to eliminate $F_m(\omega)$ gives:

$$V_k(\omega) = Z_C(\omega)I_k(\omega) + A(\omega)(V_m(\omega) + Z_m(\omega)I_m(\omega)) \quad (6.40)$$

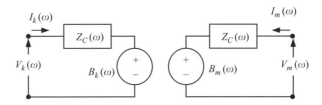

Figure 6.7   Thevenin equivalent for frequency-dependent transmission line

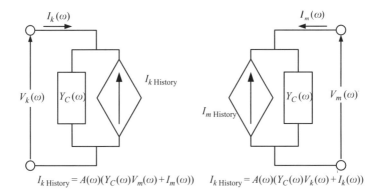

$I_{k \text{ History}} = A(\omega)(Y_C(\omega)V_m(\omega) + I_m(\omega))$ $I_{k \text{ History}} = A(\omega)(Y_C(\omega)V_k(\omega) + I_k(\omega))$

Figure 6.8   Norton equivalent for frequency-dependent transmission line

Rearranging equation 6.40 gives the Norton form of the frequency dependent transmission line, i.e.

$$I_k(\omega) = Y_C(\omega)V_k(\omega) - A(\omega)(I_m(\omega) + Y_C(\omega)V_m(\omega)) \quad (6.41)$$

and a similar expression can be written for the other end of the line.

The Norton frequency-dependent transmission line model is displayed in Figure 6.8.

6.3.1 Frequency to time domain transformation

The frequency domain equations 6.40 and 6.41 can be transformed to the time domain by using the convolution principle, i.e.

$$A(\omega)F_m(\omega) \Leftrightarrow a(t) * f_m = \int_\tau^t a(u) f_m(t-u) \, du \quad (6.42)$$

where

$$A(\omega) = e^{-\Gamma l} = e^{-\gamma(\omega)l} = e^{-\alpha(\omega)l} e^{-j\beta(\omega)l} \quad (6.43)$$

is the propagation matrix. The propagation matrix is frequency dependent and it comprises two components, the attenuation ($e^{-\alpha(\omega)l}$) and phase shift ($e^{-j\beta(\omega)l}$). The time

Transmission lines and cables 133

domain equivalent of these are $a(t)$ and β, where $a(t)$ is the time domain transform (impulse response) of $e^{-\alpha(\omega)l}$ and β is a pure time delay (travelling time). The lower limit of the integral in equation 6.42, τ, is the time (in seconds) for an impulse to travel from one end of the line to the other.

Thus converting equations 6.40 and 6.41 to the time domain yields:

$$v_k(t) = Z_C(t) * i_{km}(t) + a(t) * (v_m(t) + Z_C(t) * i_{mk}(t)) \tag{6.44}$$

$$i_k(t) = Y_C(t) * v_k(t) - a(t) * (Y_C(t) * v_m(t-\tau) - i_m(t-\tau)) \tag{6.45}$$

This process can be evaluated efficiently using recursive convolution if $a(u)$ is an exponential. This is achieved using the partial fraction expansion of a rational function to represent $A(\omega)$ in the frequency domain as the inverse Laplace transform of $k_m/(s+p_m)$ which is $k_m \cdot e^{-p_m t}$. Hence the convolution of equation 6.42 becomes:

$$y(t) = k_m \int_\tau^t e^{-p_m(T)} f_m(t-T)\,dT \tag{6.46}$$

Semlyen and Dabuleanu [8] showed that for a single time step the above equation yields:

$$y(t) = e^{-p_m \Delta t} \cdot y(t-\Delta t) + \int_0^{\Delta t} k_m e^{-p_m T} u(t-T)\,dT \tag{6.47}$$

It is a recursive process because $y(t)$ is found from $y(t-\Delta t)$ with a simple integration over one single time step. If the input is assumed constant during the time step, it can be taken outside the integral, which can then be determined analytically, i.e.

$$y(t) = e^{-p_m \Delta t} y(t-\Delta t) + u(t-\Delta t) \int_0^{\Delta t} k_m e^{-p_m T}\,dT \tag{6.48}$$

$$= e^{-p_m \Delta t} y(t-\Delta t) + \frac{k_m}{p_m}(1-e^{-p_m \Delta t}) u(t-\Delta t) \tag{6.49}$$

If the input is assumed to vary linearly, i.e.

$$u(t-T) = \frac{(u(t-\Delta t) - u(t))}{\Delta t} T + u(t) \tag{6.50}$$

the resulting recursive equation becomes:

$$y(t) = e^{-p_m \Delta t} x_{n-1} + \frac{k_m}{p_m}\left(1 - \frac{1}{p_m \Delta t}(1 - e^{-a\Delta t})\right) u(t)$$

$$+ \frac{k_m}{p_m}\left(-e^{-a\Delta t} + \frac{1}{p_m \Delta t}(1 - e^{-a\Delta t})\right) u(t-\Delta t) \tag{6.51}$$

The propagation constant can be approximated by the following rational function

$$A_{\text{approx}}(s) = e^{-s\tau} \frac{(s+z_1)(s+z_2)\cdots(s+z_n)}{(s+p_1)(s+p_2)\cdots(s+p_m)} \tag{6.52}$$

134 *Power systems electromagnetic transients simulation*

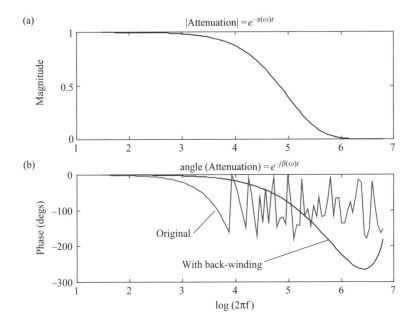

Figure 6.9 Magnitude and phase angle of propagation function

The time delay (which corresponds to a phase shift in the frequency domain) is implemented by using a buffer of previous history terms. A partial fraction expansion of the remainder of the rational function is:

$$k\frac{(s+z_1)(s+z_2)\cdots(s+z_N)}{(s+p_1)(s+p_2)\cdots(s+p_n)} = \frac{k_1}{(s+p_1)} + \frac{k_2}{(s+p_2)} + \cdots + \frac{k_n}{(s+p_n)} \quad (6.53)$$

The inverse Laplace transform gives:

$$a_{\text{approx}}(t) = e^{-p_1\tau}(k_1 \cdot e^{-p_1 t} + k_2 \cdot e^{-p_2 t} + \cdots + k_n \cdot e^{-p_n t}) \quad (6.54)$$

Because of its form as the sum of exponential terms, recursive convolution is used.

Figure 6.9 shows the magnitude and phase of the propagation function $(e^{-(\alpha(\omega)+j\beta(\omega))l})$ as a function of frequency, for a single-phase line, where l is the line length. The propagation constant is expressed as $\alpha(\omega) + j\beta(\omega)$ to emphasise that it is a function of frequency. The amplitude (shown in Figure 6.9(a)) displays a typical low-pass characteristic. Note also that, since the line length is in the exponent, the longer the line the greater is the attenuation of the travelling waves.

Figure 6.9(b) shows that the phase angle of the propagation function becomes more negative as the frequency increases. A negative phase represents a phase lag in the waveform traversing from one end of the line to the other and its counterpart in the time domain is a time delay. Although the phase angle is a continuous negative growing function, for display purposes it is constrained to the range −180 to 180 degrees. This is a difficult function to fit, and requires a high order rational function to achieve

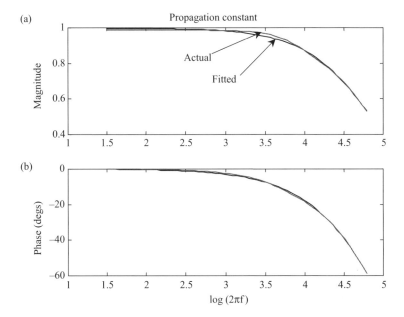

Figure 6.10   Fitted propagation function

sufficient accuracy. Multiplication by $e^{-js\tau}$, where τ represents the nominal travelling time for a wave to go from one end of the line to the other (in this case 0.33597 ms) produces the smooth function shown in Figure 6.9(b). This procedure is referred to as back-winding [9] and the resulting phase variation is easily fitted with a low order rational function. To obtain the correct response the model must counter the phase advance introduced in the frequency-domain fitting (i.e. back-winding). This is performed in the time domain implementation by incorporating a time delay τ. A buffer of past voltages and currents at each end of the line is maintained and the values delayed by τ are used. Because τ in general is not an integer multiple of the time step, interpolation between the values in the buffer is required to get the correct time delay.

Figure 6.10 shows the match obtained when applying a least squares fitting of a rational function (with numerator order 2 and denominator order 3). The number of poles is normally one more than the zeros, as the attenuation function magnitude must go to zero when the frequency approaches infinity.

Although the fitting is good, close inspection shows a slight error at the fundamental frequency. Any slight discrepancy at the fundamental frequency shows up as a steady-state error, which is undesirable. This occurs because the least squares fitting tends to smear the error across the frequency range. To control the problem, a weighting factor can be applied to specified frequency ranges (such as around d.c. or the fundamental frequency) when applying the fitting procedure. When the fitting has been completed any slight error still remaining is removed by multiplying the rational function by a constant k to give the correct value at low frequency. This sets the d.c. gain (i.e. its value when s is set to zero) of the fitted rational function. The value of k controls the d.c. gain of this rational function and is calculated from the d.c. resistance and the d.c. gain of the surge impedance, thereby ensuring that the correct d.c. resistance is exhibited by the model.

Some fitting techniques force the poles and zeros to be real and stable (i.e. in the left-hand half of the s-plane) while others allow complex poles and use other methods to ensure stable fits (either reflecting unstable poles in the y-axis or deleting them). A common approach is to assume a minimum-phase function and use real half-plane poles. Fitting can be performed either in the s-domain or z-domain, each alternative having advantages and disadvantages. The same algorithm can be used for fitting the characteristic impedance (or admittance if using the Norton form), the number of poles and zeros being the same in both cases. Hence the partial expansion of the fitted rational function is:

$$k \frac{(s+z_1)(s+z_2)\cdots(s+z_n)}{(s+p_1)(s+p_2)\cdots(s+p_n)} = k_0 + \frac{k_1}{(s+p_1)} + \frac{k_2}{(s+p_2)} + \cdots + \frac{k_n}{(s+p_n)}$$

(6.55)

It can be implemented by using a series of RC parallel blocks (the Foster I realisation), which gives $R_0 = k_0$, $R_i = k_i/p_i$ and $C_i = 1/k_i$. Either the trapezoidal rule can be applied to the RC network, or better still, recursive convolution. The shunt conductance $G'(\omega)$ is not normally known. If it is assumed zero, at low frequencies the surge impedance becomes larger as the frequency approaches zero, i.e.

$$Z_C(\omega) = \lim_{\omega \to 0} \sqrt{\frac{R'(\omega) + j\omega L'(\omega)}{j\omega C'(\omega)}} \to \infty$$

This trend can be seen in Figure 6.11 which shows the characteristic (or surge) impedance calculated by a transmission line parameter program down to 5 Hz. In practice the characteristic impedance does not tend to infinity as the frequency goes to zero; instead

$$Z_C(\omega) = \lim_{\omega \to 0} \sqrt{\frac{R'(\omega) + j\omega L'(\omega)}{G'(\omega) + j\omega C'(\omega)}} \to \sqrt{\frac{R'_{DC}}{G'_{DC}}}$$

To mitigate the problem a starting frequency is entered, which flattens the impedance curve at low frequencies and thus makes it more realistic. Entering a starting frequency is equivalent to introducing a shunt conductance G'. The higher the starting frequency the greater the shunt conductance and, hence, the shunt loss. On the other hand choosing a very low starting frequency will result in poles and zeros at low frequencies and the associated large time constants will cause long settling times to reach the steady state. The value of G' is particularly important for d.c. line models and trapped charge on a.c. lines.

6.3.2 Phase domain model

EMTDC version 3 contains a new curve-fitting technique as well as a new phase domain transmission line model [10]. In this model the propagation matrix $[A_p]$ is first

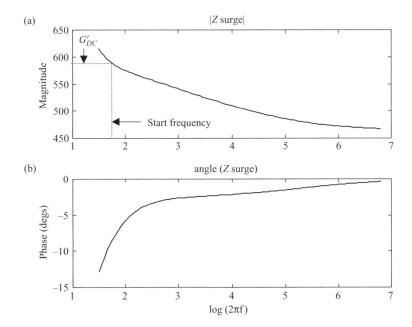

Figure 6.11   Magnitude and phase angle of characteristic impedance

fitted in the modal domain, and the resulting poles and time delays determined. Modes with similar time delays are grouped together. These poles and time delays are used for fitting the propagation matrix $[A_p]$ in the phase domain, on the assumption that all poles contribute to all elements of $[A_p]$. An over-determined linear equation involving all elements of $[A_p]$ is solved in the least-squares sense to determine the unknown residuals. As all elements in $[A_p]$ have identical poles a columnwise realisation can be used, which increases the efficiency of the time domain simulation [4].

6.4 Overhead transmission line parameters

There are a number of ways to calculate the electrical parameters from the physical geometry of a line, the most common being Carson's series equations.

To determine the shunt component Maxwell's potential coefficient matrix is first calculated from:

$$P'_{ij} = \frac{1}{2\pi \varepsilon_0} \ln\left(\frac{D_{ij}}{d_{ij}}\right) \qquad (6.56)$$

where ε_0 is the permittivity of free space and equals 8.854188×10^{-12} hence $1/2\pi \varepsilon_0 = 17.975109 \, \text{km} \, \text{F}^{-1}$.

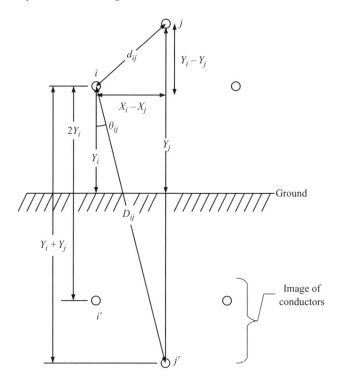

Figure 6.12   Transmission line geometry

if $i \neq j$

$$D_{ij} = \sqrt{(X_i - X_j)^2 - (Y_i + Y_j)^2}$$
$$d_{ij} = \sqrt{(X_i - X_j)^2 - (Y_i - Y_j)^2}$$

if $i = j$

$D_{ij} = 2Y_i$

$d_{ij} = $ GMR$_i$ (bundled conductor) or R_i (radius for single conductor)

In Figure 6.12 the conductor heights Y_i and Y_j are the average heights above ground which are $Y_{\text{tower}} - 2/3 Y_{\text{sag}}$.

Maxwell's potential coefficient matrix relates the voltages to the charge per unit length, i.e.

$$\mathbf{V} = [P']\mathbf{q}$$

Hence the capacitance matrix is given by

$$[C] = [P']^{-1} \tag{6.57}$$

Transmission lines and cables

The series impedance may be divided into two components, i.e. a conductor internal impedance that affects only the diagonal elements and an aerial and ground return impedance, i.e.

$$Z_{ij} = \frac{j\omega\mu_0}{2\pi} \left(\ln\left(\frac{D_{ij}}{d_{ij}}\right) + 2 \int_0^\infty \frac{e^{-\alpha \cdot \cos(\theta_{ij})} \cos(\alpha \cdot \sin(\theta_{ij}))}{\alpha + \sqrt{\alpha^2 + j \cdot r_{ij}^2}} \, d\alpha \right) \quad (6.58)$$

In equation 6.58 the first term defines the aerial reactance of the conductor assuming that the conductance of the ground is perfect. The second term is known as Carson's integral and defines the additional impedance due to the imperfect ground. In the past the evaluation of this integral required expressions either as a power or asymptotic series; however it is now possible to perform the integration numerically. The use of two Carson's series (for low and high frequencies respectively) is not suitable for frequency-dependent lines, as a discontinuity occurs where changing from one series to the other, thus complicating the fitting.

Deri et al. [11] developed the idea of complex depth of penetration by showing that:

$$2 \int_0^\infty \frac{e^{-\alpha \cdot \cos(\theta_{ij})} \cos(\alpha \cdot \sin(\theta_{ij}))}{\alpha + \sqrt{\alpha^2 + j \cdot r_{ij}^2}} \, d\alpha$$

$$\approx \frac{\sqrt{(Y_i + Y_j + 2\sqrt{\rho_g/2j\omega\mu})^2 + (X_i - X_j)^2}}{d_{ij}} \quad (6.59)$$

This has a maximum error of approximately 5 per cent, which is acceptable considering the accuracy by which earth resistivity is known.

PSCAD/EMTDC uses the following equations (which can be derived from equation 6.59):

$$Z_{ij} = \frac{j\omega\mu_0}{2\pi} \left(\ln\left(\frac{D_{ij}}{d_{ij}}\right) + \frac{1}{2} \ln\left(1 + \frac{4 \cdot D_e \cdot (Y_i + Y_j + D_e)}{D_{ij}^2}\right) \right) \Omega \, \text{m}^{-1} \quad (6.60)$$

$$Z_{ii} = \frac{j\omega\mu_0}{2\pi} \left(\ln\left(\frac{D_{ii}}{r_i}\right) + \frac{0.3565}{\pi \cdot R_C^2} + \frac{\rho_C M \coth^{-1}(0.777 R_C M)}{2\pi R_C} \right) \Omega \, \text{m}^{-1} \quad (6.61)$$

where

$$M = \sqrt{\frac{j\omega\mu_0}{\rho_C}}$$

$$D_e = \sqrt{\frac{\rho_g}{j\omega\mu_0}}$$

ρ_C = conductor resistivity(Ω m) = $R_{dc} \times$ Length/Area
ρ_g = ground resistivity(Ω m)
$\mu_0 = 4\pi \times 10^{-7}$.

6.4.1 Bundled subconductors

Bundled subconductors are often used to reduce the electric field strength at the surface of the conductors, as compared to using one large conductor. This therefore reduces the likelihood of corona. The two alternative methods of modelling bundling are:

1. Replace the bundled subconductors with an equivalent single conductor.
2. Explicitly represent subconductors and use matrix elimination of subconductors.

In method 1 the GMR (Geometric Mean Radius) of the bundled conductors is calculated and a single conductor of this GMR is used to represent the bundled conductors. Thus with only one conductor represented $\text{GMR}_{\text{equiv}} = \text{GMR}_i$.

$$\text{GMR}_{\text{equiv}} = \sqrt[n]{n \cdot \text{GMR}_{\text{conductor}} \cdot R_{\text{Bundle}}^{n-1}}$$

and

$$R_{\text{equiv}} = \sqrt[n]{n \cdot R_{\text{conductor}} \cdot R_{\text{Bundle}}^{n-1}}$$

where

n = number of conductors in bundle
R_{Bundle} = radius of bundle
$R_{\text{conductor}}$ = radius of conductor
R_{equiv} = radius of equivalent single conductor
$\text{GMR}_{\text{conductor}}$ = geometric mean radius of individual subconductor
$\text{GMR}_{\text{equiv}}$ = geometric mean radius of equivalent single conductor.

The use of GMR ignores proximity effects and hence is only valid if the subconductor spacing is much smaller than the spacing between the phases of the line.

Method 2 is a more rigorous approach and is adopted in PSCAD/EMTDC version 3. All subconductors are represented explicitly in $[Z']$ and $[P']$ (hence the order is 12×12 for a three-phase line with four subconductors). As the elimination procedure is identical for both matrices, it will be illustrated in terms of $[Z']$. If phase A comprises four subconductors A_1, A_2, A_3 and A_4, and R represents their total equivalent for phase A, then the sum of the subconductor currents equals the phase current and the change of voltage with distance is the same for all subconductors, i.e.

$$\sum_{i=1}^{n} I_{A_i} = I_R$$

$$\frac{dV_{A_1}}{dx} = \frac{dV_{A_2}}{dx} = \frac{dV_{A_3}}{dx} = \frac{dV_{A_4}}{dx} = \frac{dV_R}{dx} = \frac{dV_{\text{Phase}}}{dx}$$

Figure 6.13(a) illustrates that I_R is introduced in place of I_{A_1}. As $I_{A_1} = I_R - I_{A_2} - I_{A_3} - I_{A_4}$ column A_1 must be subtracted from columns A_2, A_3 and A_4. Since V/dx is the same for each subconductor, subtracting row A_1 from rows A_2, A_3 and A_4 (illustrated in Figure 6.13b) will give zero in the V_{A_2}/dx vector. Then partitioning as shown in Figure 6.13(c) allows Kron reduction to be performed to give the reduced equation (Figure 6.13d).

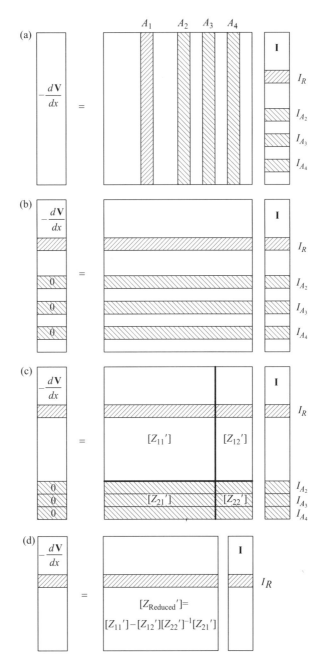

Figure 6.13   Matrix elimination of subconductors

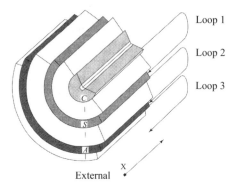

Figure 6.14   Cable cross-section

This method does include proximity effects and hence is generally more accurate; however the difference with respect to using one equivalent single conductor of appropriate GMR is very small when the phase spacing is much greater than the bundle spacing.

6.4.2   Earth wires

When earth wires are continuous and grounded at each tower then for frequencies below 250 kHz it is reasonable to assume that the earth wire potential is zero along its length. The reduction procedure for $[Z']$ and $[P']$ is the same. $[P']$ is reduced prior to inverting to give the capacitance matrix. The matrix reduction is next illustrated for the series impedance.

Assuming a continuous earth wire grounded at each tower then $dV_e/dx = 0$ and $V_e = 0$. Partitioning into conductors and earth wires gives:

$$-\begin{pmatrix}\left(\dfrac{dV_c}{dx}\right)\\ \left(\dfrac{dV_e}{dx}\right)\end{pmatrix} = -\begin{pmatrix}\left(\dfrac{dV_c}{dx}\right)\\ (0)\end{pmatrix} = \begin{bmatrix}[Z'_{cc}] & [Z'_{ce}]\\ [Z'_{ec}] & [Z'_{ee}]\end{bmatrix}\begin{pmatrix}(I_c)\\ (I_e)\end{pmatrix} \quad (6.62)$$

$$-\left(\dfrac{dV_c}{dx}\right) = [Z_{\text{Reduced}'}](I_c)$$

where $[Z_{\text{Reduced}'}] = [Z'_{cc}] - [Z'_{ce}][Z'_{ee}]^{-1}[Z'_{ec}]$.

When the earth wires are bundled the same technique used for bundled phase subconductors can be applied to them.

6.5   Underground cable parameters

A unified solution similar to that of overhead transmission is difficult for underground cables because of the great variety in their construction and layouts.

Transmission lines and cables

The cross-section of a coaxial cable, although extremely complex, can be simplified to that of Figure 6.14 and its series per unit length harmonic impedance is calculated by the following set of loop equations.

$$\begin{pmatrix} \dfrac{dV_1}{dx} \\ \dfrac{dV_2}{dx} \\ \dfrac{dV_3}{dx} \end{pmatrix} = \begin{bmatrix} Z'_{11} & Z'_{12} & 0 \\ Z'_{21} & Z'_{22} & Z'_{23} \\ 0 & Z'_{32} & Z'_{33} \end{bmatrix} \begin{pmatrix} I_1 \\ I_2 \\ I_3 \end{pmatrix} \qquad (6.63)$$

where

Z'_{11} = the sum of the following three component impedances:
$Z_{\text{core-outside}}$ = internal impedance of the core with the return path outside the core
$Z_{\text{core-insulation}}$ = impedance of the insulation surrounding the core
$Z_{\text{sheath-inside}}$ = internal impedance of the sheath with the return path inside the sheath.

Similarly

$$Z'_{22} = Z_{\text{sheath-outside}} + Z_{\text{sheath/armour-insulation}} + Z_{\text{armour-inside}} \qquad (6.64)$$

$$Z'_{33} = Z_{\text{armour-outside}} + Z_{\text{armour/earth-insulation}} + Z_{\text{earth-inside}} \qquad (6.65)$$

The coupling impedances $Z'_{12} = Z'_{21}$ and $Z'_{23} = Z'_{32}$ are negative because of opposing current directions (I_2 in negative direction in loop 1, and I_3 in negative direction in loop 2), i.e.

$$Z'_{12} = Z'_{21} = -Z_{\text{sheath-mutual}} \qquad (6.66)$$

$$Z'_{23} = Z'_{32} = -Z_{\text{armour-mutual}} \qquad (6.67)$$

where

$Z_{\text{sheath-mutual}}$ = mutual impedance (per unit length) of the tubular sheath between the inside loop 1 and the outside loop 2.
$Z_{\text{armour-mutual}}$ = mutual impedance (per unit length) of the tubular armour between the inside loop 2 and the outside loop 3.

Finally, $Z'_{13} = Z'_{31} = 0$ because loop 1 and loop 3 have no common branch. The impedances of the insulation ($\Omega\,\text{m}^{-1}$) are given by

$$Z'_{\text{insulation}} = j\omega \ln\left(\dfrac{r_{\text{outside}}}{r_{\text{inside}}}\right) \qquad (6.68)$$

where

r_{outside} = outside radius of insulation
r_{inside} = inside radius of insulation.

If there is no insulation between the armour and earth, then $Z_{\text{insulation}} = 0$. The internal impedances and the mutual impedance of a tubular conductor are a function of frequency, and can be derived from Bessel and Kelvin functions.

$$Z'_{\text{tube-inside}} = \frac{j\omega\mu}{2\pi D \cdot mq}[I_0(mq)K_1(mr) + K_0(mq)I_1(mr)] \quad (6.69)$$

$$Z'_{\text{tube-outside}} = \frac{j\omega\mu}{2\pi D \cdot mr}[I_0(mr)K_1(mq) + K_0(mr)I_1(mq)] \quad (6.70)$$

$$Z'_{\text{tube-mutual}} = \frac{j\omega\mu}{2\pi D \cdot mq \cdot mr} \quad (6.71)$$

with

μ = the permeability of insulation in H m^{-1}
$D = I_1(mr)K_1(mq) - I_1(mq)K_1(mr)$
$mr = \sqrt{K/(1-s^2)}$
$mq = \sqrt{Ks^2/(1-s^2)}$
$K = j8\pi \times 10^{-4} f\mu_r/R'_{\text{dc}}$
$s = q/r$
q = inside radius
r = outside radius
R'_{dc} = d.c. resistance in Ω km^{-1}.

The only remaining term is $Z_{\text{earth-inside}}$ in equation 6.65 which is the earth return impedance for underground cables, or the sea return impedance for submarine cables. The earth return impedance can be calculated approximately with equation 6.69 by letting the outside radius go to infinity. This approach, also used by Bianchi and Luoni [12] to find the sea return impedance, is quite acceptable considering the fact that sea resistivity and other input parameters are not known accurately. Equation 6.63 is not in a form compatible with the solution used for overhead conductors, where the voltages with respect to local ground and the actual currents in the conductors are used as variables. Equation 6.63 can easily be brought into such a form by introducing the appropriate terminal conditions, i.e.

$$V_1 = V_{\text{core}} - V_{\text{sheath}} \qquad I_1 = I_{\text{core}}$$
$$V_2 = V_{\text{sheath}} - V_{\text{armour}} \qquad I_2 = I_{\text{core}} + I_{\text{sheath}}$$
$$V_3 = V_{\text{armour}} \qquad I_3 = I_{\text{core}} + I_{\text{sheath}} + I_{\text{armour}}$$

Thus equation 6.63 can be rewritten as

$$-\begin{pmatrix} \dfrac{dV_{core}}{dx} \\ \dfrac{dV_{sheath}}{dx} \\ \dfrac{dV_{armour}}{dx} \end{pmatrix} = \begin{bmatrix} Z'_{cc} & Z'_{cs} & Z'_{ca} \\ Z'_{sc} & Z'_{ss} & Z'_{sa} \\ Z'_{ac} & Z'_{as} & Z'_{aa} \end{bmatrix} \begin{pmatrix} I_{core} \\ I_{sheath} \\ I_{armour} \end{pmatrix} \quad (6.72)$$

where

$$Z'_{cc} = Z'_{11} + 2Z'_{12} + Z'_{22} + 2Z'_{23} + Z'_{33}$$
$$Z'_{cs} = Z'_{sc} = Z'_{12} + Z'_{22} + 2Z'_{23} + Z'_{33}$$
$$Z'_{ca} = Z'_{ac} = Z'_{sa} = Z'_{as} = Z'_{23} + Z'_{33}$$
$$Z'_{ss} = Z'_{22} + 2Z'_{23} + Z'_{33}$$
$$Z'_{aa} = Z'_{33}$$

A good approximation for many cables with bonding between the sheath and the armour, and with the armour earthed to the sea, is $V_{sheath} = V_{armour} = 0$.

Therefore the model can be reduced to

$$\dfrac{-dV_{core}}{dx} = Z I_{core} \quad (6.73)$$

where Z is a reduction of the impedance matrix of equation 6.72.

Similarly, for each cable the per unit length harmonic admittance is:

$$-\begin{pmatrix} \dfrac{dI_1}{dx} \\ \dfrac{dI_2}{dx} \\ \dfrac{dI_3}{dx} \end{pmatrix} = \begin{bmatrix} j\omega C'_1 & 0 & 0 \\ 0 & j\omega C'_2 & 0 \\ 0 & 0 & j\omega C'_3 \end{bmatrix} \begin{pmatrix} V_1 \\ V_2 \\ V_3 \end{pmatrix} \quad (6.74)$$

where $C'_i = 2\pi\varepsilon_0\varepsilon_r / \ln(r/q)$. Therefore, when converted to core, sheath and armour quantities,

$$-\begin{pmatrix} \dfrac{dI_{core}}{dx} \\ \dfrac{dI_{sheath}}{dx} \\ \dfrac{dI_{armour}}{dx} \end{pmatrix} = \begin{bmatrix} Y'_1 & -Y'_1 & 0 \\ -Y'_1 & Y'_1 + Y'_2 & -Y'_2 \\ 0 & -Y'_2 & Y'_2 + Y'_3 \end{bmatrix} \begin{pmatrix} V_{core} \\ V_{sheath} \\ V_{armour} \end{pmatrix} \quad (6.75)$$

where $Y_i = j\omega l_i$. If, as before, $V_{sheath} = V_{armour} = 0$, equation 6.75 reduces to

$$-dI_{core}/dx = Y_1 V_{core} \quad (6.76)$$

146 Power systems electromagnetic transients simulation

Therefore, for the frequencies of interest, the cable per unit length impedance, Z', and admittance, Y', are calculated with both the zero and positive sequence values being equal to the Z in equation 6.73, and the Y_1 in equation 6.76, respectively. In the absence of rigorous computer models, such as described above, power companies often use approximations to the skin effect by means of correction factors.

6.6 Example

To illustrate various transmission line representations let us consider two simple lines with the parameters shown in Tables 6.1 and 6.2.

For the transmission line with the parameters shown in Table 6.1, $\gamma = 0.500000E-08$, $Z_c = 100\,\Omega$ and the line travelling delay is 0.25 ms (or 5 time steps). This delay can clearly be seen in Figures 6.15 and 6.16. Note also the lack of reflections when the line is terminated by the characteristic impedance (Figure 6.15). Reflections cause a step change every 0.5 ms, or twice the travelling time.

When the load impedance is larger than the characteristic impedance (Figure 6.16) a magnified voltage at the receiving end (of 33 per cent in this case) appears 0.25 ms after the step occurs at the sending end. This also results in a receiving end current beginning to flow at this time. The receiving end voltage and current then propagate back to the sending end, after a time delay of 0.25 ms, altering the sending end current.

Table 6.1 Parameters for transmission line example

L'	$500 \times 10^{-9}\,\mathrm{H\,m^{-1}}$
C'	$50 \times 10^{-9}\,\mathrm{F\,m^{-1}}$
L	50 km
R (source)	$0.1\,\Omega$
Δt	$50\,\mu s$

Table 6.2 Single phase test transmission line

Description	Value
Ground resistivity (Ω m)	100.0
Line length (km)	100.0
Conductor radius (cm)	2.03454
Height at tower Y (m)	30.0
Sag at mid-span (m)	10.0
d.c. resistance ($\Omega\,\mathrm{km}^{-1}$)	0.03206

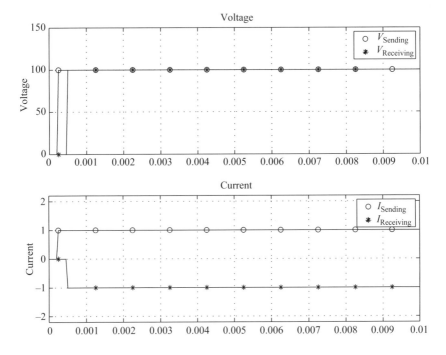

Figure 6.15 Step response of a lossless line terminated by its characteristic impedance

This change in sending end current propagates down the receiving end, influencing its voltages and currents again. Hence in the case of a higher than characteristic impedance loading the initial receiving voltage and current magnitudes are larger than the steady-state value and each subsequent reflection opposes the last, causing a decaying oscillation.

With a smaller than characteristic impedance loading (Figure 6.17) the receiving voltage and current magnitudes are smaller than their steady-state values, and each subsequent reflection reinforces the previous one, giving the damped response shown in Figure 6.17. The FORTRAN code for this example is given in Appendix H.4.

Figures 6.18–6.20 show the same simulation except that the Bergeron model has been used instead. The FORTRAN code for this case is given in Appendix H.5. The line loss is assumed to be $R' = 1.0 \times 10^{-4}\,\Omega\,\mathrm{m}^{-1}$. With characteristic impedance loading there is now a slight transient (Figure 6.18) after the step change in receiving end voltage as the voltage and current waveforms settle, taking into account the line losses. The changes occur every 0.25 ms, which is twice the travelling time of a half-line section, due to reflections from the middle of the line.

The characteristics of Figures 6.18–6.20 are very similar to those of the lossless counterparts, with the main step changes occurring due to reflections arriving in intervals of twice the travelling time of the complete line. However now there is also a small step change in between, due to reflections from the middle of the line. The

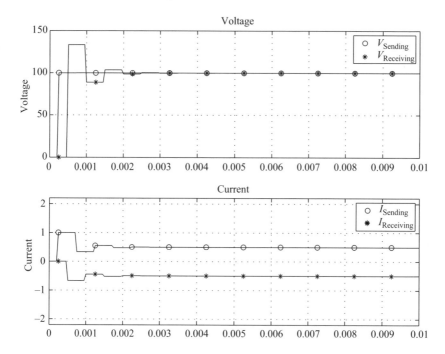

Figure 6.16 Step response of a lossless line with a loading of double characteristic impedance

voltage drop can be clearly seen, the larger voltage drop occurring when the current is greater.

To illustrate a frequency-dependent transmission line model a simple single wire transmission line with no earth wire is used next. The line parameters shown in Table 6.2 are used to obtain the electrical parameters of the line and then curve fitting is performed. There are two main ways of calculating the time convolutions required to implement a frequency-dependent transmission line. These are either recursive convolutions, which require s-domain fitting, or ARMA using z-domain fitting [13].

Figures 6.21 and 6.23 show the match for the attenuation constant and characteristic impedance respectively, while the errors associated with the fit are shown in Figures 6.22 and 6.24. The fitted rational function for the characteristic impedance is shown in Table 6.3 and the partial fraction expansion of its inverse (characteristic admittance) in Table 6.4.

The ratio of d.c. impedance (taken as the impedance at the lowest frequency, which is $614.41724\,\Omega$) over the d.c. value of the fitted function (670.1023) is 0.91690065830247, therefore this is multiplied with the residuals (k terms in equation 6.55). To ensure the transmission line exhibits the correct d.c. resistance the attenuation function must also be scaled. The surge impedance function evaluated at d.c. is $Z_C(\omega = 0)$ and R_{dc} is the line resistance per unit length. Then G' is calculated from $G' = R_{dc}/Z_C^2(\omega = 0)$ and the constant term of the attenuation function

Transmission lines and cables 149

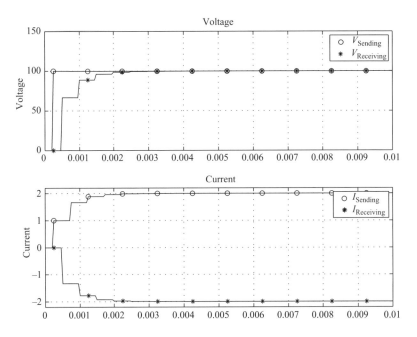

Figure 6.17   Step response of a lossless line with a loading of half its characteristic impedance

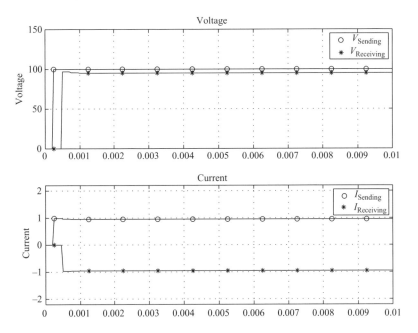

Figure 6.18   Step response of Bergeron line model for characteristic impedance termination

150 Power systems electromagnetic transients simulation

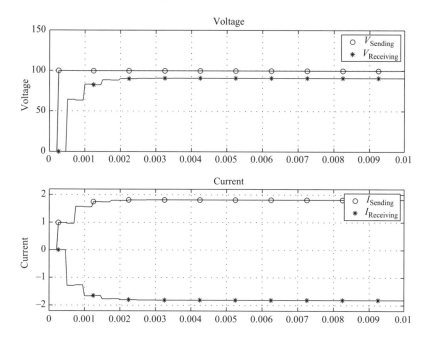

Figure 6.19 *Step response of Bergeron line model for a loading of half its characteristic impedance*

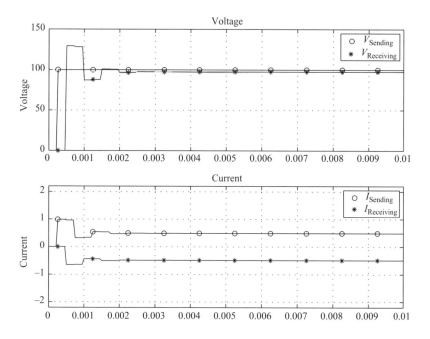

Figure 6.20 *Step response of Bergeron line model for a loading of double characteristic impedance*

Transmission lines and cables 151

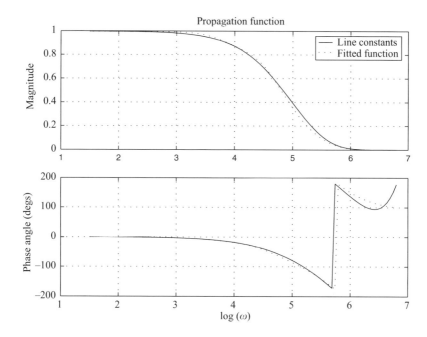

Figure 6.21   Comparison of attenuation (or propagation) constant

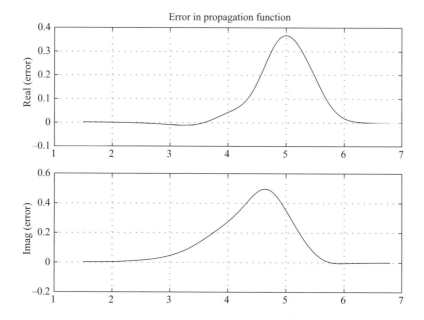

Figure 6.22   Error in fitted attenuation constant

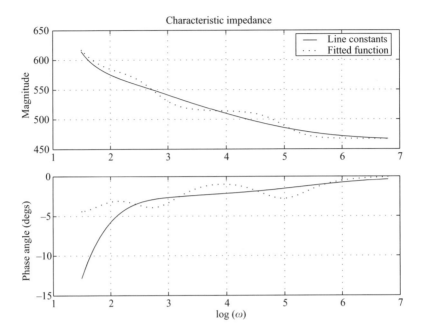

Figure 6.23 Comparison of surge impedance

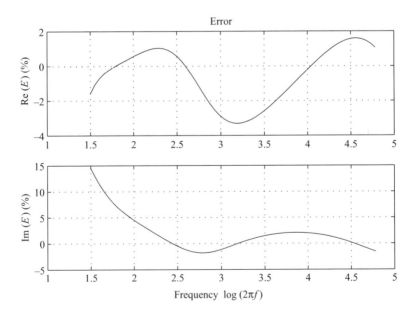

Figure 6.24 Error in fitted surge impedance

Table 6.3 s-domain fitting of characteristic impedance

	Constant	s	s^2
Numerator	−2.896074e+01	−6.250320e+02	−1.005140e+05
Denominator	−2.511680e+01	−5.532123e+02	−9.130399e+04
Constant	467.249168		

Table 6.4 Partial fraction expansion of characteristic admittance

Quantity	Constant	s	s^2
Residual	−19.72605872772154	−0.14043511946635	−0.00657234249032
Denominator	−1.005140e+05	−0.00625032e+05	−0.0002896074e+05
k_0	0.00214018572698	–	–

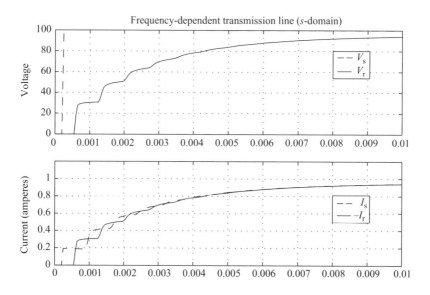

Figure 6.25 Step response of frequency-dependent transmission line model (load = 100 Ω)

is calculated from $e^{-\sqrt{R_{dc}G'}}$. The d.c. line resistance is sensitive to the constant term and the difference between using 0.99 and 0.999 is large.

The response derived from the implementation of this model is given in Figures 6.25, 6.26 and 6.27 for loads of 100, 1000 and 50 ohms respectively.

154 *Power systems electromagnetic transients simulation*

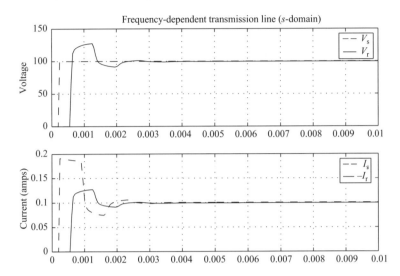

Figure 6.26 Step response of frequency-dependent transmission line model (load = 1000 Ω)

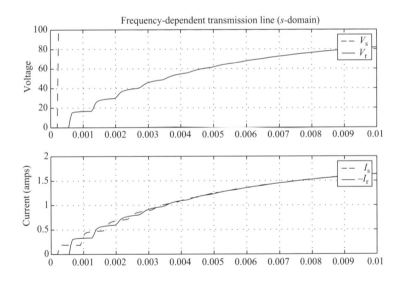

Figure 6.27 Step response of frequency-dependent transmission line model (load = 50 Ω)

Appendix H.6 contains the FORTRAN program used for the simulation of this example.

The fitted rational function for the attenuation function is shown in Table 6.5, and its partial fraction expansion in Table 6.6.

Table 6.5 Fitted attenuation function (s-domain)

	Constant	s	s^2	s^3
Numerator	−7.631562e+03	–	–	–
Denominator	−6.485341e+03	−4.761763e+04	−5.469828e+05	−5.582246e+05
Constant term	0.9952270			

Table 6.6 Partial fraction expansion of fitted attenuation function (s-domain)

Quantity	Constant	s	s^2	s^3
Residual	−2.137796e+06	−2.1858274e+06	0.046883e+06	0.001149e+06
Denominator	−5.582246e+05	−5.469828e+05	−4.761763e+04	−6.485341e+03

Table 6.7 Pole/zero information from PSCAD V2 (characteristic impedance)

Zeros	−2.896074e+01	−6.250320e+02	−1.005140e+05
Poles	−2.511680e+01	−5.532123e+02	−9.130399e+04
H	6.701023e+02		

PSCAD version 2 stores the negative of the poles ($-p_k$) and zeros ($-z_k$) as well as the constant term H, using the form:

$$H \frac{(1+s/z_1)(1+s/z_2)\cdots(1+s/z_n)}{(1+s/p_1)(1+s/p_2)\cdots(1+s/p_m)} \quad (6.77)$$

Relating this expression to equations 6.53 and 6.55 gives:

$$k = H \frac{\prod_{k=1}^{n} p_k}{\prod_{k=1}^{N} z_k}$$

The poles, zeros and constant term H for the characteristic impedance and attenuation are shown in Tables 6.7 and 6.8.

Sequence components are used for data entry (PI model) and output, particularly in the line constants of EMTP. The transformation to sequence components is

$$\begin{pmatrix} V_0 \\ V_+ \\ V_- \end{pmatrix} = \frac{1}{K_1} \begin{bmatrix} 1 & 1 & 1 \\ 1 & a & a^2 \\ 1 & a^2 & a \end{bmatrix} \cdot \begin{pmatrix} V_a \\ V_b \\ V_c \end{pmatrix} \quad (6.78)$$

Table 6.8   Pole/zero information from PSCAD V2 (attenuation function)

Zeros −7.631562e+03
Poles −6.485341e+03 −4.761763e+04 −5.469828e+05 −5.582246e+05
H   9.952270e−01

---

and the reverse transform:

$$\begin{pmatrix} V_a \\ V_b \\ V_c \end{pmatrix} = \frac{1}{K_2} \begin{bmatrix} 1 & 1 & 1 \\ 1 & a^2 & a \\ 1 & a & a^2 \end{bmatrix} \cdot \begin{pmatrix} V_0 \\ V_+ \\ V_- \end{pmatrix}$$

where $a = e^{j120} = -1/2 + j\sqrt{3}/2$.

The power industry uses values of $K_1 = 3$ and $K_2 = 1$, but in the normalised version both K_1 and K_2 are equal to $\sqrt{3}$. Although the choice of factors affect the sequence voltages and currents, the sequence impedances are unaffected by them.

6.7 Summary

For all except very short transmission lines, travelling wave transmission line models are preferable. If frequency dependence is important then a frequency transmission line dependent model will be used. Details of transmission line geometry and conductor data are then required in order to calculate accurately the frequency-dependent electrical parameters of the line. The simulation time step must be based on the shortest response time of the line.

Many variants of frequency-dependent multiconductor transmission line models exist. A widely used model is based on ignoring the frequency dependence of the transformation matrix between phase and mode domains (i.e. the J. Marti model in EMTP [14]).

At present phase-domain models are the most accurate and robust for detailed transmission line representation. Given the complexity and variety of underground cables, a rigorous unified solution similar to that of the overhead line is only possible based on a standard cross-section structure and under various simplifying assumptions. Instead, power companies often use correction factors, based on experience, for skin effect representation.

6.8 References

1 CARSON, J. R.: 'Wave propagation in overhead wires with ground return', *Bell System Technical Journal*, 1926, **5**, pp. 539–54

2 POLLACZEK, F.: 'On the field produced by an infinitely long wire carrying alternating current', *Elektrische Nachrichtentechnik*, 1926, **3**, pp. 339–59
3 POLLACZEK, F.: 'On the induction effects of a single phase ac line', *Elektrische Nachrichtentechnik*, 1927, **4**, pp. 18–30
4 GUSTAVSEN, B. and SEMLYEN, A.: 'Simulation of transmission line transients using vector fitting and modal decomposition', *IEEE Transactions on Power Delivery*, 1998, **13** (2), pp. 605–14
5 BERGERON, L.: 'Du coup de Belier en hydraulique au coup de foudre en electricite' (Dunod, 1949). (English translation: 'Water hammer in hydraulics and wave surges in electricity', ASME Committee, Wiley, New York, 1961.)
6 WEDEPOHL, L. M., NGUYEN, H. V. and IRWIN, G. D.: 'Frequency-dependent transformation matrices for untransposed transmission lines using Newton-Raphson method', *IEEE Transactions on Power Systems*, 1996, **11** (3), pp. 1538–46
7 CLARKE, E.: 'Circuit analysis of AC systems, symmetrical and related components' (General Electric Co., Schenectady, NY, 1950)
8 SEMLYEN, A. and DABULEANU, A.: 'Fast and accurate switching transient calculations on transmission lines with ground return using recursive convolutions', *IEEE Transactions on Power Apparatus and Systems*, 1975, **94** (2), pp. 561–71
9 SEMLYEN, A.: 'Contributions to the theory of calculation of electromagnetic transients on transmission lines with frequency dependent parameters', *IEEE Transactions on Power Apparatus and Systems*, 1981, **100** (2), pp. 848–56
10 MORCHED, A., GUSTAVSEN, B. and TARTIBI, M.: 'A universal model for accurate calculation of electromagnetic transients on overhead lines and underground cables', *IEEE Transactions on Power Delivery*, 1999, **14** (3), pp. 1032–8
11 DERI, A., TEVAN, G., SEMLYEN, A. and CASTANHEIRA, A.: 'The complex ground return plane, a simplified model for homogenous and multi-layer earth return', *IEEE Transactions on Power Apparatus and Systems*, 1981, **100** (8), pp. 3686–93
12 BIANCHI, G. and LUONI, G.: 'Induced currents and losses in single-core submarine cables', *IEEE Transactions on Power Apparatus and Systems*, 1976, **95**, pp. 49–58
13 NODA, T.: 'Development of a transmission-line model considering the skin and corona effects for power systems transient analysis' (Ph.D. thesis, Doshisha University, Kyoto, Japan, December 1996)
14 MARTI, J. R.: 'Accurate modelling of frequency-dependent transmission lines in electromagnetic transient simulations', *IEEE Transactions on Power Apparatus and Systems*, 1982, **101** (1), pp. 147–57

Chapter 7

Transformers and rotating plant

7.1 Introduction

The simulation of electrical machines, whether static or rotative, requires an understanding of the electromagnetic characteristics of their respective windings and cores. Due to their basically symmetrical design, rotating machines are simpler in this respect. On the other hand the latter's transient behaviour involves electromechanical as well as electromagnetic interactions. Electrical machines are discussed in this chapter with emphasis on their magnetic properties. The effects of winding capacitances are generally negligible for studies other than those involving fast fronts (such as lightning and switching).

The first part of the chapter describes the dynamic behaviour and computer simulation of single-phase, multiphase and multilimb transformers, including saturation effects [1]. Early models used with electromagnetic transient programs assumed a uniform flux throughout the core legs and yokes, the individual winding leakages were combined and the magnetising current was placed on one side of the resultant series leakage reactance. An advanced multilimb transformer model is also described, based on unified magnetic equivalent circuit recently implemented in the EMTDC program.

In the second part, the chapter develops a general dynamic model of the rotating machine, with emphasis on the synchronous generator. The model includes an accurate representation of the electrical generator behaviour as well as the mechanical characteristics of the generator and the turbine. In most cases the speed variations and torsional vibrations can be ignored and the mechanical part can be left out of the simulation.

7.2 Basic transformer model

The equivalent circuit of the basic transformer model, shown in Figure 7.1, consists of two mutually coupled coils. The voltages across these coils is expressed as:

$$\begin{pmatrix} v_1 \\ v_2 \end{pmatrix} = \begin{bmatrix} L_{11} & L_{21} \\ L_{12} & L_{22} \end{bmatrix} \frac{d}{dt} \begin{pmatrix} i_1 \\ i_2 \end{pmatrix} \quad (7.1)$$

where L_{11} and L_{22} are the self-inductance of winding 1 and 2 respectively, and L_{12} and L_{21} are the mutual inductance between the windings.

In order to solve for the winding currents the inductance matrix has to be inverted, i.e.

$$\frac{d}{dt} \begin{pmatrix} i_1 \\ i_2 \end{pmatrix} = \frac{1}{L_{11}L_{22} - L_{12}L_{21}} \begin{bmatrix} L_{22} & -L_{21} \\ -L_{12} & L_{11} \end{bmatrix} \begin{pmatrix} v_1 \\ v_2 \end{pmatrix} \quad (7.2)$$

Since the mutual coupling is bilateral, L_{12} and L_{21} are identical. The coupling coefficient between the two coils is:

$$K_{12} = \frac{L_{12}}{\sqrt{L_{11}L_{22}}} \quad (7.3)$$

Rewriting equation 7.1 using the turns ratio ($a = v_1/v_2$) gives:

$$\begin{pmatrix} v_1 \\ av_2 \end{pmatrix} = \begin{bmatrix} L_{11} & L_{21} \\ aL_{12} & a^2 L_{22} \end{bmatrix} \frac{d}{dt} \begin{pmatrix} i_1 \\ i_2/a \end{pmatrix} \quad (7.4)$$

This equation can be represented by the equivalent circuit shown in Figure 7.2, where

$$L_1 = L_{11} - a L_{12} \quad (7.5)$$

$$L_2 = a^2 L_{22} - a L_{12} \quad (7.6)$$

Consider a transformer with a 10% leakage reactance equally divided between the two windings and a magnetising current of 0.01 p.u. Then the input impedance with the second winding open circuited must be 100 p.u. (Note from equation 7.5,

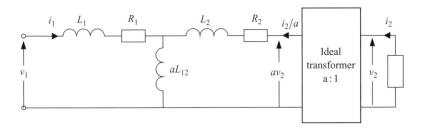

Figure 7.1 Equivalent circuit of the two-winding transformer

Transformers and rotating plant 161

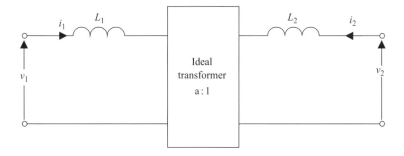

Figure 7.2   Equivalent circuit of the two-winding transformer, without the magnetising branch

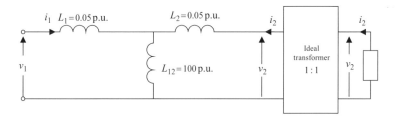

Figure 7.3   Transformer example

$L_1 + L_{12} = L_{11}$ since $a = 1$ in the per unit system.) Hence the equivalent in Figure 7.3 is obtained, the corresponding equation (in p.u.) being:

$$\begin{pmatrix} v_1 \\ v_2 \end{pmatrix} = \begin{bmatrix} 100.0 & 99.95 \\ 99.95 & 100.0 \end{bmatrix} \frac{d}{dt} \begin{pmatrix} i_1 \\ i_2 \end{pmatrix} \quad (7.7)$$

or in actual values:

$$\begin{pmatrix} v_1 \\ v_2 \end{pmatrix} = \frac{1}{S_{\text{Base}}} \begin{bmatrix} 100.0 \times v_{\text{Base\_1}}^2 & 99.95 \times v_{\text{Base\_1}} v_{\text{Base\_2}} \\ 99.95 \times v_{\text{Base\_1}} v_{\text{Base\_2}} & 100.0 \times v_{\text{Base\_2}}^2 \end{bmatrix} \frac{d}{dt} \begin{pmatrix} i_1 \\ i_2 \end{pmatrix} \text{ volts} \quad (7.8)$$

7.2.1 Numerical implementation

Separating equation 7.2 into its components:

$$\frac{di_1}{dt} = \frac{L_{22}}{L_{11}L_{22} - L_{12}L_{21}} v_1 - \frac{L_{21}}{L_{11}L_{22} - L_{12}L_{21}} v_2 \quad (7.9)$$

$$\frac{di_2}{dt} = \frac{-L_{12}}{L_{11}L_{22} - L_{12}L_{21}} v_1 + \frac{L_{11}}{L_{11}L_{22} - L_{12}L_{21}} v_2 \quad (7.10)$$

Solving equation 7.9 by trapezoidal integration yields:

$$i_1(t) = \frac{L_{22}}{L_{11}L_{22} - L_{12}L_{21}} \int_0^t v_1 \, dt - \frac{L_{21}}{L_{11}L_{22} - L_{12}L_{21}} \int_0^t v_2 \, dt$$

$$= i_1(t - \Delta t) + \frac{L_{22}}{L_{11}L_{22} - L_{12}L_{21}} \int_{t-\Delta t}^t v_1 \, dt$$

$$- \frac{L_{21}}{L_{11}L_{22} - L_{12}L_{21}} \int_{t-\Delta t}^t v_2 \, dt$$

$$= i_1(t - \Delta t) + \frac{L_{22}\Delta t}{2(L_{11}L_{22} - L_{12}L_{21})}(v_1(t - \Delta t) + v_1(t))$$

$$- \frac{L_{21}\Delta t}{2(L_{11}L_{22} - L_{12}L_{21})}(v_2(t - \Delta t) + v_2(t)) \quad (7.11)$$

Collecting together the past History and Instantaneous terms gives:

$$i_1(t) = I_h(t - \Delta t) + \left(\frac{L_{22}\Delta t}{2(L_{11}L_{22} - L_{12}L_{21})} - \frac{L_{21}\Delta t}{2(L_{11}L_{22} - L_{12}L_{21})}\right) v_1(t)$$

$$+ \frac{L_{21}\Delta t}{2(L_{11}L_{22} - L_{12}L_{21})}(v_1(t) - v_2(t)) \quad (7.12)$$

where

$$I_h(t - \Delta t) = i_1(t - \Delta t)$$

$$+ \left(\frac{L_{22}\Delta t}{2(L_{11}L_{22} - L_{12}L_{21})} - \frac{L_{21}\Delta t}{2(L_{11}L_{22} - L_{12}L_{21})}\right) v_1(t - \Delta t)$$

$$+ \frac{L_{21}\Delta t}{2(L_{11}L_{22} - L_{12}L_{21})}(v_1(t - \Delta t) - v_2(t - \Delta t)) \quad (7.13)$$

A similar expression can be written for $i_2(t)$. The model that these equations represents is shown in Figure 7.4. It should be noted that the discretisation of these models using the trapezoidal rule does not give complete isolation between its terminals for d.c. If a d.c. source is applied to winding 1 a small amount will flow in winding 2, which in practice would not occur. Simulation of the test system shown in Figure 7.5 will clearly demonstrate this problem. This test system also shows ill-conditioning in the inductance matrix when the magnetising current is reduced from 1 per cent to 0.1 per cent.

7.2.2 Parameters derivation

Transformer data is not normally given in the form used in the previous section. Either results from short-circuit and open-circuit tests are available or the magnetising current and leakage reactance are given in p.u. quantities based on machine rating.

In the circuit of Figure 7.1, shorting winding 2 and neglecting the resistance gives:

$$I_1 = \frac{V_1}{\omega(L_1 + L_2)} \quad (7.14)$$

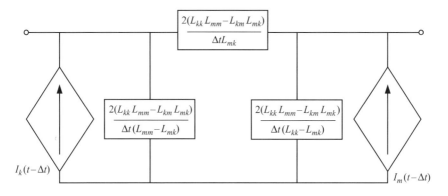

Figure 7.4 Transformer equivalent after discretisation

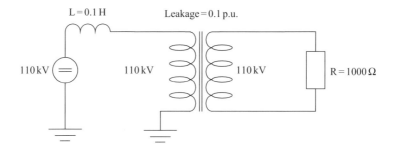

Figure 7.5 Transformer test system

Similarly, open-circuit tests with windings 2 or 1 open-circuited, respectively give:

$$I_1 = \frac{V_1}{\omega(L_1 + aL_{12})} \tag{7.15}$$

$$I_2 = \frac{a^2 V_2}{\omega(L_2 + aL_{12})} \tag{7.16}$$

Short and open circuit tests provide enough information to determine aL_{12}, L_1 and L_2. These calculations are often performed internally in the transient simulation program, and the user only needs to enter directly the leakage and magnetising reactances.

The inductance matrix contains information to derive the magnetising current and also, indirectly through the small differences between L_{11} and L_{12}, the leakage (short-circuit) reactance.

The leakage reactance is given by:

$$L_{\text{Leakage}} = L_{11} - L_{21}^2/L_{22} \tag{7.17}$$

In most studies the leakage reactance has the greatest influence on the results. Thus the values of the inductance matrix must be specified very accurately to reduce errors due to loss of significance caused by subtracting two numbers of very similar magnitude.

Mathematically the inductance becomes ill-conditioned as the magnetising current gets smaller (it is singular if the magnetising current is zero). The matrix equation expressing the relationships between the derivatives of current and voltage is:

$$\frac{d}{dx}\begin{pmatrix} i_1 \\ i_2 \end{pmatrix} = \frac{1}{L}\begin{bmatrix} 1 & a \\ -a & a^2 \end{bmatrix}\begin{pmatrix} v_1 \\ v_2 \end{pmatrix} \tag{7.18}$$

where $L = L_1 + a^2 L_2$. This represents the equivalent circuit shown in Figure 7.2.

7.2.3 Modelling of non-linearities

The magnetic non-linearity and core loss components are usually incorporated by means of a shunt current source and resistance respectively, across one winding. Since the single-phase approximation does not incorporate inter-phase magnetic coupling, the magnetising current injection is calculated at each time step independently of the other phases.

Figure 7.6 displays the modelling of saturation in mutually coupled windings. The current source representation is used, rather than varying the inductance, as the latter would require retriangulation of the matrix every time the inductance changes. During start-up it is recommended to inhibit saturation and this is achieved by using a flux limit for the result of voltage integration. This enables the steady state to be reached faster. Prior to the application of the disturbance the flux limit is removed, thus allowing the flux to go into the saturation region.

Another refinement, illustrated in Figure 7.7, is to impose a decay time on the in-rush currents, as would occur on energisation or fault recovery.

Typical studies requiring the modelling of saturation are: In-rush current on energising a transformer, steady-state overvoltage studies, core-saturation instabilities and ferro-resonance.

A three-phase bank can be modelled by the correct connection of three two-coupled windings. For example the wye/delta connection is achieved as shown in Figure 7.8, which produces the correct phase shift automatically between the

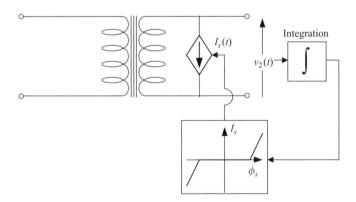

Figure 7.6   Non-linear transformer

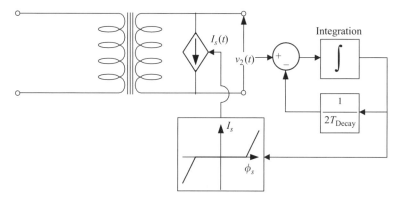

Figure 7.7 Non-linear transformer model with in-rush

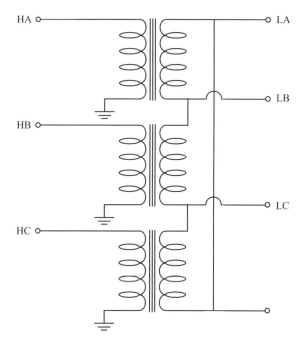

Figure 7.8 Star–delta three-phase transformer

primary and secondary windings (secondary lagging primary by 30 degrees in the case shown) [2].

7.3 Advanced transformer models

To take into account the magnetising currents and core configuration of multilimb transformers the EMTP package has developed a program based on the principle of duality [3]. The resulting duality-based equivalents involve a large number of components; for instance, 23 inductances and nine ideal transformers are required to represent the three-phase three-winding transformer. Additional components are used to isolate the true non-linear series inductors required by the duality method, as their implementation in the EMTP program is not feasible [4].

To reduce the complexity of the equivalent circuit two alternatives based on an equivalent inductance matrix have been proposed. However one of them [5] does not take into account the core non-linearity under transient conditions. In the second [6], the non-linear inductance matrix requires regular updating during the EMTP solution, thus reducing considerably the program efficiency.

Another model [7] proposes the use of a Norton equivalent representation for the transformer as a simple interface with the EMTP program. This model does not perform a direct analysis of the magnetic circuit; instead it uses a combination of the duality and leakage inductance representation.

The rest of this section describes a model also based on the Norton equivalent but derived directly from magnetic equivalent circuit analysis [8], [9]. It is called the UMEC (Unified Magnetic Equivalent Circuit) model and has been recently implemented in the EMTDC program.

The UMEC principle is first described with reference to the single-phase transformer and later extended to the multilimb case.

7.3.1 Single-phase UMEC model

The single-phase transformer, shown in Figure 7.9(a), can be represented by the UMEC of Figure 7.9(b). The m.m.f. sources $N_1 i_1(t)$ and $N_2 i_2(t)$ represent each winding individually. The primary and secondary winding voltages, $v_1(t)$ and $v_2(t)$, are used to calculate the winding limb fluxes $\phi_1(t)$ and $\phi_2(t)$, respectively. The

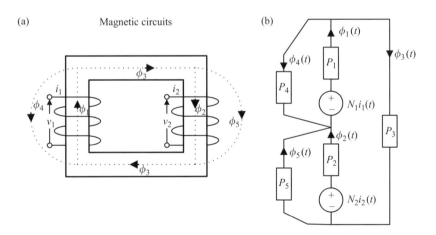

Figure 7.9 UMEC single-phase transformer model: (a) core flux paths; (b) unified magnetic equivalent circuit

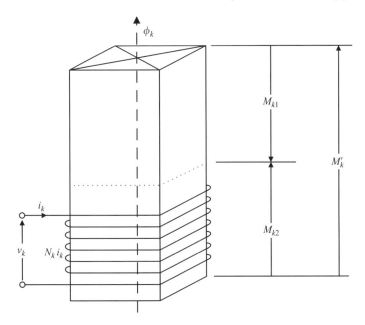

Figure 7.10 Magnetic equivalent circuit for branch

winding limb flux divides between leakage and yoke paths and, thus, a uniform core flux is not assumed.

Although single-phase transformer windings are not generally wound separately on different limbs, each winding can be separated in the UMEC. In Figure 7.9(b) P_1 and P_2 represent the permeances of transformer winding limbs and P_3 that of the transformer yokes. If the total length of core surrounded by windings L_w has a uniform cross-sectional area A_w, then $A_1 = A_2 = A_{w1}$. The upper and lower yokes are assumed to have the same length L_y and cross-sectional area A_y. Both yokes are represented by the single UMEC branch 3 of length $L_3 = 2L_y$ and area $A_3 = A_y$. Leakage information is obtained from the open and short-circuit tests and, therefore, the effective lengths and cross-sectional areas of leakage flux paths are not required to calculate the leakage permeances P_4 and P_5.

Figure 7.10 shows a transformer branch where the branch reluctance and winding magnetomotive force (m.m.f.) components have been separated.

The non-linear relationship between branch flux (ϕ_k) and branch m.m.f. drop (M_{k1}) is

$$M_{k1} = r_k(\phi_k) \tag{7.19}$$

where r_k is the magnetising characteristic (shown in Figure 7.11).

The m.m.f. of winding N_k is:

$$M_{k2} = N_k i_k \tag{7.20}$$

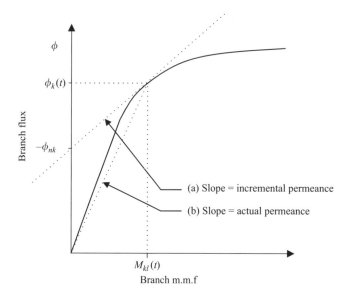

Figure 7.11   Incremental and actual permeance

The resultant branch m.m.f. $\left(M'_{k2}\right)$ is thus

$$M'_{k2} = M_{k2} - M_{k1} \tag{7.21}$$

The magnetising characteristic displayed in Figure 7.11 shows that, as the transformer core moves around the knee region, the change in incremental permeance (P_k) is much larger and more sudden (especially in the case of highly efficient cores) than the change in actual permeance $\left(P_k^*\right)$. Although the incremental permeance forms the basis of steady-state transformer modelling, the use of the actual permeance is favoured for the transformer representation in dynamic simulation.

In the UMEC branch the flux is expressed using the actual permeance $\left(P_k^*\right)$, i.e.

$$\phi_k(t) = P_k^* M_{k1}(t) \tag{7.22}$$

From Figure 7.11, ϕ_k can be expressed as

$$\phi_k = P_k^* \left(N_k i_k - M'_k\right) \tag{7.23}$$

which written in vector form

$$\boldsymbol{\phi} = \left[P_k^*\right]\left([N_k]\mathbf{i}_k - \mathbf{M}'_k\right) \tag{7.24}$$

represents all the branches of a multilimb transformer.

Transformers and rotating plant 169

7.3.1.1 UMEC Norton equivalent

The linearised relationship between winding current and branch flux can be extended to incorporate the magnetic equivalent-circuit branch connections. Let the node–branch connection matrix of the magnetic circuit be $[A]$ and the vector of nodal magnetic drops ϕ_{Node}. At each node the flux must sum to zero, i.e.

$$[A]\boldsymbol{\phi}_{\text{Node}} = \mathbf{0} \tag{7.25}$$

Application of the branch–node connection matrix to the vector of nodal magnetic drops gives the branch m.m.f.

$$[A]\mathbf{M}_{\text{Node}} = \mathbf{M}' \tag{7.26}$$

Combining equations 7.24, 7.25 and 7.26 finally yields:

$$\boldsymbol{\phi} = [Q^*][P^*][N]\mathbf{i} \tag{7.27}$$

where

$$[Q^*] = [I] - [P^*][A]\left([A]^T[P^*][A]\right)^{-1}[A]^T \tag{7.28}$$

The winding voltage v_k is related to the branch flux ϕ_k by:

$$v_k = N_k \frac{d\phi_k}{dt} \tag{7.29}$$

Using the trapezoidal integration rule to discretise equation 7.29 gives:

$$\boldsymbol{\phi}_s(t) = \boldsymbol{\phi}_s(t - \Delta t) + \frac{\Delta t}{2}[N_s]^{-1}(\mathbf{v}_s(t) + \mathbf{v}_s(t - \Delta t)) \tag{7.30}$$

where

$$\boldsymbol{\phi}_s(t - \Delta t) = \boldsymbol{\phi}_s(t - 2\Delta t) + \frac{\Delta t}{2}[N_s]^{-1}(\mathbf{v}_s(t - \Delta t) + \mathbf{v}_s(t - 2\Delta t)) \tag{7.31}$$

Partitioning the vector of branch flux $\boldsymbol{\phi}$ into branches associated with each transformer winding $\boldsymbol{\phi}_s$ and using equation 7.30 leads to the Norton equivalent:

$$\mathbf{i}_s(t) = \left[Y^*_{ss}\right]\mathbf{v}_s(t) + \mathbf{i}^*_{ns}(t) \tag{7.32}$$

where

$$[Y^*_{ss}] = ([Q^*_{ss}][P^*_s][N_s])^{-1}\frac{\Delta t}{2}[N_s]^{-1} \tag{7.33}$$

and

$$\mathbf{i}^*_{ns}(t) = ([Q^*_{ss}][P^*_s][N_s])^{-1}\left(\frac{\Delta t}{2}[N_s]^{-1}\mathbf{v}_s(t - \Delta t) + \boldsymbol{\phi}(t - \Delta t)\right) \tag{7.34}$$

Calculation of the UMEC branch flux ϕ_k requires the expansion of the linearised equation 7.27

$$\begin{pmatrix} \boldsymbol{\phi}_s \\ - \\ \boldsymbol{\phi}_r \end{pmatrix} = \begin{bmatrix} [Q^*_{ss}] & | & [Q^*_{sr}] \\ - & - & - \\ [Q^*_{rs}] & | & [Q^*_{rr}] \end{bmatrix} \begin{bmatrix} [P^*_s] & | & [0] \\ - & - & - \\ [0] & | & [P^*_r] \end{bmatrix} \begin{pmatrix} [N_s]\mathbf{i}_s \\ - \\ 0 \end{pmatrix} \qquad (7.35)$$

The winding-limb flux $\phi_s(t - \Delta t)$ is calculated from the winding current by using the upper partition of equation 7.35, i.e.

$$(\boldsymbol{\phi}_s) = [Q^*_{ss}][P^*_s][N_s](\mathbf{i}_s) \qquad (7.36)$$

The yoke and leakage path flux $\phi_r(t - \Delta t)$ is calculated from the winding current by using the lower partition of equation 7.35, i.e.

$$(\boldsymbol{\phi}_r) = [Q^*_{rs}][P^*_s][N_s](\mathbf{i}_s) \qquad (7.37)$$

The branch actual permeance (P^*_k) is calculated directly from a hyperbola approximation of the saturated magnetising characteristic using the solved branch flux $\phi_k(t-\Delta t)$. Once $[P^*_k]$ is known the per-unit admittance matrix $[Y^*_{ss}]$ and current source i^*_{ns} can be obtained. For the UMEC of Figure 7.9, equation 7.32 becomes:

$$\begin{pmatrix} \mathbf{i}_1(t) \\ \mathbf{i}_2(t) \end{pmatrix} = \begin{bmatrix} y_{11} & y_{12} \\ y_{12} & y_{22} \end{bmatrix} \begin{pmatrix} \mathbf{v}_1(t) \\ \mathbf{v}_2(t) \end{pmatrix} + \begin{pmatrix} \mathbf{i}_{ns1}(t) \\ \mathbf{i}_{ns2}(t) \end{pmatrix} \qquad (7.38)$$

which can be represented by the Norton equivalent circuit shown in Figure 7.12.

The Norton equivalent circuit is in an ideal form for dynamic simulation of the EMTDC type. The symmetric admittance matrix $[Y^*_{ss}]$ is non-diagonal, and thus includes mutual couplings. All the equations derived above are general and apply to any magnetic-equivalent circuit consisting of a finite number of branches, such as that shown in Figure 7.10.

If required, the winding copper loss can be represented by placing series resistances at the terminals of the Norton equivalent.

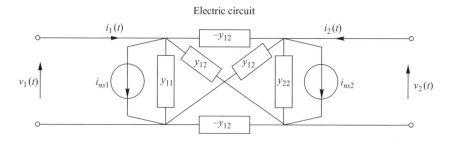

Figure 7.12   *UMEC Norton equivalent*

7.3.2 UMEC implementation in PSCAD/EMTDC

Figure 7.13 illustrates the transformer implementation of the above formulation in PSCAD/EMTDC. An exact solution of the magnetic/electrical circuit at each time step requires a Newton-type iterative process since the system is non-linear. The iterative process finds a solution for the branch fluxes such that nodal flux and loop m.m.f. sums are zero, and with the branch permeances consistent with the flux through them. With small simulation steps of the order of 50 μs, acceptable results can still be obtained in a non-iterative solution if the branch permeances are calculated with the flux solution from the previous time step. The resulting errors are small and confined to the zero sequence of the magnetising currents.

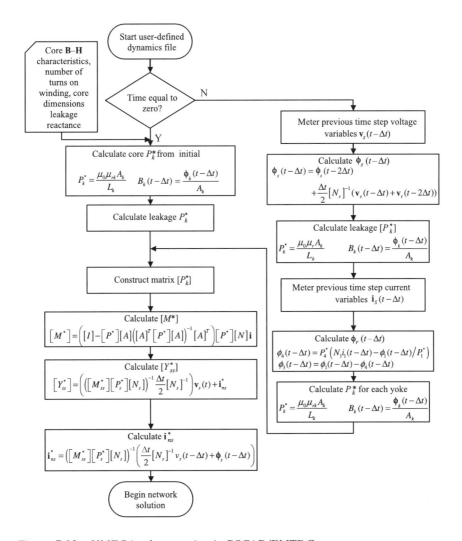

Figure 7.13 UMEC implementation in PSCAD/EMTDC

The leakage-flux branch permeances are constant and the core branch-saturation characteristic is the steel flux density magnetising force (**B–H**) curve. Individual branch per unit $\phi-i$ characteristics are not a conventional specification but, if required, these can be provided by the manufacturer.

Core dimensions, branch length L_k and cross-sectional area A_k, are required to calculate real value permeances from

$$P_k^* = \frac{\mu_0 \mu_{rk} A_k}{L_k} \tag{7.39}$$

The branch flux $\phi_k(t - \Delta t)$ is converted to branch flux density by

$$B_k(t - \Delta t) = \frac{\phi_k(t - \Delta t)}{A_k} \tag{7.40}$$

The branch permeability $\mu_0 \mu_{rk}$ is then calculated from the core **B–H** characteristic.

Figure 7.13 also shows that the winding-limb flux $\boldsymbol{\phi}_s(t - \Delta t)$ is calculated using trapezoidal integration rather than the linearised equation 7.27. Trapezoidal integration requires storage of vectors $\boldsymbol{\phi}_s(t - 2\Delta t)$ and $\mathbf{v}_s(t - 2\Delta t)$. In equation 7.27 matrices $[Q_{ss}^*]$ and $[P_{ss}^*]$ must be stored and, although $[P_{ss}^*]$ is diagonal, $[Q_{ss}^*]$ is full; therefore in this method element storage increases with the square of the UMEC winding-limb branch number.

The elements of $\boldsymbol{\phi}_r(t - \Delta t)$ can be calculated using magnetic circuit theory, whereby the m.m.f. around the primary winding limb and leakage branch loop must sum to zero, i.e. with reference to Figure 7.9(b),

$$\phi_4(t - \Delta t) = P_4^* \left(N_1 i_1(t - \Delta t) - \phi_1(t - \Delta t)/P_1^* \right) \tag{7.41}$$

Also, the m.m.f. around the secondary winding limb and leakage branch loop must sum to zero

$$\phi_5(t - \Delta t) = P_5^* \left(N_2 i_2(t - \Delta t) - \phi_2(t - \Delta t)/P_2^* \right) \tag{7.42}$$

and, finally, the flux at node N_1 must sum to zero:

$$\phi_3(t - \Delta t) = \phi_1(t - \Delta t)) - \phi_4(t - \Delta t) \tag{7.43}$$

The yoke branch actual permeance P_k^* is calculated directly from the solved branch flux $\phi_k(t - \Delta t)$ using equations 7.39 and 7.40. Once $[P^*]$ is known, the real-valued admittance matrix $[Y_{ss}^*]$ and current source vector $\mathbf{i}_{ns}^*$ can be obtained.

7.3.3 Three-limb three-phase UMEC

An extension of the single-phase UMEC concept to the three-phase transformer, shown in Figure 7.14(a), leads to the UMEC of Figure 7.14(b). There is no need to specify in advance the distribution of magnetising current components, which have been shown to be determined by the transformer internal and external circuit parameters.

Transformers and rotating plant 173

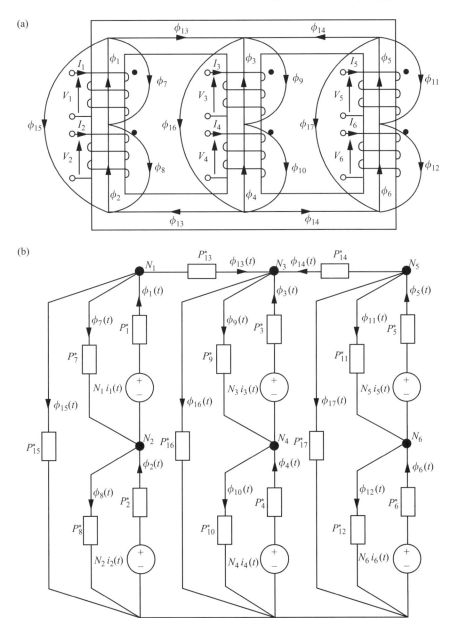

Figure 7.14   UMEC PSCAD/EMTDC three-limb three-phase transformer model: (a) core; (b) electrical equivalents of core flux paths

The m.m.f. sources $N_1i_1(t)$–$N_6i_6(t)$ represent each transformer winding individually, and the winding voltages $v_1(t)$–$v_6(t)$ are used to calculate the winding-limb fluxes $\phi_1(t)$–$\phi_6(t)$, respectively.

P_1^*–P_6^* represent the permeances of transformer winding limbs. If the total length of each phase-winding limb L_w has uniform cross-sectional area A_w, the UMEC branches 1–6 have length $L_w/2$ and cross-sectional area A_w. P_{13}^* and P_{14}^* represent the permeances of the transformer left and right hand yokes, respectively. The upper and lower yokes are assumed to have the same length L_y and cross-sectional area A_y. Both left and right-hand yokes are represented by UMEC branches 13 and 14 of length $L_{13} = L_{14} = 2L_y$ and area $A_{13} = A_{14} = 2A_y$. Zero-sequence permeances P_{15}–P_{17} are obtained from in-phase excitation of all three primary or secondary windings.

Leakage permeances are obtained from open and short-circuit tests and, therefore, the effective length and cross-sectional areas of UMEC leakage branches 7–12 are not required to calculate P_7–P_{12}.

The UMEC circuit of Figure 7.14(b) places the actual permeance formulation in the real-value form

$$\begin{pmatrix} i_1(t) \\ i_2(t) \\ i_3(t) \\ i_4(t) \\ i_5(t) \\ i_6(t) \end{pmatrix} = \begin{bmatrix} y_{11} & y_{12} & y_{13} & y_{14} & y_{15} & y_{16} \\ y_{21} & y_{22} & y_{23} & y_{24} & y_{25} & y_{26} \\ y_{31} & y_{32} & y_{33} & y_{34} & y_{35} & y_{36} \\ y_{41} & y_{42} & y_{43} & y_{44} & y_{45} & y_{46} \\ y_{51} & y_{52} & y_{53} & y_{54} & y_{55} & y_{56} \\ y_{61} & y_{62} & y_{63} & y_{64} & y_{65} & y_{66} \end{bmatrix} \begin{pmatrix} v_1(t) \\ v_2(t) \\ v_3(t) \\ v_4(t) \\ v_5(t) \\ v_6(t) \end{pmatrix} + \begin{pmatrix} i_{ns1} \\ i_{ns2} \\ i_{ns3} \\ i_{ns4} \\ i_{ns5} \\ i_{ns6} \end{pmatrix} \quad (7.44)$$

The matrix $[Y_{ss}]$ is symmetric and this Norton equivalent is implemented in PSCAD/EMTDC as shown in Figure 7.15, where only the blue-phase network of a star-grounded/star-grounded transformer is shown.

The flow diagram of Figure 7.13 also describes the three-limb three-phase UMEC implementation in PSCAD/EMTDC with only slight modifications. The trapezoidal integration equation is applied to the six transformer windings to calculate the winding-limb flux vector $\boldsymbol{\phi}_s(t - \Delta t)$. Equations 7.39 and 7.40 are used to calculate the permeances of the winding branches. Once the previous time step winding-current vector $i_s(t - \Delta t)$ is formed, the flux leakage elements of $\phi_r(t - \Delta t)$ can be calculated using

$$\begin{aligned}
\phi_7(t - \Delta t) &= P_7^* \left(N_1 i_1(t - \Delta t) - \phi_1(t - \Delta t)/P_1^* \right) \\
\phi_8(t - \Delta t) &= P_8^* \left(N_2 i_2(t - \Delta t) - \phi_2(t - \Delta t)/P_2^* \right) \\
\phi_9(t - \Delta t) &= P_9^* \left(N_1 i_3(t - \Delta t) - \phi_3(t - \Delta t)/P_3^* \right) \\
\phi_{10}(t - \Delta t) &= P_{10}^* \left(N_2 i_4(t - \Delta t) - \phi_4(t - \Delta t)/P_4^* \right) \\
\phi_{11}(t - \Delta t) &= P_{11}^* \left(N_1 i_5(t - \Delta t) - \phi_5(t - \Delta t)/P_5^* \right) \\
\phi_{12}(t - \Delta t) &= P_{12}^* \left(N_2 i_6(t - \Delta t) - \phi_6(t - \Delta t)/P_6^* \right)
\end{aligned} \quad (7.45)$$

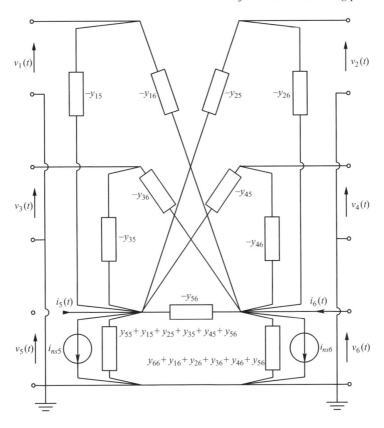

Figure 7.15 UMEC three-limb three-phase Norton equivalent for blue phase (Y-g/Y-g)

The zero-sequence elements of $\boldsymbol{\phi}_r(t - \Delta t)$ are calculated using the m.m.f. loop sum around the primary and secondary winding-limb and zero-sequence branch

$$\phi_{15}(t - \Delta t) = P_{15}^*\big(N_1 i_1(t - \Delta t) + N_2 i_2(t - \Delta t) - \phi_1(t - \Delta t)/P_1^* \\ - \phi_2(t - \Delta t)/P_2^*\big) \qquad (7.46)$$

$$\phi_{16}(t - \Delta t) = P_{16}^*\big(N_1 i_3(t - \Delta t) + N_2 i_4(t - \Delta t) - \phi_3(t - \Delta t)/P_3^* \\ - \phi_4(t - \Delta t)/P_4^*\big) \qquad (7.47)$$

$$\phi_{17}(t - \Delta t) = P_{17}^*\big(N_1 i_5(t - \Delta t) + N_2 i_6(t - \Delta t) - \phi_5(t - \Delta t)/P_5^* \\ - \phi_6(t - \Delta t)/P_6^*\big) \qquad (7.48)$$

Finally, the yoke flux is obtained using the flux summation at nodes N_1 and N_2

$$\phi_{13}(t - \Delta t) = \phi_1(t - \Delta t) - \phi_7(t - \Delta t) - \phi_{15}(t - \Delta t) \\ \phi_{14}(t - \Delta t) = \phi_5(t - \Delta t) - \phi_{11}(t - \Delta t) - \phi_{17}(t - \Delta t) \qquad (7.49)$$

The yoke-branch permeances P_{13}^* and P_{14}^* are again calculated directly from solved branch fluxes ϕ_{13} and ϕ_{14} using equations 7.39 and 7.40. Once $[P^*]$ is known the real-valued admittance matrix $[Y]$ can be obtained.

7.3.4 Fast transient models

The inter-turn capacitance is normally ignored at low frequencies, but for high-frequency events this capacitance becomes significant. When subjected to impulse test the capacitance determines the voltage distribution across the internal windings of the transformer. Moreover the inter-turn capacitance and winding inductance have a resonant frequency that may be excited. Hence transformer failures can be caused by high-frequency overvoltages due to internal resonances [10], [11]. These internal winding resonances (typically in the 5–200 kHz range), are initiated by fast transients and may not cause an immediate breakdown, but partial discharges may occur, thereby accelerating ageing of the transformer winding [12]. To determine the voltage levels across the internal transformer insulation during a specific external transient requires the use of a detailed high-frequency transformer model. Though the general high-frequency models are very accurate and detailed, they are usually too large to be incorporated in a general model of the power system [13]. Hence reduced order models, representing the transformer's terminal behaviour, are normally developed and used in the system study [14]. These reduced order models need to be custom models developed by the user for the EMTP-type program available. There is a multitude of modelling techniques. The resulting transient can be used as the external transient into a more detailed high-frequency transformer model, some of which can calculate down to turn-to-turn voltages.

The difficulty in modelling transformers in detail stems from the fact that some transformer parameters are both non-linear and frequency dependent. The iron core losses and inductances are non-linear due to saturation and hysteresis. They are also frequency dependent due to eddy currents in the laminations. During resonance phenomena the resistances greatly influence the maximum winding voltages. These resistances represent both the copper and iron losses and are strongly frequency dependent [15]–[17]. Parameters for these models are extracted from laboratory testing and are only valid for the transformer type and frequency range of the tests performed.

7.4 The synchronous machine

The synchronous machine model to be used in each case depends on the time span of interest. For example the internal e.m.f. behind subtransient reactance is perfectly adequate for electromagnetic transients studies of only a few cycles, such as the assessment of switching oscillations. At the other extreme, transient studies involving speed variations and/or torsional vibrations need to model adequately the generator rotor and turbine rotor masses. Thus a general-purpose model should include the generator electrical parameters as well as the generator and turbine mechanical parameters.

7.4.1 Electromagnetic model

All the models used in the various versions of the EMTP method are based on Park's transformation from phase to $dq0$ components [18], a frame of reference in which the self and mutual machine inductances are constant. Although a state variable formulation of the equations is used, their solution is carried out using the numerical integrator substitution method.

In the EMTDC program the machine d and d axis currents are used as state variables, whereas fluxes are used instead in the EMTP program [19].

Figure 7.16 depicts a synchronous machine with three fixed windings and one rotating winding (at this point the damping windings are not included). Let $\theta(t)$ be the angle between the field winding and winding a at time t. From Faraday's law:

$$\begin{pmatrix} V_a - i_a R_a \\ V_b - i_b R_b \\ V_c - i_c R_c \end{pmatrix} = \frac{d}{dt} \begin{pmatrix} \psi_a \\ \psi_b \\ \psi_c \end{pmatrix} \qquad (7.50)$$

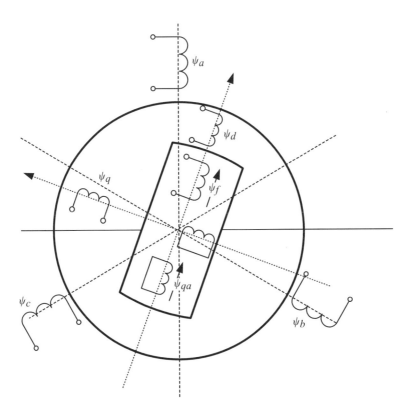

Figure 7.16   Cross-section of a salient pole machine

where

$$\begin{pmatrix} \psi_a \\ \psi_b \\ \psi_c \end{pmatrix} = \begin{bmatrix} L_{aa} & L_{ab} & L_{ac} & L_{af} \\ L_{ba} & L_{bb} & L_{bc} & L_{bf} \\ L_{ca} & L_{cb} & L_{cc} & L_{cf} \end{bmatrix} \begin{pmatrix} i_a \\ i_b \\ i_c \\ i_f \end{pmatrix}$$

The inductances are of a time varying nature, e.g.

$$\begin{aligned} L_{aa} &= L_a + L_m \cos(\theta) \\ L_{bb} &= L_a + L_m \cos(2(\theta - 2\pi/3)) \\ L_{cc} &= L_a + L_m \cos(2(\theta - 4\pi/3)) \end{aligned} \quad (7.51)$$

Assuming a sinusoidal winding distribution then the mutual inductances are:

$$\begin{aligned} L_{ab} &= L_{ba} = -M_s - L_m \cos(2(\theta - \pi/6)) \\ L_{bc} &= L_{cb} = -M_s - L_m \cos(2(\theta - \pi/2)) \\ L_{ca} &= L_{ac} = -M_s - L_m \cos(2(\theta + \pi/2)) \end{aligned} \quad (7.52)$$

and the inductances of the field winding:

$$\begin{aligned} L_{af} &= L_{fa} = M_f \cos(\theta) \\ L_{bf} &= L_{fb} = M_f \cos(\theta - 2\pi/3) \\ L_{cf} &= L_{fc} = M_f \cos(\theta - 4\pi/3) \end{aligned} \quad (7.53)$$

In compact notation

$$\begin{pmatrix} \psi_a \\ \psi_b \\ \psi_c \\ \psi_f \end{pmatrix} = \left[\begin{array}{c|c} L_{abc} & L_{abcf} \\ \hline L_{fabc} & L_{ff} \end{array} \right] \begin{pmatrix} i_a \\ i_b \\ i_c \\ i_f \end{pmatrix} \quad (7.54)$$

or

$$\begin{pmatrix} \mathbf{\psi}_{abc} \\ \psi_f \end{pmatrix} = \left[\begin{array}{c|c} L_{abc} & L_{abcf} \\ \hline L_{fabc} & L_{ff} \end{array} \right] \begin{pmatrix} \mathbf{i}_{abc} \\ i_f \end{pmatrix} \quad (7.55)$$

Taking the top partition

$$\mathbf{\psi}_{abc} = [L_{abc}]\mathbf{i}_{abc} + [L_{abcf}]\mathbf{i}_f \quad (7.56)$$

Choosing a matrix $[T(\theta)]$ that diagonalises $[L_{abc}]$ gives:

$$\mathbf{i}_{dq0} = [T(\theta)]\mathbf{i}_{abc} \quad (7.57)$$

$$\mathbf{v}_{dq0} = [T(\theta)]\mathbf{v}_{abc} \quad (7.58)$$

$$\mathbf{\psi}_{dq0} = [T(\theta)]\mathbf{\psi}_{abc} \quad (7.59)$$

and substituting in equation 7.56 gives

$$[T(\theta)]^{-1}\Psi_{dq0} = [L_{abc}][T(\theta)]^{-1}i_{dq0} + [L_{abcf}]i_f \qquad (7.60)$$

Thus

$$\Psi_{dq0} = [T(\theta)][L_{abc}][T(\theta)]^{-1}i_{dq0} + [T(\theta)][L_{abcf}]i_f$$

A common choice of $[T(\theta)]$ is:

$$[T(\theta)] = \frac{2}{3}\begin{bmatrix} \cos(\theta) & \cos(\theta - 2\pi/3) & \cos(\theta + 2\pi/3) \\ \sin(\theta) & \sin(\theta - 2\pi/3) & \sin(\theta + 2\pi/3) \\ 1/2 & 1/2 & 1/2 \end{bmatrix} \qquad (7.61)$$

thus

$$[T(\theta)]^{-1} = \begin{bmatrix} \cos(\theta) & \sin(\theta) & 1 \\ \cos(\theta - 2\pi/3) & \sin(\theta - 2\pi/3) & 1 \\ \cos(\theta + 2\pi/3) & \sin(\theta + 2\pi/3) & 1 \end{bmatrix} \qquad (7.62)$$

This matrix is known as Park's transformation. Therefore the following expression results in $dq0$ coordinates:

$$\begin{pmatrix} \psi_d \\ \psi_q \\ \psi_0 \end{pmatrix} = \begin{bmatrix} L_a + M_s + \frac{3}{2}L_m & 0 & 0 \\ 0 & L_a + M_s - \frac{3}{2}L_m & 0 \\ 0 & 0 & L_a - 2M_s \end{bmatrix} \begin{pmatrix} i_d \\ i_q \\ i_0 \end{pmatrix} + \begin{pmatrix} \sqrt{\frac{3}{2}}M_f \\ 0 \\ 0 \end{pmatrix} (i_f) \qquad (7.63)$$

or

$$\begin{pmatrix} \psi_d \\ \psi_q \\ \psi_0 \end{pmatrix} = \begin{bmatrix} L_a + L_{md} & 0 & 0 \\ 0 & L_a + L_{md} & 0 \\ 0 & 0 & L_0 \end{bmatrix} \begin{pmatrix} i_d \\ i_q \\ i_0 \end{pmatrix} + \begin{pmatrix} L_{md} \\ 0 \\ 0 \end{pmatrix} (i'_f) \qquad (7.64)$$

where

$$L_{md} = M_s + \frac{3}{2}L_m$$

$$i'_f = \frac{\sqrt{3/2}M_f}{L_{md}} i_f$$

The equation for the field flux now becomes time dependent, i.e.

$$(\psi_f) = (M_f \cos(\theta) \quad M_f \cos(\theta - 2\pi/3) \quad M_f \cos(\theta - 4\pi/3))$$

$$\times \begin{bmatrix} \cos(\theta) & \sin(\theta) & 1 \\ \cos(\theta - 2\pi/3) & \sin(\theta - 2\pi/3) & 1 \\ \cos(\theta + 2\pi/3) & \sin(\theta + 2\pi/3) & 1 \end{bmatrix} \begin{pmatrix} i_d \\ i_q \\ i_0 \end{pmatrix} + [L_{ff}] \cdot i_f$$

$$= (\tfrac{3}{2}M_f \quad 0 \quad 0) \begin{pmatrix} i_d \\ i_q \\ i_0 \end{pmatrix} + [L_{ff}] \cdot i_f$$

$$= \tfrac{3}{2}M_f i_d + [L_{ff}] \cdot i_f \qquad (7.65)$$

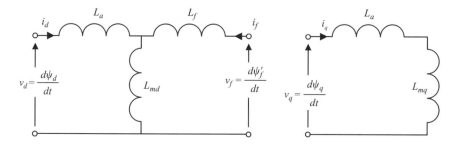

Figure 7.17 Equivalent circuit for synchronous machine equations

Similarly the field circuit equation can be expressed as:

$$\psi'_f = [L_{md}]i_d + [L_{md} + L_f]i'_f \tag{7.66}$$

This can be thought of as transforming the field current to the same base as the stator currents. Figure 7.17 depicts the equivalent circuit based on these equations. From Kirchhoff's current law:

$$\begin{pmatrix} v_a - i_a R_a \\ v_b - i_b R_b \\ v_c - i_c R_c \end{pmatrix} = \frac{d}{dt}\begin{pmatrix} \psi_a \\ \psi_b \\ \psi_c \end{pmatrix} \tag{7.67}$$

and

$$v'_f - i'_f R'_f = \frac{d\psi'_f}{dt} \tag{7.68}$$

Applying Park's transformation gives:

$$[T(\theta)]\begin{pmatrix} v_a - i_a R_a \\ v_b - i_b R_b \\ v_c - i_c R_c \end{pmatrix} = [T(\theta)]\frac{d}{dt}\left([T(\theta)]^{-1}\begin{pmatrix} \psi_d \\ \psi_q \\ \psi_0 \end{pmatrix}\right) \tag{7.69}$$

or

$$[T(\theta)]\begin{pmatrix} v_d - i_d R_d \\ v_q - i_q R_q \\ v_0 - i_0 R_0 \end{pmatrix} = [T(\theta)]\left(\frac{d}{dt}[T(\theta)]^{-1}\begin{pmatrix} \psi_d \\ \psi_q \\ \psi_0 \end{pmatrix} + [T(\theta)]^{-1}\frac{d}{dt}\begin{pmatrix} \psi_d \\ \psi_q \\ \psi_0 \end{pmatrix}\right)$$

$$= [T(\theta)]\left(\frac{d}{d\theta}[T(\theta)]^{-1}\frac{d\theta}{dt}\begin{pmatrix} \psi_d \\ \psi_q \\ \psi_0 \end{pmatrix} + [T(\theta)]^{-1}\frac{d}{dt}\begin{pmatrix} \psi_d \\ \psi_q \\ \psi_0 \end{pmatrix}\right)$$

$$= \underbrace{[T(\theta)]\frac{d}{d\theta}[T(\theta)]^{-1}\omega\begin{pmatrix} \psi_d \\ \psi_q \\ \psi_0 \end{pmatrix}}_{\text{speed emf}} + \underbrace{\frac{d}{dt}\begin{pmatrix} \psi_d \\ \psi_q \\ \psi_0 \end{pmatrix}}_{\text{transformer emf}} \tag{7.70}$$

where ω is the angular speed.

Evaluating the speed term:

$$\frac{d}{d\theta}[T(\theta)]^{-1} = \frac{d}{d\theta}\begin{bmatrix} \cos(\theta) & \sin(\theta) & 1 \\ \cos(\theta - 2\pi/3) & \sin(\theta - 2\pi/3) & 1 \\ \cos(\theta - 4\pi/3) & \sin(\theta - 4\pi/3) & 1 \end{bmatrix} \quad (7.71)$$

$$= \begin{bmatrix} -\sin(\theta) & \cos(\theta) & 0 \\ -\sin(\theta - 2\pi/3) & \cos(\theta - 2\pi/3) & 0 \\ -\sin(\theta - 4\pi/3) & \cos(\theta - 4\pi/3) & 0 \end{bmatrix} \quad (7.72)$$

and

$$[T(\theta)]\frac{d}{d\theta}[T(\theta)]^{-1} = \frac{2}{3}\begin{bmatrix} \cos(\theta) & \cos(\theta - 2\pi/3) & \cos(\theta - 4\pi/3) \\ \sin(\theta) & \sin(\theta - 2\pi/3) & \sin(\theta - 4\pi/3) \\ 1/2 & 1/2 & 1/2 \end{bmatrix}$$

$$\times \begin{bmatrix} -\sin(\theta) & \cos(\theta) & 0 \\ -\sin(\theta - 2\pi/3) & \cos(\theta - 2\pi/3) & 0 \\ -\sin(\theta - 4\pi/3) & \cos(\theta - 4\pi/3) & 0 \end{bmatrix}$$

$$= \frac{2}{3}\begin{bmatrix} 0 & 2/3 & 0 \\ -2/3 & 0 & 0 \\ 0 & 0 & 0 \end{bmatrix} = \begin{bmatrix} 0 & 1 & 0 \\ -1 & 0 & 0 \\ 0 & 0 & 0 \end{bmatrix} \quad (7.73)$$

Hence equation 7.67 becomes:

$$\begin{aligned} v_d - i_d R_a &= \omega\psi_q + \frac{d\psi_d}{dt} \\ v_q - i_q R_a &= -\omega\psi_d + \frac{d\psi_q}{dt} \\ v_0 - i_0 R_0 &= \frac{d\psi_0}{dt} \end{aligned} \quad (7.74)$$

while the field circuit remains unchanged, i.e.

$$v'_f - i'_f R'_f = \frac{d\psi'_f}{dt} \quad (7.75)$$

If, as is normally the case, the winding connection is ungrounded star then $i_0 = 0$ and the third equation in 7.74 disappears.

Adkins' [20] equivalent circuit, shown in Figure 7.18, consists of a machine with three coils on the d-axis and two on the q-axis, although the model can easily be

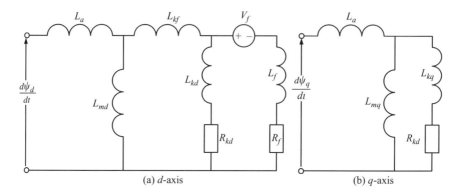

Figure 7.18 The a.c. machine equivalent circuit

extended to include further coils. The following equations can be written:

$$\begin{pmatrix} V_d - \omega \Psi_q - R_a i_d \\ V_f - R_f i_f \\ -R_{kd} i_{kd} \end{pmatrix}$$

$$= \begin{bmatrix} L_{md} + L_a & L_{md} & L_{md} \\ L_{md} & L_{md} + L_f + L_{kf} & L_{md} + L_{kf} \\ L_{md} & L_{md} + L_{kf} & L_{md} + L_{kf} + L_{kd} \end{bmatrix} \frac{d}{dt} \begin{pmatrix} i_d \\ i_f \\ i_{kd} \end{pmatrix}$$

$$= [\mathbf{L}_d] \frac{d}{dt} \begin{pmatrix} i_d \\ i_f \\ i_{kd} \end{pmatrix} \tag{7.76}$$

and

$$\begin{pmatrix} V_q + \omega \Psi_d - R_a i_q \\ -R_{kq} i_{kq} \end{pmatrix} = \begin{bmatrix} L_{mq} + L_a & L_{mq} \\ L_{mq} & L_{mq} + L_{kq} \end{bmatrix} \frac{d}{dt} \begin{pmatrix} i_q \\ i_{kq} \end{pmatrix}$$

$$= [\mathbf{L}_q] \frac{d}{dt} \begin{pmatrix} i_q \\ i_{kq} \end{pmatrix} \tag{7.77}$$

The flux paths associated with the various d-axis inductances is shown in Figure 7.19.

The additional inductance L_{kd} represents the mutual flux linking only the damper and field windings (not the stator windings); this addition has been shown to be necessary for the accurate representation of the transient currents in the rotor circuits. Saturation is taken into account by making inductances L_{md} and L_f functions of the magnetising current and this information is derived from the machine open-circuit characteristics.

Solving equations 7.76 and 7.77 for the currents yields:

$$\frac{d}{dt} \begin{pmatrix} i_d \\ i_f \\ i_{kd} \end{pmatrix} = [\mathbf{L}_d]^{-1} \begin{pmatrix} -\omega \Psi_d - R_a i_q \\ -R_f i_f \\ -R_{kd} i_{kd} \end{pmatrix} + [\mathbf{L}_d]^{-1} \begin{pmatrix} V_d \\ V_f \\ 0 \end{pmatrix} \tag{7.78}$$

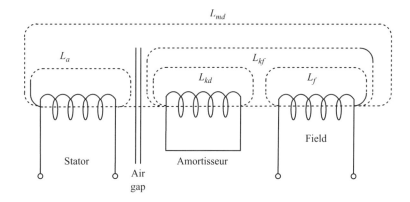

Figure 7.19 d-axis flux paths

$$\frac{d}{dt}\begin{pmatrix} i_q \\ i_{kq} \end{pmatrix} = [\mathbf{L}_q]^{-1}\begin{pmatrix} \omega\Psi_d - R_a i_q \\ -R_{kq} i_{kq} \end{pmatrix} + [\mathbf{L}_q]^{-1}\begin{pmatrix} V_q \\ 0 \end{pmatrix} \quad (7.79)$$

which are in the standard form of the state variable formulation, i.e.

$$\dot{\mathbf{x}} = [A]\mathbf{x} + [B]\mathbf{u} \quad (7.80)$$

where the state vector **x** represents the currents and the input vector **u** the applied voltages.

7.4.2 Electromechanical model

The accelerating torque is the difference between the mechanical and electrical torque, hence:

$$J\frac{d\omega}{dt} = T_{\text{mech}} - T_{\text{elec}} - D\omega \quad (7.81)$$

where
  J is the angular moment of inertia
  D is the damping constant
  ω is the angular speed $(d\theta/dt)$

In matrix form this is:

$$\frac{d}{dt}\begin{pmatrix} \theta \\ \omega \end{pmatrix} = \begin{bmatrix} 0 & 1 \\ 0 & -D/J \end{bmatrix}\begin{pmatrix} \theta \\ \omega \end{pmatrix} + \begin{pmatrix} 0 \\ (T_{\text{mech}} - T_{\text{elec}})/J \end{pmatrix} \quad (7.82)$$

This equation is numerically integrated to calculate the rotor position θ. Multimass systems can be modelled by building up mass-inertia models. Often only ω is passed as input to the machine model. A model for the governor can be interfaced, which accepts ω and calculates T_{mech}.

The electromagnetic torque can be expressed as:

$$T_{\text{elec}} = \frac{\psi_d i_q - \psi_q i_d}{2} \quad (7.83)$$

184 Power systems electromagnetic transients simulation

The large time constants associated with governors and exciters would require long simulation times to reach steady state if the initial conditions are not accurate. Therefore techniques such as artificially lowering the inertia or increasing damping for start-up have been developed.

7.4.2.1 Per unit system

PSCAD/EMTDC uses a per unit system for electrical machines. The associated base quantities are:

ω_0 – rated angular frequency (314.1592654 for 50 Hz system or 376.99111184 for 60 Hz system)
$P_0 = 3V_b I_b$ – base power
$t_0 = 1/\omega_0$ – time base
$\psi_0 = V_b/\omega_0$ – base flux linkage
$n_{b\_Mech} = \omega_0/p$ – mechanical angular speed
$\theta_{b\_Mech} = \theta/p$ – mechanical angle

where

p is the number of pole pairs
V_b is the base voltage (RMS phase voltage)
I_b is the base current (RMS phase current)

Hence Δt is multiplied by ω_0 to give per unit incremental time.

7.4.2.2 Multimass representation

Although a single-mass representation is usually adequate for hydroturbines, this is not the case for thermal turbines where torsional vibrations can occur due to subsynchronous resonance. In such cases a lumped mass representation is commonly used, as depicted in Figure 7.20.

The moments of inertia and the stiffness coefficients (K) are normally available from design data. D are the damping coefficients and represent two damping effects, i.e. the self-damping of a mass (frictional and windage losses) and the damping created in the shaft between masses k and $k - 1$ when twisted with speed.

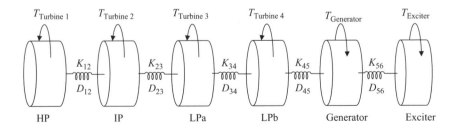

Figure 7.20   Multimass model

The resulting damping torque is:

$$T_{\text{Damping}\_k} = D_k \frac{d\theta_k}{dt} + D_{k-1,k}\left(\frac{d\theta_k}{dt} - \frac{d\theta_{k-1}}{dt}\right) + D_{k,k+1}\left(\frac{d\theta_k}{dt} - \frac{d\theta_{k+1}}{dt}\right) \quad (7.84)$$

The damping matrix is thus:

$$[D] = \begin{bmatrix} D_1 + D_{12} & -D_{12} & 0 & \cdots & 0 & 0 \\ -D_{12} & D_2 + D_{12} + D_{23} & -D_{23} & \cdots & 0 & 0 \\ \vdots & \vdots & \vdots & \ddots & & \vdots \\ 0 & 0 & -D_{n-2,n-2} & D_{n-1} + D_{n-2,n-1} - D_{n-1,n} & -D_{n-1,n} \\ 0 & 0 & 0 & & -D_{n-1,n} & D_{n-1,n} + D_n \end{bmatrix} \quad (7.85)$$

The spring action of the shaft section between the kth and $(k-1)^{\text{th}}$ mass creates a force proportional to the angular twist, i.e.

$$T_{\text{Spring}\_k-1} - T_{\text{Spring}\_k} = K_{k-1,k}(\theta_{k-1} - \theta_k) \quad (7.86)$$

Hence these add the term $[K]\,d\theta/dt$ to equation 7.81, where

$$[K] = \begin{bmatrix} K_{12} & -K_{12} & 0 & \cdots & 0 & 0 \\ -K_{12} & K_{12} + K_{23} & -K_{23} & \cdots & 0 & 0 \\ \vdots & \vdots & \vdots & \ddots & & \vdots \\ 0 & 0 & -K_{n-2,n-2} & K_{n-2,n-2} + K_{n-1,n} & -K_{n-1,n} \\ 0 & 0 & 0 & & -K_{n-1,n} & K_{n-1,n} \end{bmatrix} \quad (7.87)$$

Therefore equation 7.81 becomes:

$$[J]\frac{d\omega}{dt} = T_{\text{mech}} - T_{\text{elec}} - [D]\omega - [K]\frac{d\theta}{dt} \quad (7.88)$$

7.4.3 Interfacing machine to network

Although the equations for the detailed synchronous machine model have been more or less the same in the various synchronous machine models already developed, different approaches have been used for their incorporation into the overall algorithm. The EMTP uses V. Brandwajn's model, where the synchronous machine is represented as an internal voltage source behind some impedance, with the voltage source being recomputed at each time step and the impedance incorporated in the system nodal conductance matrix. This solution requires the prediction of some variables and could cause numerical instability; however it has been refined sufficiently to be reasonably stable and reliable.

The EMTDC program uses the current injection method. To prevent numerical instabilities this model is complemented by an interfacing resistor, as shown in Figure 7.21, which provides a more robust solution. The conductance is $\Delta t / 2L_d''$ and

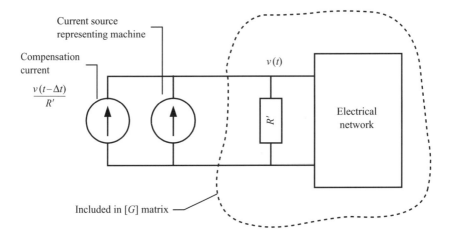

Figure 7.21 Interfacing electrical machines

the current source $i(t - \Delta t) + G_{eq}v(t - \Delta t)$ (where L_d'' is the subtransient reactance of a synchronous machine or leakage reactance if interfacing an induction machine). However, there is a one time step delay in the interface (i.e. $G_{eq}v(t - \Delta t)$ is used), thus leading to potential instabilities (especially during open circuit). This instability due to the time step delay manifests itself as a high-frequency oscillation. To stabilise this, a series RC circuit has been added to the above interface. R is set to $20Z_{base}$ and C is chosen such that $RC = 10\Delta t$. This selection of values provides high-frequency damping without adding significant fundamental or low-frequency losses.

The following steps are used to interface machines to the electrical network:

1. Assume v_a, v_b, v_c from the previous time step and calculate v_d, v_q, v_0 using the transformation matrix $[T(\theta)]$
2. Choose ψ_d, ψ_q, ψ_f (and ψ_0 if a zero sequence is to be considered) as state variables

$$\begin{pmatrix} i_d \\ i_f' \end{pmatrix} = \begin{bmatrix} L_{md} + L_a & L_{md} \\ L_{md} & L_{md} + L_a \end{bmatrix}^{-1} \begin{pmatrix} \psi_d \\ \psi_f' \end{pmatrix} d\text{-axis} \quad (7.89)$$

$$(i_q) = [L_{mq} + L_a]^{-1} (\psi_q) \, q\text{-axis} \quad (7.90)$$

3. The following equations are solved by numerical integration.

$$\frac{d\psi_d}{dt} = v_d - i_d R_a - \omega \psi_q \quad (7.91)$$

$$\frac{d\psi_f'}{dt} = v_f' - i_f' R_f' \quad (7.92)$$

$$\frac{d\psi_q}{dt} = v_q - i_q R_a + \omega \psi_d \quad (7.93)$$

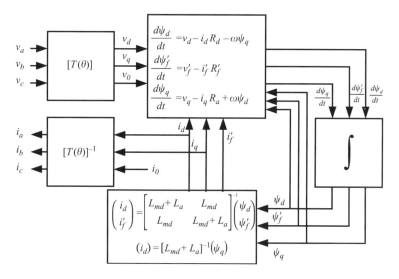

Figure 7.22 Electrical machine solution procedure

All quantities (e.g. v_d, v'_f, etc.) on the right-hand side of these equations are known from the previous time step. The currents i_d, i'_f and i_q are linear combinations of the state variables and therefore they have a state variable formulation $\dot{x} = [A]x + [B]u$ (if ω is assumed constant).

4. After solving for ψ_d, ψ'_f and ψ_q then i_d, i'_f and i_q (and i_0) are calculated and transformed back to obtain i_a, i_b and i_c. Also the equation $d\theta/dt = \omega$ is used to update θ in the transformation. If the machine's mechanical transients are ignored then $\theta = \omega t + \phi$ and the integration for θ is not needed.

The above implementation is depicted in Figure 7.22.

Additional rotor windings need to be represented due to the presence of damper (amortisseur) windings. In this case more than two circuits exist on the d-axis and more than one on the q-axis. Equation 7.89 is easily extended to cope with this. For example for one damper winding equation 7.89 becomes:

$$\begin{pmatrix} i_d \\ i'_f \\ i_{kd} \end{pmatrix} = \begin{bmatrix} L_{md} + L_a & L_{md} & L_{md} \\ L_{md} & L_{md} & L_{md} + L_f \\ L_{md} & L_{md} + L_{kd} & L_{md} \end{bmatrix}^{-1} \begin{pmatrix} \psi_d \\ \psi'_f \\ \psi_{kd} \end{pmatrix} \quad (7.94)$$

where the subscripts kd and kq denote the damper winding on the d and q-axis, respectively.

The damper windings are short circuited and hence the voltage is zero. The extended version of equations 7.91–7.93 is:

$$\frac{d\psi_d}{dt} = v_d - i_d R_a - \omega \psi_q$$
$$\frac{d\psi_{kd}}{dt} = 0 - i_{kd} R_{kd}$$
$$\frac{d\psi_q}{dt} = v_q - i_q R_a - \omega \psi_d \qquad (7.95)$$
$$\frac{d\psi_{kq}}{dt} = 0 - i_{kq} R_{kq}$$
$$\frac{d\psi'_f}{dt} = v'_f - i'_f R'_f$$

Saturation can be introduced in many ways but one popular method is to make L_{md} and L_{mq} functions of the magnetising current (i.e. $i_d + i_{kd} + i_f$ in the d-axis). For round rotor machines it is necessary to saturate both L_{md} and L_{mq}. For start-up transients sometimes it is also necessary to saturate L_a.

The place of the governor and exciter blocks in relation to the machine equations is shown in Figure 7.23. Figure 7.24 shows a block diagram of the complete a.c. machine model, including the electromagnetic and mechanical components, as well as their interface with the rest of the network.

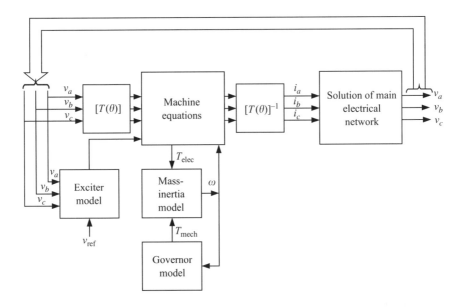

Figure 7.23 The a.c. machine system

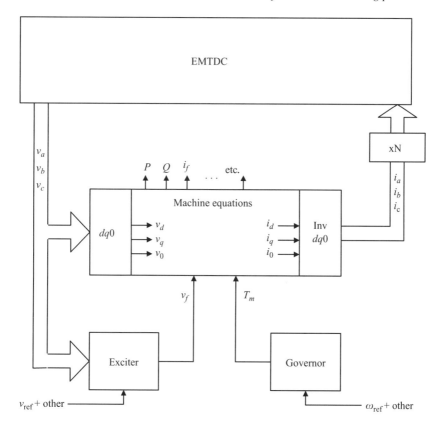

Figure 7.24 Block diagram synchronous machine model

7.4.4 Types of rotating machine available

Besides the synchronous machine developed by V. Brandwajn, the universal machine model was added to EMTP by H.K. Lauw and W.S. Meyer to enable various types of machines to be studied using the same model [21], [22]. Under the universal machine model the EMTP package includes the following different types of rotating machines:

1. synchronous machine with a three-phase armature
2. synchronous machine with a two-phase armature
3. induction machine with a three-phase armature
4. induction machine with a three-phase armature and three-phase rotor
5. induction machine with a two-phase armature
6. single-phase induction or synchronous machine with one-phase excitation
7. single-phase induction or synchronous machine with two-phase excitation
8. d.c. machine separately excited
9. d.c. machine with series compound field
10. d.c. machine series excited

11. d.c. machine with parallel compound field
12. d.c. machine shunt excited

There are two types of windings, i.e. armature and field windings. Which of them rotates and which are stationary is irrelevent as only the relative motion is important. The armature or power windings are on the stator in the synchronous and induction machines and on the rotor in the d.c. machines, with the commutator controlling the flow of current through them.

The user must choose the interface method for the universal machine. There are only minor differences between the electrical models of the synchronous and universal machines but their mechanical part differs significantly, and a lumped RLC equivalent electrical network is used to represent the mechanical part.

Like EMTP the EMTDC program also uses general machine theory for the different machines, but with separate FORTRAN subroutines for each machine type to simplify the parameter entry. For example, the salient pole machine has different parameters on both axes but only the d-axis mutual saturation is significant; on the other hand the induction machine has the same equivalent circuit in both axes and experiences saturation of both mutual and leakage reactances. The state variable formulation and interfacing detailed in this chapter are the same for all the machines. The machine subroutines can use a different time step and integration procedure from the main program. Four machine models are available in EMTDC; the original salient pole synchronous machine (MAC100 Model), a new round rotor synchronous machine model (SNC375), the squirrel cage induction machine (SQC100) and the wound rotor induction motor.

The MAC100 model has been superseded by SNC375, which can model both salient and round rotor machines due to the extra damper winding on the q-axis. The squirrel cage induction machine is modelled as a double cage and the motor convention is used; that is terminal power and shaft torque are positive when power is going into the machine and the shaft is driving a load.

7.5 Summary

The basic theory of the single-phase transformer has been described, including the derivation of parameters, the modelling of magnetisation non-linearities and its numerical implementation in a form acceptable for electromagnetic transients programs.

The need for advanced models has been justified, and a detailed description made of UMEC (the Unified Magnetic Equivalent Circuit), a general transformer model developed for the accurate representation of multiphase, multiwinding arrangements. UMEC is a standard transformer model in the latest version of PSCAD/EMTDC.

Rotating machine modelling based on Park's transformation is reasonably standard, the different implementations relating to the way of interfacing the machine to the system. A state variable formulation of the equations is used but the solution, in line with EMTP philosophy, is carried out by numerical integrator substitution.

The great variety of rotating machines in existence precludes the use of a common model and so the electromagnetic transient programs include models for the main types in current use, most of these based on general machine theory.

7.6 References

1. BRANDWAJN, V., DOMMEL, H. W. and DOMMEL, I. I.: 'Matrix representation of three-phase n-winding transformers for steady-state and transient studies', *IEEE Transactions on Power Apparatus and Systems*, 1982, **101** (6), pp. 1369–78
2. CHEN, M. S. and DILLON, W. E.: 'Power systems modelling', *Proceedings of IEEE*, 1974, **162** (7), pp. 901–15
3. ARTURI, C. M.: 'Transient simulation and analysis of a three-phase five-limb step-up transformer following out-of-phase synchronisation', *IEEE Transactions on Power Delivery*, 1991, **6** (1), pp. 196–207
4. DOMMELEN, D. V.: 'Alternative transients program rule book', technical report, Leuven EMTP Center, 1987
5. HATZIARGYRIOU, N. D., PROUSALIDIS, J. M. and PAPADIAS, B. C.: 'Generalised transformer model based on analysis of its magnetic circuit', *IEE Proceedings Part C*, 1993, **140** (4), pp. 269–78
6. CHEN, X. S. and NEUDORFER, P.: 'The development of a three-phase multi-legged transformer model for use with EMTP', technical report, Dept. of Energy Contract DE-AC79-92BP26702, USA, 1993
7. LEON, F. D. and SEMLYEN, A.: 'Complete transformer model for electromagnetic transients', *IEEE Transactions on Power Delivery*, 1994, **9** (1), pp. 231–9
8. ENRIGHT, W. G.: 'Transformer models for electromagnetic transient studies with particular reference to HVdc transmission' (Ph.D. thesis, University of Canterbury, New Zealand, Private Bag 4800, Christchurch, New Zealand, 1996)
9. ENRIGHT, W. G., WATSON, N. R. and ARRILLAGA, J.: 'Improved simulation of HVdc converter transformers in electromagnetic transients programs', *Proceedings of IEE, Part-C*, 1997, **144** (2), pp. 100–6
10. BAYLESS, R. S., SELMAN, J. D., TRUAX, D. E. and REID, W. E.: 'Capacitor switching and transformer transients', *IEEE Transactions on Power Delivery*, 1988, **3** (1), pp. 349–57
11. MOMBELLO, E. E. and MOLLER, K.: 'New power transformer model for the calculation of electromagnetic resonant transient phenomena including frequency-dependent losses', *IEEE Transactions on Power Delivery*, 2000, **15** (1), pp. 167–74
12. VAN CRAENENBROECK, T., DE CEUSTER, J., MARLY, J. P., DE HERDT, H., BROUWERS, B. and VAN DOMMELEN, D.: 'Experimental and numerical analysis of fast transient phenomena in distribution transformers', *Proceedings of Power Engineering Society Winter Meeting*, 2000, **3**, pp. 2193–8
13. DE HERDT, H., DECLERCQ, J., SEIS, T., VAN CRAENENBROECK, T. and VAN DOMMELEN, D.: 'Fast transients and their effect on transformer

insulation: simulation and measurements', *Electricity Distribution, 2001. Part 1: Contributions. CIRED. 16th International Conference and Exhibition (IEE Conf. Publ No. 482)*, 2001, Vol. 1, No. 1.6

14 DE LEON, F. and SEMLYEN, A.: 'Reduced order model for transformer transients', *IEEE Transactions on Power Delivery*, 1992, **7** (1), pp. 361–9

15 VAKILIAN, M. and DEGENEFF, R. C.: 'A method for modeling nonlinear core characteristics of transformers during transients', *IEEE Transactions on Power Delivery*, 1994, **9** (4), pp. 1916–25

16 TARASIEWICZ, E. J., MORCHED, A. S., NARANG, A. and DICK, E. P.: 'Frequency dependent eddy current models for nonlinear iron cores', *IEEE Transactions on Power Systems*, 1993, **8** (2), pp. 588–96

17 SEMLYEN, A. and DE LEON, F.: 'Eddy current add-on for frequency dependent representation of winding losses in transformer models used in computing electromagnetic transients', *Proceedings of IEE on Generation, Transmission and Distribution (Part C)*, 1994, **141** (3), pp. 209–14

18 KIMBARK, E. W.: Power system stability; synchronous machines (Dover Publications, New York, 1965)

19 WOODFORD, D. A., GOLE, A. M. and MENZIES, R. W.: 'Digital simulation of DC links and AC machines', *IEEE Transactions on Power Apparatus and Systems*, 1983, **102** (6), pp. 1616–23

20 ADKINS, B. and HARLEY, R. G.: *'General theory of AC machines'* (Chapman & Hall, London, 1975)

21 LAUW, H. K. and MEYER W. S.: 'Universal machine modelling for the representation of rotating electric machinery in an electromagnetic transients program', *IEEE Transactions on Power Apparatus and Systems*, 1982, **101** (6), pp. 1342–51

22 LAUW, H. K.: 'Interfacing for universal multi-machine system modelling in an electromagnetic transients program', *IEEE Transactions on Power Apparatus and Systems*, 1985, **104** (9), pp. 2367–73

Chapter 8
Control and protection

8.1 Introduction

As well as accurate modelling of the power components, effective transient simulation requires detailed representation of their control and protection processes.

A variety of network signals need to be generated as inputs to the control system, such as active and reactive powers, r.m.s. voltages and currents, phase angles, harmonic frequencies, etc. The output of the control functions are then used to control voltage and current sources as well as provide switching signals and firing pulses to the power electronic devices. These signals can also be used to dynamically control the values of resistors, inductors and capacitors.

A concise description of the control functions attached to the state variable solution has been made in Chapter 3. The purpose of this chapter is to discuss the implementation of control and protection systems in electromagnetic transient programs.

The control blocks, such as integrators, multipliers, etc. need to be translated into a discrete form for digital computer simulation. Thus the controller itself must also be represented by difference equations. Although the control equations could be solved simultaneously with the main circuit in one large set of linear equations [1], [2], considering the large size of the main circuit, such an approach would result in loss of symmetry and increased computation. Therefore electromagnetic transient programs solve the control equations separately, even though this introduces time-step delays in the algorithm. For analogue controls a combined electrical/control solution is also possible, but laborious, by modelling in detail the analogue components (e.g. op-amps, . . .) as electrical components.

The modelling of protection systems in electromagnetic transients programs lies behind those of other system components. At this stage only a limited number of relay types have been modelled and the reliability and accuracy of these models is still being assessed. Therefore this chapter only provides general guidelines on their eventual implementation.

8.2 Transient analysis of control systems (TACS)

Originally developed to represent HVDC converter controls, the TACS facility of the EMTP package is currently used to model any devices or phenomena which cannot be directly represented by the basic network components. Examples of application are HVDC converter controls, excitation systems of synchronous machines, current limiting gaps in surge arresters, arcs in circuit breakers, etc.

The control systems, devices and phenomena modelled in TACS and the electric network are solved separately. Output quantities from the latter are used as input signals in TACS over the same time step, while the output quantities from TACS become input signals to the network solution over the next time step [3].

As illustrated in Figure 8.1, the network solution is first advanced from $t - \Delta t$ without TACS direct involvement; there is, of course, an indirect link between them as the network will use voltage and current sources defined between $t - \Delta t$ and t, derived in the preceding step (i.e. between $t - 2\Delta t$ to $t - \Delta t$). NETWORK also receives orders for the opening and closing of switches at time t which were calculated by TACS in the preceding step.

However, in the latter case, the error caused in the network solution by the Δt time delay is usually negligible. This is partly due to the small value of Δt (of the order of 50 μs) and partly because the delay in closing a thyristor switch is compensated by the converter control; in the case of HVDC transmission, the controller alternately

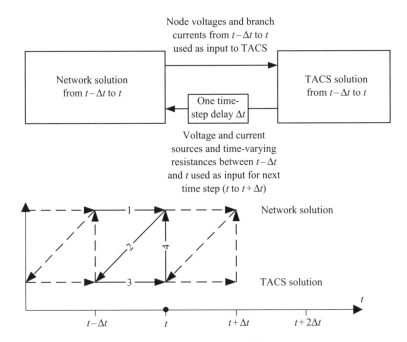

Figure 8.1 Interface between network and TACS solution

advances and retards the firing of thyristor switches to keep the power constant during steady-state operation.

When a NETWORK solution has been completed, the specified voltage and current signals are used to derive the TACS solution from $t - \Delta t$ to t; in this part of the interface the only time delays are those due to the way in which the control blocks are solved. Any delays can manifest themselves in a steady-state firing error as well as causing numerical instabilities, particularly if non-linear devices are present.

Besides modelling transfer functions, adders and limiters, TACS allows components to be modelled using FORTRAN-like functions and expressions. The user can also supply FORTRAN code as a TACS element [3].

Two approaches can be used to alleviate the effects of time-step delays. One involves a phase advance technique, using extrapolation to estimate the output of the control solution for the network solution; this method performs poorly during abrupt changes (such as fault inception). The second method reduces the number of internal variables in the control equations and uses a simultaneous solution of the network and reduced control equations [1].

MODELS [4] is a general-purpose description language supported by an extensive set of simulation tools for the representation and study of time-dependent systems. It was first developed in 1985 by Dube and was interfaced to BPA's EMTP in 1989. MODELS provides an alternative approach to the representation and simulation of new components and of control systems in EMTP. However TACS continues to be used for representing simple control systems which can be described using the existing TACS building blocks.

8.3 Control modelling in PSCAD/EMTDC

To help the user to build a complex control system, PSCAD/EMTDC contains a Control Block Library (the Continuous System Model Functions CSMF shown in Figure 8.2). The voltages and currents, used as input to the control blocks, are integrated locally (i.e. within each block) to provide flexibility. As explained previously this causes a time-step delay in the feedback paths [5]. The delay can be removed by modelling the complete transfer function encompassing the feedback paths, rather than using the multiple block representation, however this lacks flexibility.

The trapezoidal discretisation of the control blocks is discussed in Chapter 2.

PSCAD simulation is further enhanced by using the exponential form of the difference equations, as described in Chapter 5, to simulate the various control blocks.

With the use of digital controls the time-step delay in the control interface is becoming less of an issue. Unlike analogue controls, which give an instantaneous response, digital controllers sample and hold the continuous voltages and currents, which is effectively what the simulation performs. Modern digital controls have multiple-time-steps and incorporate event-driven controls, all of which can be modelled. EMTDC version 3 has an algorithm which controls the position of time-delay in feedback paths. Rather than using the control library to build a representation of the control system, some manufacturers use a direct translation of real control code,

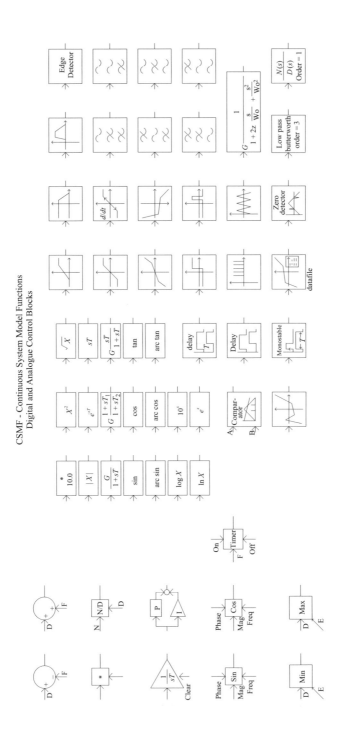

Control and protection 197

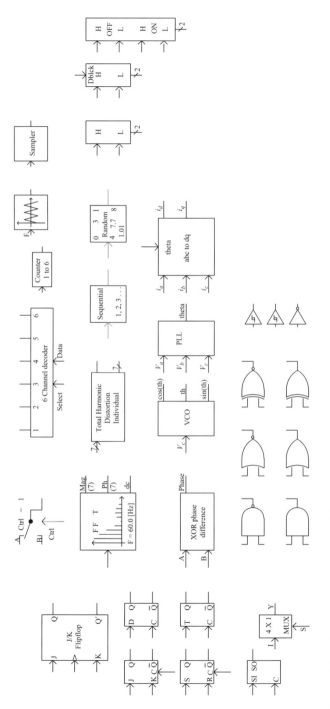

Figure 8.2 *Continuous system model function library (PSCAD/EMTDC)*

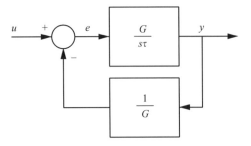

Figure 8.3 First-order lag

written in FORTRAN, C, C++, MATLAB or BASIC, and link this with EMTDC. Interpolation of controls is also an important issue for accurate results [6].

8.3.1 Example

The first-order lag control system, depicted in Figure 8.3, is used to demonstrate the use of the z-domain for the prediction of instabilities.

The corresponding transfer function is:

$$\frac{y}{u} = \frac{1}{1+fg} = \frac{G}{1+s\tau} \qquad (8.1)$$

where
f = feedback path = $1/G$
g = forward path = $G/(s\tau)$
τ = time lag.

The equations for the two blocks are:

$$e = u - \frac{1}{G}y \qquad (8.2)$$

$$y = \frac{Ge}{s\tau} \qquad (8.3)$$

Substitution of the trapezoidal rule to form difference equations gives:

$$e_k = u_k - \frac{1}{G}y_k \qquad (8.4)$$

$$y_k = \frac{\Delta t(e_k + e_{k-1})G}{2\tau} \qquad (8.5)$$

The difference in data paths becomes apparent. If solved as two separate difference equations, then e_k must be calculated from y at the previous time step as y_k is not available and this introduces one time-step delay in the y data path. Swapping the order of equations will result in the same problem for the e data path. Substituting one equation into the other and rearranging, results in a difference equation with no delay

Control and protection 199

in data path. This is equivalent to performing integrator substitution on the transfer function for the complete controller

Time-step delay in data path
If there is a time-step delay in the feedback path due to the way the difference equation for each block is simulated, then

$$e_k = \left(u_k - \frac{1}{G}y_{k-1}\right)$$

$$y = \frac{G}{s\tau}e$$
(8.6)

Applying trapezoidal integration gives:

$$\begin{aligned}
y_k &= y_{k-1} + \frac{\Delta t G}{2\tau}(e_k + e_{k-1}) \\
&= y_{k-1} + \frac{\Delta t G}{2\tau}\left(u_k - \frac{1}{G}y_{k-1} + u_{k-1} - \frac{1}{G}y_{k-2}\right) \\
&= \left(y_{k-1} - \frac{\Delta t}{2\tau}y_{k-1} - \frac{\Delta t}{2\tau}y_{k-2}\right) + \frac{\Delta t G}{2\tau}(u_k + u_{k-1})
\end{aligned}$$
(8.7)

Transforming equation 8.7 into the z-plane yields:

$$Y\left(1 - z^{-1}\left(1 - \frac{\Delta t}{2\tau}\right) + z^{-2}\frac{\Delta t}{2\tau}\right) = \frac{\Delta t G}{2\tau}(1 + z^{-1})U$$
(8.8)

Rearranging gives:

$$\begin{aligned}
\frac{Y}{U} &= \frac{(\Delta t G/(2\tau))(1 + z^{-1})}{\left(1 - z^{-1}(1 - \Delta t/(2\tau)) + z^{-2}\Delta t/(2\tau)\right)} \\
&= \frac{(\Delta t G/2\tau)z(z+1)}{\left(z^2 - z^1(1 - \Delta t/(2\tau)) + \Delta t/(2\tau)\right)}
\end{aligned}$$
(8.9)

The roots are given by:

$$z_1, z_2 = \frac{-b \pm \sqrt{b^2 - 4ac}}{2a}$$
(8.10)

$$\begin{aligned}
z_1, z_2 &= \frac{1}{2}\left[\left(1 - \frac{\Delta t}{2\tau}\right) \pm \sqrt{\left(1 - \frac{\Delta t}{2\tau}\right)^2 - 4\frac{\Delta t}{2\tau}}\right] \\
&= \frac{1}{2}\left[\left(1 - \frac{\Delta t}{2\tau}\right) \pm \sqrt{\left(1 - 3\frac{\Delta t}{\tau}\right) + \frac{\Delta t^2}{4\tau^2}}\right]
\end{aligned}$$
(8.11)

Stability is assured so long as the roots are within the unit circle $|z| \leq 1$.

No time-step delay in data path
If there is no delay in the feedback path implementation then $e_k = (u_k - (1/G)y_k)$
Applying trapezoidal integration gives:

$$y_k = y_{k-1} + \frac{\Delta t G}{2\tau}(e_k + e_{k-1})$$

$$= y_{k-1} + \frac{\Delta t G}{2\tau}\left(u_k - \frac{1}{G}y_k + u_{k-1} - \frac{1}{G}y_{k-1}\right)$$

$$= \left(y_{k-1} - \frac{\Delta t}{2\tau}y_k - \frac{\Delta t}{2\tau}y_{k-1}\right) + \frac{\Delta t G}{2\tau}(u_k + u_{k-1}) \qquad (8.12)$$

Transforming equation 8.12 into the z-plane yields:

$$Y\left(\left(1 + \frac{\Delta t}{2\tau}\right) + z^{-1}\left(\frac{\Delta t}{2\tau} - 1\right)\right) = \frac{\Delta t G}{2\tau}(1 + z^{-1})U \qquad (8.13)$$

Rearranging gives:

$$\frac{Y}{U} = \frac{(\Delta t G/(2\tau))(1 + z^{-1})}{(1 + \Delta t/(2\tau)) + z^{-1}(\Delta t/(2\tau) - 1)}$$

$$= \frac{(\Delta t G/(2\tau))(z + 1)}{z(1 + \Delta t/(2\tau)) + (\Delta t/(2\tau) - 1)} \qquad (8.14)$$

The pole (root of characteristic equation) is:

$$z = \frac{(1 - \Delta t/(2\tau))}{(1 + \Delta t/(2\tau))} \qquad (8.15)$$

Note that $|z_{\text{pole}}| \leq 1$ for all $\Delta t/2\tau > 0$, therefore this method is always stable. However this does not mean that numerical oscillations will not occur due to errors in the trapezoidal integration.

Root-matching technique
Applying the root-matching technique to this control system (represented mathematically by equation 8.1) gives the difference equation:

$$\frac{Y(z)}{U(z)} = \frac{G(1 - e^{-\Delta t/\tau})}{(1 - z^{-1}e^{-\Delta t/\tau})} \qquad (8.16)$$

hence multiplying both side of equation 8.16 by $U(z)(1 - z^{-1}e^{-\Delta t/\tau})$

$$Y(z) = e^{-\Delta t/\tau}z^{-1}Y(z) + G(1 - e^{-\Delta t/\tau})U(z) \qquad (8.17)$$

Transforming to the time domain yields the difference equation:

$$y_k = e^{-\Delta t/\tau}y_{k-1} + G\left(1 - e^{-\Delta t/\tau}\right)u_k \qquad (8.18)$$

Control and protection 201

The pole in the z-plane is:

$$z_{\text{pole}} = e^{-\Delta t/\tau} \qquad (8.19)$$

Note that $|z_{\text{pole}}| \leq 1$ for all $e^{-\Delta t/\tau} \leq 1$ hence for all $\Delta t/\tau \geq 0$.

Numerical illustration

The first-order lag system of Figure 8.3 is analysed using the three difference equations developed previously, i.e. the trapezoidal rule with no feedback (data path) delay, the trapezoidal rule with data path delay and the exponential form using the root-matching technique. The step response is considered using three different time-steps, $\Delta t = \tau/10, \tau, 10\tau$ ($\tau = 50\,\mu s$) and the corresponding results are shown in Figures 8.4–8.6.

When $\Delta t/\tau = 1/10$ the poles for trapezoidal integration with delay in the data path are obtained by solving equation 8.11; these are:

$$z_1, z_2 = \frac{19}{20} \pm \sqrt{\left(\left(\frac{19}{20}\right)^2 - \frac{4}{20}\right)} = 0.0559 \text{ and } 0.8941$$

Since two real roots exist ($z_1 = 0.0559$ and $z_2 = 0.894$) and both are smaller than one, the resulting difference equation is stable. This can clearly be seen in Figure 8.4, which also shows that the exponential form (pole $= 0.9048$) and the trapezoidal rule (pole $= 0.9048$) with no data path delay are indistinguishable, while the error introduced by the trapezoidal rule with data path delay is noticeable.

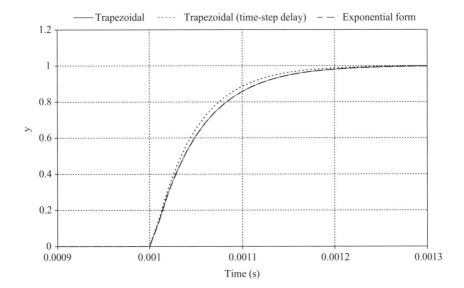

Figure 8.4 Simulation results for a time step of 5 μs

202 *Power systems electromagnetic transients simulation*

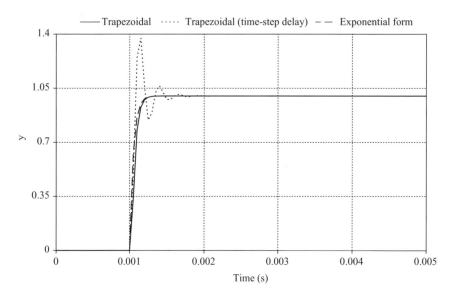

Figure 8.5 Simulation results for a time step of 50 μs

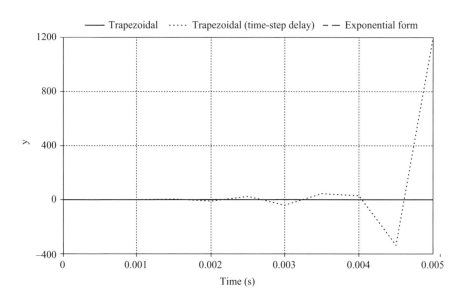

Figure 8.6 Simulation results for a time step of 500 μs

The poles when $\Delta t/\tau = 1$ are given by:

$$z_1, z_2 = \frac{1}{2}\left(\frac{1}{2} \pm \sqrt{\left(\left(\frac{1}{2}\right)^2 - 4\frac{1}{2}\right)}\right) = 0.25 \pm j0.6614$$

Hence a pair of complex conjugate roots result ($z_1, z_2 = 0.25 \pm j0.6614$). They lie inside the unit circle ($|z_1| = |z_2| = 0.7071 < 1$) which indicates stability. Figure 8.5 shows that, although considerable overshoot has been introduced by the time-step delay in the data path, this error dies down in approximately 20 time steps and the difference equations are stable. A slight difference can be seen between the trapezoidal integrator with no data delay (pole $= 0.3333$) and the exponential form (pole $= 0.3679$).

Finally when $\Delta t/\tau = 10$ the poles for the trapezoidal rule with time delay in the feedback path are:

$$z_1, z_2 = \frac{-4}{2} \pm \frac{\sqrt{((-4)^2 - 4 \times 5)}}{2} = -2.0 \pm j1.0$$

hence two complex poles exist, however they lie outside the unit circle in the z-plane and therefore the system of difference equations is unstable. This is shown in the simulation results in Figure 8.6.

The poles for the trapezoidal method with no time delay and exponential form are -0.6667 and $4.5400\mathrm{e}{-005}$ respectively. As predicted by equation 8.15, the difference equation with no data path delay is always stable but close examination of an expanded view (displayed in Figure 8.6) shows a numerical oscillation in this case. Moreover, this numerical oscillation will increase with the step length. Figure 8.6 also shows the theoretical curve and exponential form of the difference equation. The latter gives the exact answer at every point it is evaluated. The exponential form has been derived for the overall transfer function (i.e. without time delays in the data paths).

If a modular building block approach is adopted, the exponential form of difference equation can be applied to the various blocks, and the system of difference equations is solved in the same way as for the trapezoidal integrator. However, errors due to data path delays will occur. This detrimental effect results from using a modular approach to controller representation.

This example has illustrated the use of the z-domain in analysing the difference equations and data path delays and shown that with z-domain analysis instabilities can be accurately predicted. Modelling the complete controller transfer function is preferable to a modular building block approach, as it avoids the data path delays and inherent error associated with it, which can lead to instabilities. However the error introduced by the trapezoidal integrator still exists and the best solution is to use instead the exponential form of difference equation derived from root-matching techniques.

As well as control blocks, switches and latches, the PSCAD/EMTDC CSMF library contains an on-line Fourier component that is used to derive the frequency components of signals. It uses a Discrete Fourier Transform rather than an FFT. This

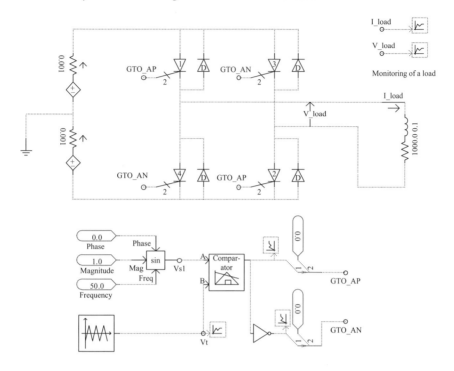

Figure 8.7 Simple bipolar PWM inverter

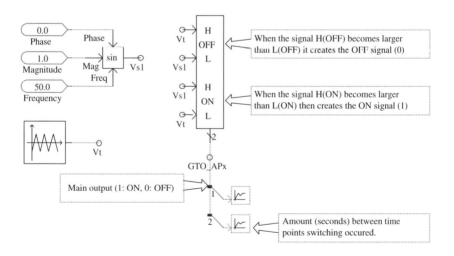

Figure 8.8 Simple bipolar PWM inverter with interpolated turn ON and OFF

enables a recursive formulation to be used, which is very efficient computationally, especially when a small number of frequency components are required (the maximum limit is 31 for the on-line Fourier component).

Figure 8.7 displays a simple bipolar inverter, where the valve is turned on at the first time point after the crossover of the triangular and control signals. To take advantage of interpolated switching the *interpolated firing pulses* block is used as shown in Figure 8.8.

8.4 Modelling of protective systems

A protective system consists of three main components, i.e. transducers, relays and circuit breakers, all of which require adequate representation in electromagnetic transients programs.

The modelling of the relays is at present the least advanced, due to insufficient design information from manufacturers. This is particularly a problem in the case of modern microprocessor-based relays, because a full description of the software involved would give the design secrets away! Manufacturers' manuals contain mostly relay behaviour in the form of operating characteristics in terms of phasor parameters. Such information, however, is not sufficient to model the relay behaviour under transient conditions. PSCAD/EMTDC does allow manufacturers to provide binary library files of their custom models which can be used as a 'black-box', thus keeping their design secret.

Even if all the design details were available, it would be an extremely complex exercise to model all the electronic, electromechanical and software components of the relay. A more practical approach is to develop models that match the behaviour of the actual relays under specified operating conditions.

A practical way of developing suitable models is by means of comprehensive validation, via physical testing of the actual relays and their intended models. The Real-Time Playback (RTP) system or Real-Time Digital Simulator (RTDS), described in Chapter 13, provide the ideal tools for that purpose. The RTDS performs electromagnetic transient simulation to provide the fault current and voltage waveforms in digital form. These are converted to analogue signals by means of A/D converters and then amplified to the appropriate levels required by the relay.

Recognising the importance of relay modelling, a Working Group of the IEEE Power System Relaying Committee has recently published a paper reviewing the present state of the art in relay modelling and recommending guidelines for further work [7].

8.4.1 Transducers

The performance of high-speed protection is closely related to the response of the instrumentation transformers to the transient generated by the power system. Therefore, to be effective, electromagnetic transient programs require adequate modelling of the current transformers, magnetic voltage transformers and capacitor voltage transformers. A Working Group (WG C-5) of the Systems Protection Subcommittee of the IEEE Power System Relaying Committee has recently published [8] a comprehensive report on the physical elements of instrumentation transformers that are important to the modelling of electromagnetic transients. The report contains valuable information on the mathematical modelling of magnetic core transducers with specific details of their implementation in all the main electromagnetic transient packages. More detailed references on the specific models being discussed are [9]–[16].

CT modelling

The transient performance of current transformers (CTs) is influenced by various factors, especially the exponential decaying d.c. component of the primary current following a disturbance. This component affects the build-up of the core flux causing saturation which will introduce errors in the magnitude and phase angle of the generated signals. The core flux consists of an alternating and a unidirectional component corresponding to the a.c. and d.c. content of the primary current. Also a high level of remanence flux may be left in the core after the fault has been cleared. This flux may either aid or oppose the build-up of core flux and could contribute to CT saturation during subsequent faults, such as high-speed autoreclosing into a permanent fault, depending on the relative polarities of the primary d.c. component and the remanent flux. Moreover, after primary fault interruption, the CT can still produce a decaying d.c. current due to the magnetic energy.

The EMTP and ATP programs contain two classes of non-linear models: one of them explicitly defines the non-linearity as the full $\psi = f(i)$ function, whereas the other defines it as a piecewise linear approximation. These programs support two additional routines. One converts the r.m.s. v–i saturation curve data into peak ψ–i and the second adds representation of the hysteresis loop to the model.

In the EMTDC program the magnetising branch of the CT is represented as a non-linear inductor in parallel with a non-linear resistor. This combination, shown in the CT equivalent circuit of Figure 8.9, produces a smooth continuous B–H loop representation similar to that of EMTP/ATP models. Moreover, as the model uses the piecewise representation, to avoid re-evaluation of the overall system conduction matrix at any time when the solution calls for a change from one section to the next, the non-linearity of both the inductive and resistive parallel branches are combined into a voltage/current relationship. These voltage and current components cannot

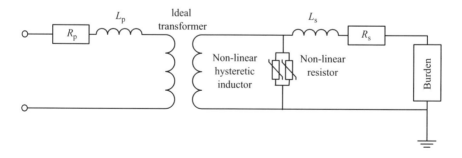

Figure 8.9   Detailed model of a current transformer

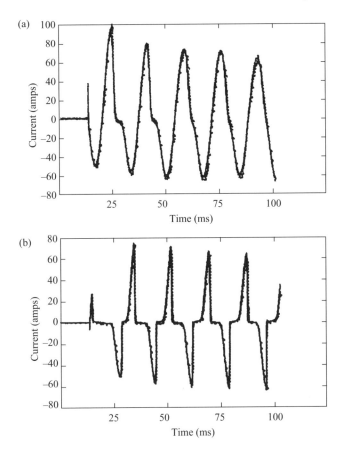

Figure 8.10 Comparison of EMTP simulation (solid line) and laboratory data (dotted line) with high secondary burden. (a) without remanence in the CT; (b) with remanence in the CT.

be calculated independently of each other, and an exact solution would require an iterative solution. However, various techniques have been suggested to reduce the simulation time in this respect [9].

Other important CT alternatives described by the Working Group document are the Seetee [15] and the Jiles–Atherton [13] models. All these models have been tested in the laboratory and shown to produce reasonable and practically identical results. An example of the comparison between the EMTP simulation and laboratory data is shown in Figure 8.10. These results must be interpreted with caution when trying to duplicate them, because there is no information on the level of CT remanence, if any, present in the laboratory tests.

CVT modelling
In the CVT the voltage transformation is achieved by a combination of a capacitive divider, which achieves the main step-down in voltage, and a small wound voltage

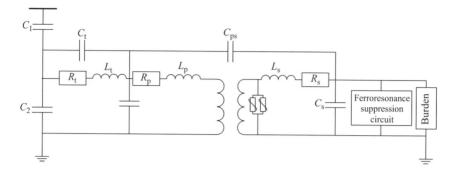

Figure 8.11   Detailed model of a capacitive voltage transformer

transformer. A detailed model equivalent of the CVT is shown in Figure 8.11. A compensating reactor (normally placed on the primary side of the voltage transformer) is used to minimise the equivalent source impedance at the fundamental frequency by tuning it to the capacitance of the $C_1//C_2$ combination; this reduces the fundamental frequency voltage drop, which can otherwise cause a large error when the burden is an electromechanical relay which draws a relatively high current. Figure 8.11 also shows a ferroresonance suppression circuit to protect against a possible resonance between the capacitors and a particular value of the combined inductance of the tuning non-linear reactor and the magnetising inductance of the transformer. The incorporation of winding capacitances C_t and C_{ps} increases the frequency range of the model (these are not needed when the frequency range of interest is below about 500 Hz). Although not shown in the figure, the CVT may contain spark gaps, saturating inductors or metal oxide varistors across the compensating inductor or ferroresonance suppression circuit and their operation during possible resonant conditions must be represented in the model.

VT modelling

The modelling of magnetic voltage transformers is similar to that of other instrumentation transformers. However, the large inductance of the primary winding and the importance of heavy saturation and hysteresis loop require special attention.

8.4.2   Electromechanical relays

The constituent parts of electromechanical relays, i.e. electrical, mechanical and magnetic, can be separated when developing a mathematical model. Figure 8.12 shows a diagram of the components of a model suitable for EMTP simulation [17], [18]. The relay burden is represented as a function of frequency and magnitude of the input current and includes the saturation non-linearities; this part is easily represented by the electrical circuit components of the EMTP method. The mechanical and magnetic parts involve mass, spring and dashpot functions all of which are available in the TACS section of the EMTP programs.

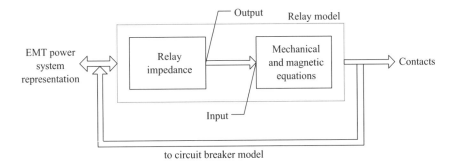

Figure 8.12 Diagram of relay model showing the combination of electrical, magnetic and mechanical parts

An alternative approach is to represent the dynamic behaviour of the relay by a differential equation [19] of the form:

$$F = a\ddot{x} + b\dot{x} + cx \tag{8.20}$$

where
  F is the difference between the applied and restrain forces
  x is the distance travelled by the relay moving contact
  a, b, c are empirically derived constants.

When the distance x is equal to the contact separation the relay operates.

8.4.3 Electronic relays

Electronic relays consist purely of static components, such as transistors, gates, flip-flops, comparators, counters, level detectors, integrators, etc. and are considerably more complex to model than electromechanical relays.

However, the TACS section of the EMTP can be used to represent all these components; simple FORTRAN statements used for logical operations can also be modelled in TACS. A brief prefault simulation (say one or two cycles) is needed prior to transient initialisation.

A detailed description of two specific distance protection relays is given in references [20], [21].

8.4.4 Microprocessor-based relays

Digital relays normally use conventional distance measuring principles, such as the phasor-based mho-circle. The required voltages and currents have to be sampled at discrete points and the resulting information is used to derive their phase values. The main components required to extract the fundamental frequency information are an anti-aliasing input filter, an ADC (analogue to digital convertor) and a Fourier detector as shown in the diagram of Figure 8.13.

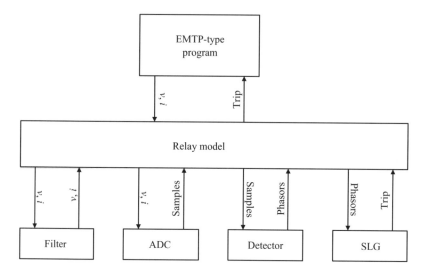

Figure 8.13 Main components of digital relay

The complete model would involve obtaining the circuit diagram and using the EMTP method to represent all the individual components, i.e. resistors, capacitors, inductors and operational amplifiers. A more practical alternative is to obtain the characteristics of the input filter, with the number of stages and signal level loss, etc. With this information a reasonable model can be produced using the s-plane in the TACS section of the EMTP program.

8.4.5 Circuit breakers

The simulation of transient phenomena caused, or affected, by circuit-breaker operations involves two related issues. One is the representation of the non-linear characteristics of the breaker, and the other the accurate placement of the switching instants.

The electrical behaviour of the arc has been represented with different levels of complexity, depending on the phenomena under investigation. In the simplest case the circuit breaker is modelled as an ideal switch that operates when the current changes sign (i.e. at the zero crossing); no attempt is made to represent the arc/system interaction.

A more realistic approach is to model the arc as a time-varying resistance, the prediction of which is based on the circuit-breaker characteristic, i.e. the effect of the system on the arc must be pre-specified.

In the most accurate models the arc resistance dynamic variation is derived from a differential or integral equation, e.g.

$$F = \int_{t_1}^{t_2} (v(t) - v_0(t))^k \, dt \tag{8.21}$$

Control and protection

where v_0 and k are constants, and t_2 is the instant corresponding to voltage breakdown, which occurs when the value of F reaches a user-defined value.

In the BPA version the voltage–time characteristic is simulated by an auxiliary switch in which the breakdown is controlled by a firing signal received from the TACS part of the EMTP.

The above considerations refer to circuit breaking. The modelling requirements are different for the circuit-making action. In the latter case the main factor affecting the transient overvoltage peak is the closing instant. Since that instant (which is different in each phase) is not normally controllable, transient programs tend to use statistical distributions of the switching overvoltages.

Considering the infrequent occurrence of power system faults, the switchings that follow protection action add little overhead to the EMTP simulation process.

8.4.6 Surge arresters

Power system protection also includes insulation coordination, mostly carried out by means of surge arresters [22].

Most arresters in present use are of the silicon carbide and metal oxide types. The former type uses a silicon carbide resistor in series with a spark gap. When the overvoltage exceeds the spark-over level (Figure 8.14) the spark gap connects the arrester to the network; the resistor, which has a non-linear voltage/current characteristic (such as shown in Figure 8.15) then limits the current through the arrester.

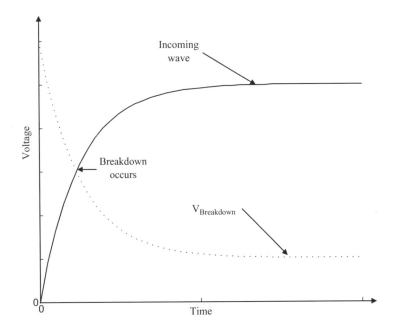

Figure 8.14 *Voltage–time characteristic of a gap*

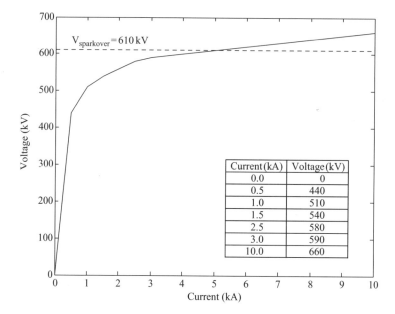

Figure 8.15 Voltage–time characteristic of silicon carbide arrestor

In the EMTP the silicon carbide arrester is modelled as a non-linear resistance in series with a gap (of constant spark-over voltage). In practice the spark-over voltage is dependent on the steepness of the income voltage waveshape; this is difficult to implement, given the irregular shape of the surges.

The non-linear resistance in series with the gap can be solved either by compensation techniques [22] or via piecewise linear models.

Metal oxide surge arresters contain highly non-linear resistances, with practically infinite slope in the normal voltage region and an almost horizontal slope in the protected region. Such characteristics, shown typically in Figure 8.16, are not amenable to a piecewise linear representation. Therefore in the EMTP programs metal oxide arresters are usually solved using the compensation method.

Interpolation is important in modelling arresters to determine the time point where the characteristic of the arrester changes. The energy calculation in the EMTDC program is interpolated to ensure a realistic result. Special care is needed in the low-current region when carrying out trapped charge studies.

Metal oxide arresters are frequency-dependent devices (i.e. the voltage across the arrester is a function of both the rate of rise and the magnitude of the current) and therefore the model must be consistent with the frequency or time-to-crest of the voltage and current expected from the disturbance. Figure 8.17 shows the frequency-dependent model of the metal oxide arrester proposed by the IEEE [22]. In the absence of a frequency-dependent model the use of simple non-linear $V-I$ characteristics, derived from test data with appropriate time-to-crest waveforms, is adequate.

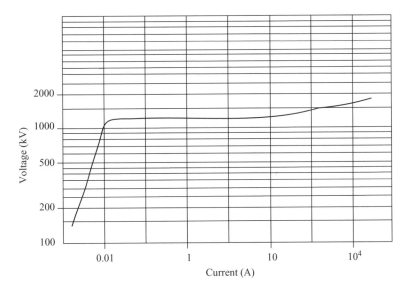

Figure 8.16   Voltage–time characteristic of metal oxide arrestor

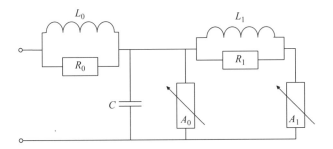

Figure 8.17   Frequency-dependent model of metal oxide arrestor

8.5   Summary

The control equations are solved separately from the power system equations though still using the EMTP philosophy, thereby maintaining the symmetry of the conductance matrix. The main facilities developed to segment the control, as well as devices or phenomena which cannot be directly modelled by the basic network components, are TACS and MODELS (in the original EMTP package) and a CMSF library (in the PSCAD/EMTDC package).

The separate solution of control and power system introduces a time-step delay, however with the sample and hold used in digital control this is becoming less of an issue. Modern digital controls, with multiple time steps, are more the norm and can be adequately represented in EMT programs.

The use of a modular approach to build up a control system, although it gives greater flexibility, introduces time-step delays in data paths, which can have a detrimental effect on the simulation results. The use of the z-domain for analysing the difference equations either generated using NIS, with and without time-step delay, or the root-matching technique, has been demonstrated.

Interpolation is important for modelling controls as well as for the non-linear surge arrester, if numerical errors and possible instabilities are to be avoided.

A description of the present state of protective system implementation has been given, indicating the difficulty of modelling individual devices in detail. Instead, the emphasis is on the use of real-time digital simulators interfaced with the actual protection hardware via digital-to-analogue conversion.

8.6 References

1 ARAUJO, A. E. A., DOMMEL, H. W. and MARTI, J. R.: 'Converter simulations with the EMTP: simultaneous solution and backtracking technique', IEEE/NTUA Athens Power Tech Conference *Planning, Operation and Control of Today's Electric Power Systems*, September 5–8, 1993, **2**, pp. 941–5

2 ARAUJO, A. E. A., DOMMEL, H. W. and MARTI, J. R.: 'Simultaneous solution of power and control system equations', *IEEE Transactions on Power Systems*, 1993, **8** (4), pp. 1483–9

3 LASSETER, R. H. and ZHOU, J.: 'TACS enhancements for the electromagnetic transient program', *IEEE Transactions on Power Systems*, 1994, **9** (2), pp. 736–42

4 DUBE, L. and BONFANTI, I.: 'MODELS: a new simulation tool in EMTP', 1992, ETEP **2** (1), pp. 45–50

5 WATSON, N. R. and IRWIN, G. D.: 'Accurate and stable electromagnetic transient simulation using root-matching techniques', *International Journal of Electrical Power & Energy Systems*, Elsevier Science Ltd, 1999, **21** (3), pp. 225–34

6 GOLE, A. M. and NORDSTROM, J. E.: 'A fully interpolated controls library for electromagnetic transients simulation of power electronic systems', Proceedings of International Conference on *Power system transients (IPST'2001)*, June 2001, pp. 669–74

7 McLAREN, P. G., MUSTAPHI, K., BENMOUYAL, G. *et al.*: 'Software models for relays', *IEEE Transactions on Power Delivery*, 2001, **16** (2), pp. 238–45

8 Working Group C5 of the Systems Protection subcommittee of the IEEE Power System Relaying Committee: 'Mathematical models for current, voltage, and coupling capacitor voltage transformer', *IEEE Transactions on Power Delivery*, 2000, **15** (1), pp. 62–72

9 LUCAS, J. R., McLAREN, P. G. and JAYASINGHE, R. P.: 'Improved simulation models for current and voltage transformers in relay studies', *IEEE Trans. on Power Delivery*, 1992, **7** (1), p. 152

10 WISEMAN, M. J.: 'CVT transient behavior during shunt capacitor switching', Ontario Hydro study no. W401, April 1993

11 McLAREN, P. G., LUCAS, J. R. and KEERTHIPALA, W. W. L.: 'A digital simulation model for CCVT in relay studies', Proceedings International Power Engineering Conference (IPEC), March 1993
12 KOJOVIC, L. A., KEZUNOVIC, M. and NILSSON, S. L.: 'Computer simulation of a ferroresonance suppression circuit for digital modeling of coupling capacitor voltage transformers', ISMM International Conference, Orlando, Florida, 1992
13 JILES, D. C. and ATHERTON, D. L.: 'Theory of ferromagnetic hysteresis', *Journal of Magnetism and Magnetic Materials*, 1986, **61**, pp. 48–60
14 JILES, D. C., THOELKE, J. B. and DEVINE, M. K.: 'Numerical determination of hysteresis parameters for modeling of magnetic properties using the theory of ferromagnetic hysteresis', *IEEE Transactions on Magnetics*, 1992, **28** (1), pp. 27–334
15 GARRET, R., KOTHEIMER, W. C. and ZOCHOLL, S. E.: 'Computer simulation of current transformers and relays', Proceedings of 41st Annual Conference for Protective Relay Engineers, 1988, Texas A&M University
16 KEZUNOVIC, M., KOJOVIC, L. J., ABUR, A., FROMEN, C. W. and SEVCIK, D. R.: 'Experimental evaluation of EMTP-based current transformer models for protective relay transient study', *IEEE Transactions on Power Delivery*, 1994, **9** (1), pp. 405–13
17 GLINKOWSKI, M. T. and ESZTRGALYOS, J.: 'Transient modeling of electromechanical relays. Part 1: armature type relay', *IEEE Transactions on Power Delivery*, 1996, **11** (2), pp. 763–70
18 GLINKOWSKI, M. T. and ESZTRGALYOS, J.: 'Transient modeling of electromechanical relays. Part 2: plunger type 50 relays', *IEEE Transactions on Power Delivery*, 1996, **11** (2), pp. 771–82
19 CHAUDARY, A. K. S, ANICH, J. B. and WISNIEWSKI, A.: 'Influence of transient response of instrument transformers on protection systems', Proceedings of Sargent and Lundy, 12th biennial Transmission and Substation Conference, 1992
20 GARRETT, B. W.: 'Digital simulation of power system protection under transient conditions' (Ph.D. thesis, University of British Columbia, 1987)
21 CHAUDHARY, A. K. S., TAM, K.-S. and PHADKE, A. G.: 'Protection system representation in the electromagnetic transients program', *IEEE Transactions on Power Delivery*, 1994, **9** (2), pp. 700–11
22 IEEE Working Group on Surge Arrester Modeling: 'Modeling of metal oxide surge arresters', *IEEE Transactions on Power Delivery*, 1992, **1** (1), pp. 302–9

Chapter 9
Power electronic systems

9.1 Introduction

The computer implementation of power electronic devices in electromagnetic transient programs has taken much of the development effort in recent years, aiming at preserving the elegance and efficiency of the EMTP algorithm. The main feature that characterises power electronic devices is the use of frequent periodic switching of the power components under their control.

The incorporation of power electronics in EMT simulation is discussed in this chapter with reference to the EMTDC version but appropriate references are made, as required, to other EMTP-based algorithms. This is partly due to the fact that the EMTDC program was specifically developed for the simulation of HVDC transmission and partly to the authors' involvement in the development of some of its recent components. A concise description of the PSCAD/EMTDC program structure is given in Appendix A.

This chapter also describes the state variable implementation of a.c.–d.c. converters and systems, which offers some advantages over the EMTP solution, as well as a hybrid algorithm involving both the state variable and EMTP methods.

9.2 Valve representation in EMTDC

In a complex power electronic system, such as HVDC transmission, valves consist of one or more series strings of thyristors. Each thyristor is equipped with a resistor–capacitor damping or snubber circuit. One or more di/dt limiting inductors are included in series with the thyristors and their snubber circuits. It is assumed that for most simulation purposes, one equivalent thyristor, snubber circuit and di/dt limiting inductor will suffice for a valve model. The di/dt limiting inductor can usually be neglected when attempting transient time domain simulations up to about 1.5–2.0 kHz frequency response. In version 3 of the EMTDC program the snubber is kept as a separate branch to allow chatter removal to be effective.

218 Power systems electromagnetic transients simulation

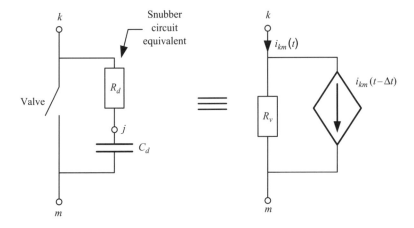

Figure 9.1 Equivalencing and reduction of a converter valve

EMTDC V2 utilised the fact that network branches of inductors and capacitors are represented as resistors with an associated current source, which allowed a valve in a converter bridge to be represented by the Norton equivalent of Figure 9.1.

With the valve blocked (not conducting), the equivalent resistor R_v is just derived from the snubber circuit. With the di/dt limiting inductor ignored, then from reference [1] this becomes:

$$R_v = R_d + \frac{\Delta t}{2C_d} \tag{9.1}$$

where

Δt = time-step length
R_d = snubber resistance
C_d = snubber capacitance

With the valve de-blocked and conducting in the forward direction, the equivalent resistor R_v is changed to a low value, e.g. $R_v = 1\,\Omega$. The equivalent current source $I_{km}(t - \Delta t)$ shown in Figure 9.1 between nodes k and m is determined by first defining the ratio Y as:

$$Y = \frac{\Delta t/(2C_d)}{R_d + \Delta t/(2C_d)} \tag{9.2}$$

From equations 4.11 and 4.13 of Chapter 4.

$$i_{km}(t) = \frac{(e_k(t) - e_m(t))}{R_d + \Delta t/(2C_d)} + I_{km}(t - \Delta t) \tag{9.3}$$

then

$$I_{km}(t - \Delta t) = -Y \left[i_{km}(t - \Delta t) + 2C_d \frac{(e_j(t - \Delta t) - e_m(t - \Delta t))}{\Delta t} \right] \tag{9.4}$$

where
$$e_j(t - \Delta t) = e_k(t - \Delta t) - R_d i_{km}(t - \Delta t) \quad (9.5)$$

For greater accuracy the above model can be extended to include the di/dt limiting inductor into the equivalent resistor and current source.

9.3 Placement and location of switching instants

The efficiency and elegance of the EMTP method relies on the use of a constant integration step. This increases the complexity of the model in the presence of frequently switching components, such as HVDC converters. The basic EMTP-type algorithm requires modification in order to accurately and efficiently model the switching actions associated with HVDC, thyristors, FACTS devices, or any other piecewise linear circuit. The simplest approach is to simulate normally until a switching is detected and then update the system topology and/or conductance matrix. The system conductance matrix must be reformed and triangulated after each change in conduction state. This increases the computational requirements of the simulation in proportion to the number of switching actions (so as to keep the conductance matrix constant to avoid retriangulation). Nevertheless, for HVDC and most FACTS applications, the switching rate is only several kHz, so that the overall simulation is still fast.

The CIGRE test system (see Appendix D) used as an example here is representative, since larger systems are likely to be broken into several subsystems, so that the ratio of switchings to system size are likely to be small. This system has been simulated (using EMTDC V2) with all the valves blocked to assess the processing overheads associated with the triangulation of the conductance matrix. The results, presented in Table 9.1, indicate that in this case the overheads are modest.

The reason for the small difference in computation time is the ordering of the system nodes. Nodes involving frequently switched elements (such as thyristors, IGBTs, etc.) are ordered last. However in version 2 of the EMTDC program infrequently switching branches (such as fault branches and CBs) are also included in the submatrix that is retriangulated. This increases the processing at every switching even though they switch infrequently.

Table 9.1 Overheads associated with repeated conductance matrix refactorisation

	Time step	Number of refactorisations	Simulation time
Unblocked	10 µs	2570	4 min 41 s
	50 µ s	2480	1 min 21 s
Blocked	10 µ s	1	4 min 24 s
	50 µ s	1	1 min 9 s

In virtually all cases switching action, or other point discontinuities, will not fall exactly on a time point, thus causing a substantial error in the simulation.

Data is stored on a subsystem basis in EMTDC and in a non-sparse format (i.e. zero elements are stored). However, in the integer arrays that are used for the calculations only the addresses of the non-zero elements are stored, i.e. no calculations are performed on the zero elements. Although keeping the storage sequential is not memory efficient, it may have performance advantages, since data transfer can be streamed more efficiently by the FORTRAN compiler than the pseudo-random allocation of elements of a sparse matrix in vectors. The column significant storage in FORTRAN (the opposite of C or C++) results in faster column indexing and this is utilized wherever possible.

Subsystem splitting reduces the amount of storage required, as only each block in the block diagonal conductance matrix is stored. For example the conductance matrix is stored in GDC(n, n, s), where n is the maximum number of nodes per subsystem and s the number of subsystems. If a circuit contains a total of approximately 10,000 nodes split over five subsystems then the memory storage is 2×10^6, compared to 100×10^6 without subsystem splitting. Another advantage of the subsystems approach is the performance gains achieved during interpolation and switching operations. These operations are performed only on one subsystem, instead of having to interpolate or switch the entire system of equations.

Depending on the number of nodes, the optimal order uses either Tinney's level II or III [2]. If the number of nodes is less that 500 then level III is used to produce faster running code, however, level II is used for larger systems as the optimal ordering would take too long. The *.map file created by PSCAD gives information on the mapping of local node numbers to optimally ordered nodes in a subsystem.

As previously mentioned, nodes connected to frequently switching components are placed at the bottom of the conductance matrix. When a branch is switched, the smallest node number to which the component is connected is determined and the conductance matrix is retriangularised from that node on. The optimal ordering is performed in two stages, first for the nodes which are not connected to frequently switching branches and then for the remaining nodes, i.e. those that have frequently switching branches connected.

9.4 Spikes and numerical oscillations (chatter)

The use of a constant step length presents some problems when modelling switching elements. If a switching occurs in between the time points it can only be represented at the next time-step point. This results in firing errors when turning the valves ON and OFF. Two problems can occur under such condition, i.e. spikes and numerical oscillations (or chatter). Voltage spikes, high Ldi/dt, in inductive circuits can occur due to current chopping (numerically this takes place when setting a non-zero current to zero).

Numerical oscillations are initiated by a disturbance of some kind and result in $v(t)$ or $i(t)$ oscillating around the true solution.

Power electronic systems 221

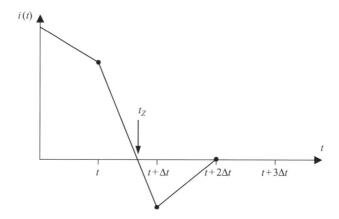

Figure 9.2   Current chopping

Voltage chatter is triggered by disturbances in circuits with nodes having only inductive and current sources connected. Similarly, current chatter occurs in circuits with loops of capacitors and voltage sources. This is a similar problem to that of using dependent state variables in the state variable analysis discussed in Chapter 3. Chatter is not only caused by current interruption (in an inductor) at a non-zero point; it also occurs even if the current zero in inductive circuits falls exactly on a time-point, due to the errors associated with the trapezoidal rule.

This effect is illustrated in Figure 9.2 where the current in a diode has reduced to zero between t and $t + \Delta t$. Because of the fixed time step the impedance of the device can only be modified (diode turns off) at $t + \Delta t$. The new conductance matrix can then be used to step from $t + \Delta t$ to $t + 2\Delta t$. Using small time steps reduces the error, as the switching occurs closer to the true turn-off. Therefore dividing the step into submultiples on detection of a discontinuity is a possible method of reducing this problem [3].

To illustrate that voltage chatter occurs even if the switching takes place exactly at the current zero, consider the current in a diode-fed RL. The differential equation for the inductor is:

$$v_L(t) = L \frac{di(t)}{dt} \tag{9.6}$$

Rearranging and applying the trapezoidal rule gives:

$$i(t + \Delta t) = i(t) + \frac{1}{2L}(v_L(t + \Delta t) + v_L(t)) \tag{9.7}$$

If the diode is turned off when the current is zero then stepping from $t + \Delta t$ to $t + 2\Delta t$ gives:

$$\frac{1}{2L}(v_L(t + 2\Delta t) + v_L(t + \Delta t)) = 0 \tag{9.8}$$

i.e.

$$v_L(t + 2\Delta t) = -v_L(t + \Delta t)$$

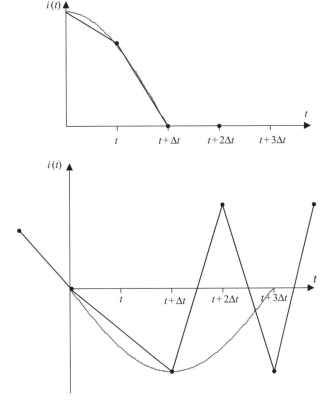

Figure 9.3 Illustration of numerical chatter

Hence there will be a sustained oscillation in voltage, as depicted in Figure 9.3. The damping of these oscillations is sensitive to the OFF resistance of the switch. A complete simulation of this effect is shown in Figure 9.4, for a diode-fed RL load with switch ON and OFF resistances of 10^{-10} Ω and 10^{10} Ω respectively. The FORTRAN and MATLAB code used in this example are given in Appendices H.3 and F.2 respectively.

9.4.1 Interpolation and chatter removal

The circuit of Figure 9.5 shows the simplest form of forced commutation. When the gate turn-OFF thyristor (GTO) turns OFF, the current from the source will go to zero. The current in the inductor cannot change instantaneously, however, so a negative voltage (due to $L di/dt$) is generated which results in the free-wheeling diode turning on immediately and maintaining the current in the inductor. With fixed time step programs however, the diode will not turn on until the end of the time step, and therefore the current in the inductor is reduced to zero, producing a large voltage spike (of one time step duration). The EMTDC program uses interpolation, so that

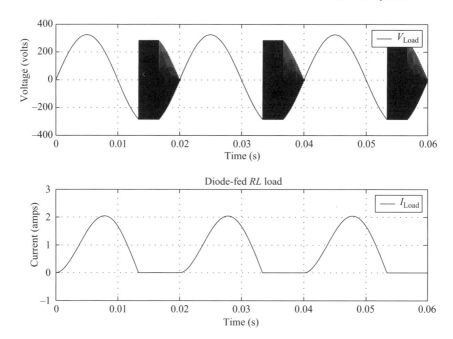

Figure 9.4 Numerical chatter in a diode-fed RL load $\left(R_{ON} = 10^{-10}, R_{OFF} = 10^{10}\right)$

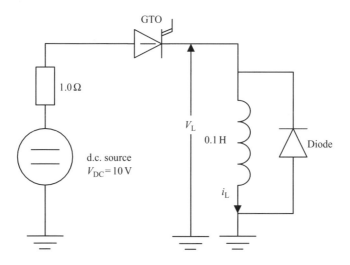

Figure 9.5 Forced commutation benchmark system

224 *Power systems electromagnetic transients simulation*

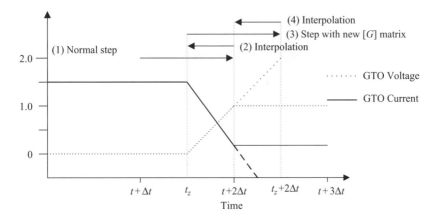

Figure 9.6 Interpolation for GTO turn-OFF (switching and integration in one step)

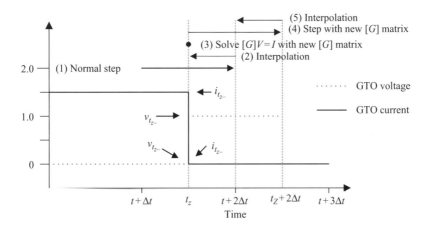

Figure 9.7 Interpolation for GTO turn-OFF (using instantaneous solution)

the diode turns ON at exactly zero voltage, not at the end of the time step. The result is that the inductor current continues to flow in the diode without interruption.

With the techniques described so far the switching and integration are effectively one step. The solution is interpolated to the point of discontinuity, the conductance matrix modified to reflect the switching and an integration step taken. This causes a fictitious power loss in forced turn-OFF devices due to the current and voltage being non-zero simultaneously [4], as illustrated in Figure 9.6. A new instantaneous solution interpolation method is now used in the PSCAD/EMTDC program (V3.07 and above) which separates the switching and integration steps, as illustrated in Figure 9.7. The node voltages, branch currents and history terms are linearly interpolated back to the switching instant giving the state at t_{z-} immediately before switching. The conductance matrix is changed to reflect the switching and $[G]V = I$ solved at t_{z+} again for the instant immediately after switching. From this point the normal integration step proceeds. Essentially there are two solutions at every point in which switching is performed, however these solution points are not written out. Moreover the solution can be interpolated numerous times in the same time step to accommodate the multiple switchings that may occur in the same time step. If a non-linear surge arrester changes state between t_{z-} and t_{z+} then the solution is interpolated to the discontinuity of the non-linear device, say t_{z-+}. The non-linear device characteristics are changed and then a new t_{z+} solution obtained, giving three solutions all at time t_z.

Ideally what should be kept constant from t_{z-} to t_{z+} are the inductor current and capacitor voltage. However, this would require changing the conductance matrix. Instead, the present scheme keeps the current source associated with inductors and capacitors constant, as the error associated with this method is very small.

Early techniques for overcoming these numerical problems was the insertion of additional damping, either in the form of external fictitious resistors (or snubber networks) or by the integration rule itself. The former is often justified by the argument that in reality the components are not ideal.

The alternative is to use a different integration rule at points of discontinuity. The most widely used technique is critical damping adjustment (CDA), in which the integration method is changed to the backward Euler for two time steps (of $\Delta t/2$) after the discontinuity. By using a step size of $\Delta t/2$ with the backward Euler the conductance matrix is the same as for the trapezoidal rule [5], [6]. The difference equations for the inductor and capacitor become:

$$i_L \left(t_Z + \frac{\Delta t}{2} \right) = \frac{\Delta t}{2L} v_L \left(t_Z + \frac{\Delta t}{2} \right) + i_L(t_{Z-})$$
$$v_C \left(t_Z + \frac{\Delta t}{2} \right) = \frac{2C}{\Delta t} v_C \left(t_Z + \frac{\Delta t}{2} \right) - \frac{2C}{\Delta t} v_C(t_{Z-})$$
(9.9)

This approach is used in the NETOMAC program [7], [8]. With reference to Figure 9.8 below the zero-crossing instant is determined by linear interpolation. All the variables (including the history terms) are interpolated back to point t_Z. Distinguishing between the instants immediately before t_{Z-} and immediately after t_{Z-} switching, the inductive current and capacitor voltages must be continuous across t_Z. However, as illustrated in Figure 9.9, the inductor voltage or capacitor current will exhibit jumps. In general the history terms are discontinuous across time. Interpolation is used to find the voltages and currents as well as the associated history terms. Strictly speaking two time points should be generated for time t_Z one immediately before switching, which is achieved by this interpolation step, and one immediately after to catch correctly this jump in voltage and/or current. However, unlike state variable analysis, this is not performed here. With these values the step is made from t_Z to $t_Z + \Delta t/2$ using the backward Euler rule. The advantage of using the backward Euler integration step is that inductor voltages or capacitor currents at t_{Z+} are not needed. NETOMAC then uses the calculated inductor voltages or capacitor currents calculated with the half

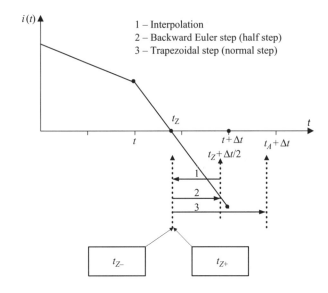

Figure 9.8 Interpolating to point of switching

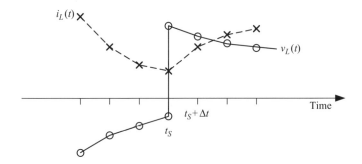

Figure 9.9 Jumps in variables

step as the values at t_{Z+} i.e.

$$v_L(t_{Z+}) = v_L\left(t_Z + \frac{\Delta t}{2}\right)$$
$$i_C(t_{Z+}) = i_C\left(t_Z + \frac{\Delta t}{2}\right)$$
(9.10)

Using these values at time point t_{Z+}, the history terms for a normal full step can be calculated by the trapezoidal rule, and a step taken. This procedure results in a shifted time grid (i.e. the time points are not equally spaced) as illustrated in Figure 9.8.

PSCAD/EMTDC also interpolates back to the zero crossing, but then takes a full time step using the trapezoidal rule. It then interpolates back on to $t + \Delta t$ so as to

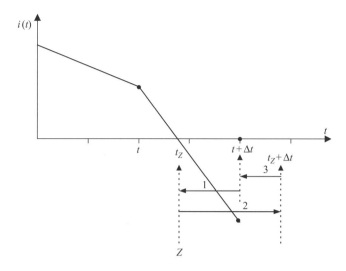

Figure 9.10   Double interpolation method (interpolating back to the switching instant)

keep the same time grid, as the post-processing programs expect equally spaced time points. This method is illustrated in Figure 9.10 and is known as double interpolation because it uses two interpolation steps.

Interpolation has been discussed so far as a method of removing spikes due, for example, to inductor current chopping. PSCAD/EMTDC also uses interpolation to remove numerical chatter. Chatter manifests itself as a symmetrical oscillation around the true solution; therefore, interpolating back half a time step will give the correct result and simulation can proceed from this point. Voltage across inductors and current in capacitors both exhibit numerical chatter. Figure 9.11 illustrates a case where the inductor current becoming zero coincides with a time point (i.e. there is no current chopping in the inductive circuit). Step 1 is a normal step and step 2 is a half time step interpolation to the true solution for $v(t)$. Step 3 is a normal step and Step 4 is another half time step interpolation to get back on to the same time grid.

The two interpolation procedures, to find the switching instant and chatter removal, are combined into one, as shown in Figure 9.12; this allows the connection of any number of switching devices in any configuration. If the zero crossing occurs in the second half of the time step (not shown in the figure) this procedure has to be slightly modified. A double interpolation is first performed to return on to the regular time grid (at $t + \Delta t$) and then a half time step interpolation performed after the next time step (to $t + 2\Delta t$) is taken. The extra solution points are kept internal to EMTDC (not written out) so that only equal spaced data points are in the output file.

PSCAD/EMTDC invokes the chatter removal algorithm immediately whenever there is a switching operation. Moreover the chatter removal detection looks for oscillation in the slope of the voltages and currents for three time steps and, if detected, implements a half time-step interpolation. This detection is needed, as chatter can be

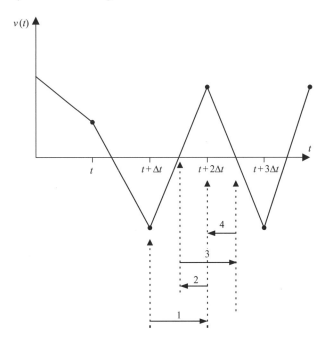

Figure 9.11   Chatter removal by interpolation

initiated by step changes in current injection or voltage sources in addition to switching actions.

The use of interpolation to backtrack to a point of discontinuity has also been adopted in the MicroTran version of EMTP [9]. MicroTran performs two half time steps forward of the backward Euler rule from the point of discontinuity to properly initialise the history terms of all components.

The ability to write a FORTRAN dynamic file gives the PSCAD/EMTDC user great flexibility and power, however these files are written assuming that they are called at every time step. To maintain compatibility this means that the sources must be interpolated and extrapolated for half time step points, which can produce significant errors if the sources are changing abruptly. Figure 9.13 illustrates this problem with a step input.

Step 1 is a normal step from $t + \Delta t$ to $t + 2\Delta t$, where the user-defined dynamic file is called to update source values at $t + 2\Delta t$.

Step 2, a half-step interpolation, is performed by the chatter removal algorithm. As the user-defined dynamic file is called only at increments the source value at $t + \Delta t/2$ has to be interpolated.

Step 3 is a normal time step (from $t + \Delta t/2$ to $t + 3\Delta t/2$) using the trapezoidal rule. This requires the source values at $t + 3\Delta t/2$, which is obtained by extrapolation from the known values at $t + \Delta t$ to $t + 2\Delta t$.

Step 4 is another half time step interpolation to get back to $t + 2\Delta t$.

Power electronic systems 229

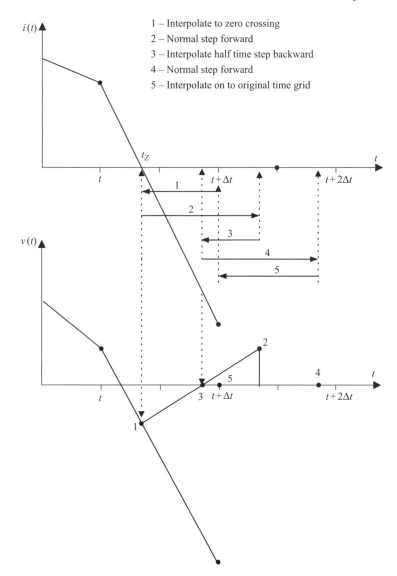

Figure 9.12 Combined zero-crossing and chatter removal by interpolation

The purpose of the methods used so far is to overcome the problem associated with the numerical error in the trapezoidal rule (or any integration rule for that matter). A better approach is to replace numerical integrator substitution by root-matching modelling techniques. As shown in Chapter 5, the root-matching technique does not exhibit chatter, and so a removal process is not required for these components. Root-matching is always numerically stable and is more efficient numerically than trapezoidal integration. Root-matching can only be formulated with branches containing

230 *Power systems electromagnetic transients simulation*

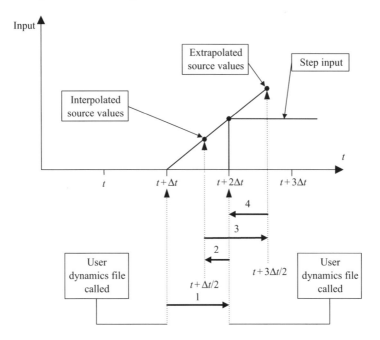

Figure 9.13 *Interpolated/extrapolated source values due to chatter removal algorithm*

two or more elements (i.e. RL, RC, RLC, LC, ...) but these branches can be intermixed in the same solution with branches solved with other integration techniques.

9.5 HVDC converters

PSCAD/EMTDC provides as a single component a six-pulse valve group, shown in Figure 9.14(a), with its associate PLO (Phase Locked Oscillator) firing control and sequencing logic. Each valve is modelled as an off/on resistance, with forward voltage drop and parallel snubber, as shown in Figure 9.14(b). The combination of on-resistance and forward-voltage drop can be viewed as a two-piece linear approximation to the conduction characteristic. The interpolated switching scheme, described in section 9.4.1 (Figure 9.10), is used for each valve.

The LDU factorisation scheme used in EMTDC is optimised for the type of conductance matrix found in power systems in the presence of frequently switched elements. The block diagonal structure of the conductance matrix, caused by a travelling-wave transmission line and cable models, is exploited by processing each associated subsystem separately and sequentially. Within each subsystem, nodes to which frequently switched elements are attached are ordered last, so that the matrix refactorisation after switching need only proceed from the switched node to the end. Nodes involving circuit breakers and faults are not ordered last, however, since they

Power electronic systems 231

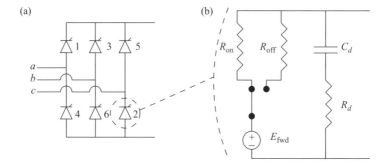

Figure 9.14   (a) The six-pulse group converter, (b) thyristor and snubber equivalent circuit

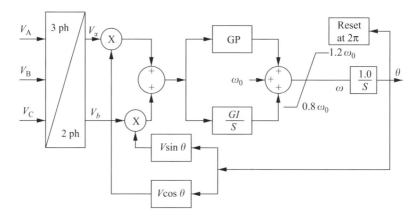

Figure 9.15   Phase-vector phase-locked oscillator

switch only once or twice in the course of a simulation. This means that the matrix refactorisation time is affected mainly by the total number of switched elements in a subsystem, and not by the total size of the subsystem. Sparse matrix indexing methods are used to process only the non-zero elements in each subsystem. A further speed improvement, and reduction in algorithmic complexity, are achieved by storing the conductance matrix for each subsystem in full form, including the zero elements. This avoids the need for indirect indexing of the conductance matrix elements by means of pointers.

Although the user has the option of building up a valve group from individual thyristor components, the use of the complete valve group including sequencing and firing control logic is a better proposition.

The firing controller implemented is of the phase-vector type, shown in Figure 9.15, which employs trigonometric identities to operate on an error signal following the phase of the positive sequence component of the commutating voltage. The output of the PLO is a ramp, phase shifted to account for the transformer phase

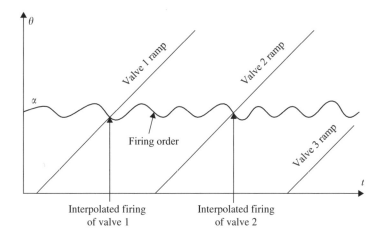

Figure 9.16   Firing control for the PSCAD/EMTDC valve group model

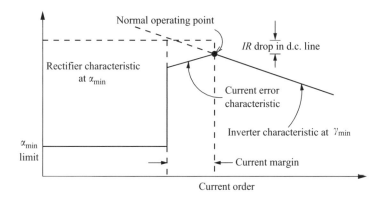

Figure 9.17   Classic V–I converter control characteristic

shift. A firing occurs for valve 1 when the ramp intersects the instantaneous value of the alpha order from the link controller. Ramps for the other five valves are obtained by adding increments of 60 degrees to the valve 1 ramp. This process is illustrated in Figure 9.16.

As for the six-pulse valve group, where the user has the option of constructing it from discrete component models, HVDC link controls can be modelled by synthesis from simple control blocks or from specific HVDC control blocks. The d.c. link controls provided are a gamma or extinction angle control and current control with voltage-dependent current limits. Power control must be implemented from general-purpose control blocks. The general extinction angle and current controllers provided with PSCAD readily enable the implementation of the classic $V-I$ characteristic for a d.c. link, illustrated in Figure 9.17.

General controller modelling is made possible by the provision of a large number of control building blocks including integrators with limits, real pole, PI control, second-order complex pole, differential pole, derivative block, delay, limit, timer and ramp. The control blocks are interfaced to the electrical circuit by a variety of metering components and controlled sources.

A comprehensive report on the control arrangements, strategies and parameters used in existing HVDC schemes has been prepared by CIGRE WG 14-02 [10]. All these facilities can easily be represented in electromagnetic transient programs.

9.6 Example of HVDC simulation

A useful test system for the simulation of a complete d.c. link is the CIGRE benchmark model [10] (described in Appendix D). This model integrates simple a.c. and d.c. systems, filters, link control, bridge models and a linear transformer model. The benchmark system was entered using the PSCAD/draft software package, as illustrated in Figure 9.18. The controller modelled in Figure 9.19 is of the proportional/integral type in both current and extinction angle control.

The test system was first simulated for 1 s to achieve the steady state, whereupon a snapshot was taken of the system state. Figure 9.20 illustrates selected waveforms of the response to a five-cycle three-phase fault applied to the inverter commutating bus. The simulation was started from the snapshot taken at the one second point. A clear advantage of starting from snapshots is that many transient simulations, for the purpose of control design, can be initiated from the same steady-state condition.

9.7 FACTS devices

The simulation techniques developed for HVDC systems are also suitable for the FACTS technology. Two approaches are currently used to that effect: the FACTS devices are either modelled from a synthesis of individual power electronic components or by developing a unified model of the complete FACTS device. The former method entails the connection of thyristors or GTOs, phase-locked loop, firing controller and control circuitry into a complicated simulation. By grouping electrical components and firing control into a single model, the latter method is more efficient, simpler to use, and more versatile. Two examples of FACTS applications, using thyristor and turn-off switching devices, are described next.

9.7.1 The static VAr compensator

An early FACTS device, based on conventional thyristor switching technology, is the SVC (Static Var Compensator), consisting of thyristor switched capacitor (TSC) banks and a thyristor controlled reactor (TCR). In terms of modelling, the TCR is the FACTS technology more similar to the six-pulse thyristor bridge. The firing instants are determined by a firing controller acting in accordance with a delay angle

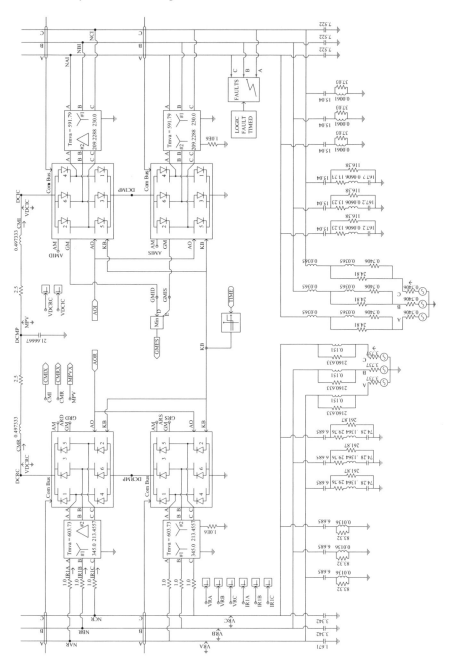

Figure 9.18 CIGRE benchmark model as entered into the PSCAD draft software

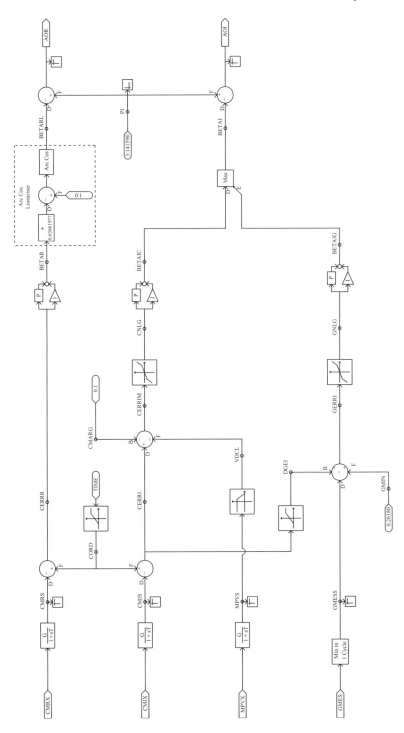

Figure 9.19 Controller for the PSCAD/EMTDC simulation of the CIGRE benchmark model

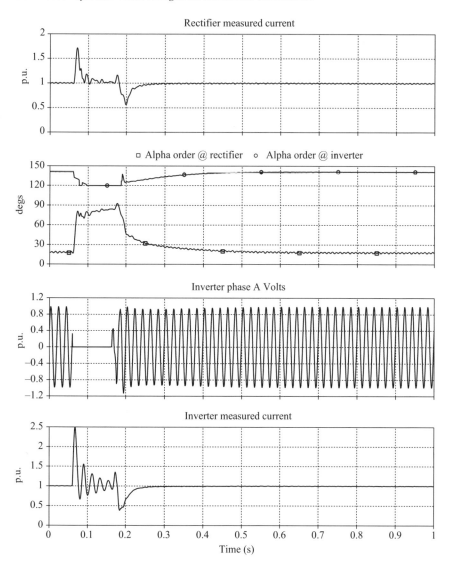

Figure 9.20 Response of the CIGRE model to five-cycle three-phase fault at the inverter bus

passed from an external controller. The end of conduction of a thyristor is unknown beforehand, and can be viewed as a similar process to the commutation in a six-pulse converter bridge.

PSCAD contains an in-built SVC model which employs the state variable formulation (but not state variable analysis) [3]. The circuit, illustrated in Figure 9.21, encompasses the electrical components of a twelve-pulse TCR, phase-shifting

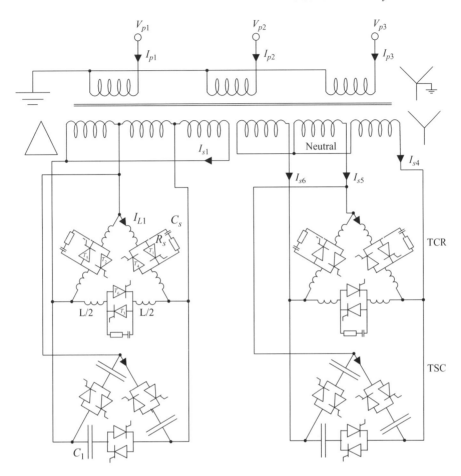

Figure 9.21 SVC circuit diagram

transformer banks and up to ten TSC banks. Signals to add or remove a TSC bank, and the TCR firing delay, must be provided from the external general-purpose control system component models. The SVC model includes a phase-locked oscillator and firing controller model. The TSC bank is represented by a single capacitor, and when a bank is switched the capacitance value and initial voltage are adjusted accordingly. This simplification requires that the current-limiting inductor in series with each capacitor should not be explicitly represented. RC snubbers are included with each thyristor.

The SVC transformer is modelled as nine mutually coupled windings on a common core, and saturation is represented by an additional current injection obtained from a flux/magnetising current relationship. The flux is determined by integration of the terminal voltage.

A total of 21 state variables are required to represent the circuit of Figure 9.21. These are the three currents in the delta-connected SVC secondary winding, two of

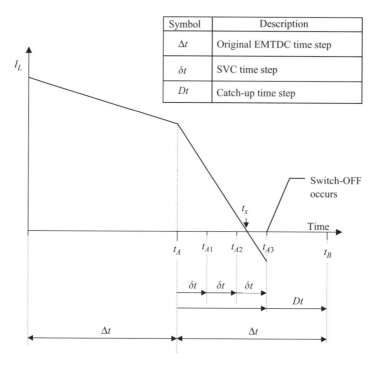

Figure 9.22 Thyristor switch-OFF with variable time step

the currents in the ungrounded star-connected secondary, two capacitor voltages in each of the two delta-connected TSCs (four variables) and the capacitor voltage on each of the back-to-back thyristor snubbers ($4 \times 3 = 12$ state variables).

The system matrix must be reformed whenever a thyristor switches. Accurate determination of the switching instants is obtained by employing an integration step length which is a submultiple of that employed in the EMTDC main loop. The detection of switchings proceeds as in Figure 9.22. Initially the step length is the same as that employed in EMTDC. Upon satisfying an inequality that indicates that a switching has occurred, the SVC model steps back a time step and integrates with a smaller time step, until the inequality is satisfied again. At this point the switching is bracketed by a smaller interval, and the system matrix for the SVC is reformed with the new topology. A catch-up step is then taken to resynchronise the SVC model with EMTDC, and the step length is increased back to the original.

The interface between the EMTDC and SVC models is by Norton and Thevenin equivalents as shown in Figure 9.23. The EMTDC network sees the SVC as a current source in parallel with a linearising resistance R_c. The linearising resistance is necessary, since the SVC current injection is calculated by the model on the basis of the terminal voltage at the previous time step. R_c is then an approximation to how the SVC current injection will vary as a function of the terminal voltage value to be calculated at the current time step. The total current flowing in this resistance may be

Power electronic systems

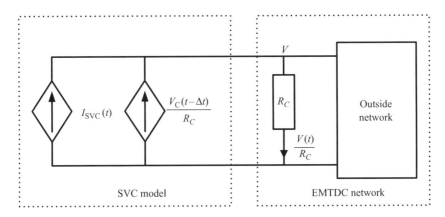

Figure 9.23 Interfacing between the SVC model and the EMTDC program

large, and unrelated to the absolute value of current flowing into the SVC. A correction offset current is therefore added to the SVC Norton current source to compensate for the current flowing in the linearising resistor. This current is calculated using the terminal voltage from the previous time step. The overall effect is that R_c acts as a linearising incremental resistance. Because of this Norton source compensation for R_c, its value need not be particularly accurate, and the transformer zero sequence leakage reactance is used.

The EMTDC system is represented in the SVC model by a time-dependent source, for example the phase A voltage is calculated as

$$V'_a = V_a + \omega \Delta t \, (V_c - V_b) \frac{(1 - (\omega \Delta t)^2)}{\sqrt{3}} \qquad (9.11)$$

which has the effect of reducing errors due to the one time-step delay between the SVC model and EMTDC.

The firing control of the SVC model is very similar to that implemented in the HVDC six-pulse bridge model. A firing occurs when the elapsed angle derived from a PLO ramp is equal to the instantaneous firing-angle order obtained from the external controller model. The phase locked oscillator is of the phase-vector type illustrated in Figure 9.15. The three-phase to two-phase dq transformation is defined by

$$V_\alpha = \left(\frac{2}{3}\right) V_a - \left(\frac{1}{3}\right) V_b - \left(\frac{1}{3}\right) V_c \qquad (9.12)$$

$$V_\beta = \left(\frac{1}{\sqrt{3}}\right) (V_b - V_c) \qquad (9.13)$$

The SVC controller is implemented using general-purpose control components, an example being that of Figure 9.24. This controller is based on that installed at Chateauguay [11]. The signals I_a, I_b, I_c and V_a, V_b, V_c are instantaneous current and voltage at the SVC terminals. These are processed to yield the reactive power

240 Power systems electromagnetic transients simulation

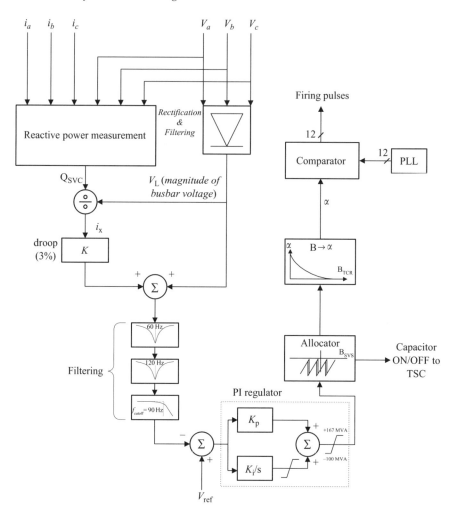

Figure 9.24 SVC controls

generation of the SVC and the terminal voltage measurement, from which a reactive current measurement is obtained. The SVC current is used to calculate a current-dependent voltage droop, which is added to the measured voltage. The measured voltage with droop is then filtered and subtracted from the voltage reference to yield a voltage error, which is acted upon by a PI controller. The PI controller output is a reactive power order for the SVC, which is split into a component from the TSC banks by means of an allocator, and a vernier component from the TCR (BTCR). A non-linear reference is used to convert the BTCR reactive power demand into a firing order for the TCR firing controller. A hysteresis TSC bank overlap of ten per cent is included in the SVC specification.

The use of the SVC model described above is illustrated in Figure 11.11 (Chapter 11) to provide voltage compensation for an arc furnace. A more accurate but laborious approach is to build up a model of the SVC using individual components (i.e. thyristors, transformers, . . . etc).

9.7.2 The static compensator (STATCOM)

The STATCOM is a power electronic controller constructed from voltage sourced converters (VSCs) [12]. Unlike the thyristors, the solid state switches used by VSCs can force current off against forward voltage through the application of a negative gate pulse. Insulated gate insulated junction transistors (IGBTs) and gate turn-off thyristors (GTOs) are two switching devices currently applied for this purpose.

The EMTDC Master Library contains interpolated firing pulse components that generate as output the two-dimensional firing-pulse array for the switching of solid-state devices. These components return the firing pulse and the interpolation time required for the ON and OFF switchings. Thus the output signal is a two-element real array, its first element being the firing pulse and the second is the time between the current computing instant and the firing pulse transition for interpolated turn-on of the switching devices.

The basic STATCOM configuration, shown in Figure 9.25, is a two-level, six-pulse VSC under pulse width modulation (PWM) control. PWM causes the valves to switch at high frequency (e.g. 2000 Hz or higher). A phase locked oscillator (PLL) plays a key role in synchronising the valve switchings to the a.c. system voltage. The two PLL functions are:

(i) The use of a single 0–360 ramp locked to phase A at fundamental frequency that produces a triangular carrier signal, as shown in Figure 9.26, whose amplitude is fixed between −1 and +1. By making the PWM frequency divisible by three, it can be applied to each IGBT valve in the two-level converter.

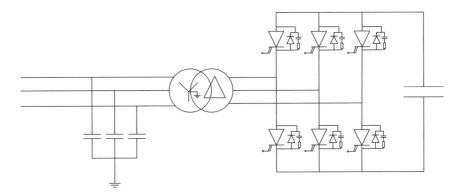

Figure 9.25   Basic STATCOM circuit

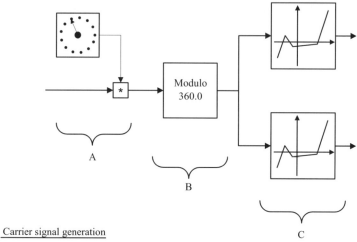

Carrier signal generation

➢ A – Increases PLL ramp slope to that required by carrier frequency

➢ B – Restrains ramps to between 0 and 360 degrees at carrier frequency

➢ C – Converts carrier ramps to carrier signals

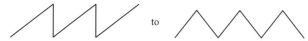

Figure 9.26 Basic STATCOM controller

(ii) The 0–360 ramp signals generated by the six-pulse PLL are applied to generate sine curves at the designated fundamental frequency. With reference to Figure 9.27, the two degrees of freedom for direct control are achieved by
– phase-shifting the ramp signals which in turn phase-shift the sine curves (signal shift), and
– varying the magnitude of the sine curves (signal M_a).

It is the control of signals Shift and M_a that define the performance of a voltage source converter connected to an active a.c. system.

The PWM technique requires mixing the carrier signal with the fundamental frequency signal defining the a.c. waveshape. PSCAD/EMTDC models both switch on and switch off pulses with interpolated firing to achieve the exact switching instants between calculation steps, thus avoiding the use of very small time steps. The PWM carrier signal is compared with the sine wave signals and generates the turn-on and turn-off pulses for the switching interpolation.

The STATCOM model described above is used in Chapter 11 to compensate the unbalance and distortion caused by an electric arc furnace; the resulting waveforms for the uncompensated and compensated cases are shown in Figures 11.10 and 11.12 respectively.

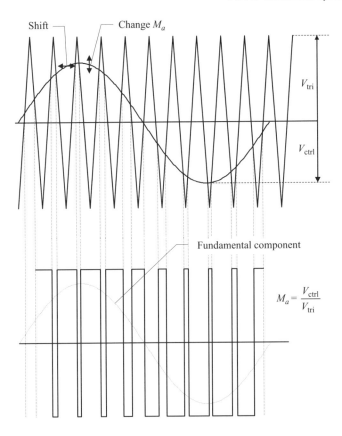

Figure 9.27   Pulse width modulation

9.8   State variable models

The behaviour of power electronic devices is clearly dominated by frequent unspecifiable switching discontinuities with intervals in the millisecond region. As their occurrence does not coincide with the discrete time intervals used by the efficient fixed-step trapezoidal technique, the latter is being 'continuously' disrupted and therefore rendered less effective. Thus the use of a unified model of a large power system with multiple power electronic devices and accurate detection of each discontinuity is impractical.

As explained in Chapter 3, state space modelling, with the system solved as a set of non-linear differential equations, can be used as an alternative to the individual component discretisation of the EMTP method. This alternative permits the use of variable step length integration, capable of locating the exact instants of switching and altering dynamically the time step to fit in with those instants. All firing control system variables are calculated at these instants together with the power circuit variables. The solution of the system is iterated at every time step, until convergence is reached with an acceptable tolerance.

Although the state space formulation can handle any topology, the automatic generation of the system matrices and state equations is a complex and time-consuming process, which needs to be done every time a switching occurs. Thus the sole use of the state variable method for a large power system is not a practical proposition. Chapter 3 has described TCS [13], a state variable program specially developed for power electronic systems. This program has provision to include all the non-linearities of a converter station (such as transformer magnetisation) and generate automatically the comprehensive connection matrices and state space equations of the multicomponent system, to produce a continuous state space subsystem. The state variable based power electronics subsystems can then be combined with the electromagnetic transients program to provide the hybrid solution discussed in the following section. Others have also followed this approach [14].

9.8.1 EMTDC/TCS interface implementation

The system has to be subdivided to represent the components requiring the use of the state variable formulation [15]. The key to a successful interface is the exclusive use of 'stable' information from each side of the subdivided system, e.g. the voltage across a capacitor and the current through an inductor [16]. Conventional HVDC converters are ideally suited for interfacing as they possess a stable commutating busbar voltage (a function of the a.c. filter capacitors) and a smooth current injection (a function of the smoothing reactor current).

A single-phase example is used next to illustrate the interface technique, which can easily be extended to a three-phase case.

The system shown in Figure 9.28 is broken into two subsystems at node M. The stable quantities in this case are the inductor current for system S_1 and the capacitor voltage for system S_2. An interface is achieved through the following relationships

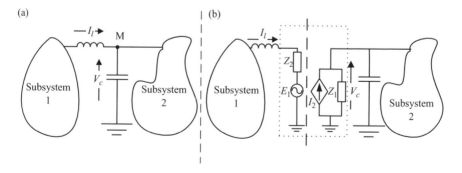

Figure 9.28 Division of a network: (a) network to be divided; (b) divided system

for the Thevenin and Norton source equivalents E_1 and I_2, respectively.

$$I_2(t) = I_l(t - \Delta t) + \frac{V_C(t - \Delta t)}{Z_1} \tag{9.14}$$

$$E_1(t) = V_C(t - \Delta t) - I_l(t - \Delta t) Z_2 \tag{9.15}$$

In equation 9.14 the value of Z_1 is the equivalent Norton resistance of the system looking from the interface point through the reactor and beyond. Similarly, the value of Z_2 in equation 9.15 is the equivalent Thevenin resistance from the interface point looking in the other direction. The interface impedances can be derived by disabling all external voltage and current sources in the system and applying a pulse of current to each reduced system at the interface point. The calculated injection node voltage, in the same time step as the current injection occurs, divided by the magnitude of the input current will yield the equivalent impedance to be used for interfacing with the next subsystem.

With reference to the d.c. converter system shown in Figure 9.29, the tearing is done at the converter busbar as shown in Figure 9.30 for the hybrid representation.

The interface between subdivided systems, as in the EMTDC solution, uses Thevenin and Norton equivalent sources. If the d.c. link is represented as a continuous state variable based system, like in the case of a back-to-back HVDC interconnection, only a three-phase two-port interface is required. A point to point interconnection can also be modelled as a continuous system if the line is represented by lumped parameters. Alternatively, the d.c. line can be represented by a distributed parameter model, in which case an extra single-phase interface is required on the d.c. side.

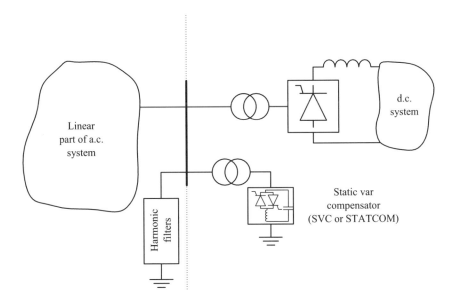

Figure 9.29 *The converter system to be divided*

246 *Power systems electromagnetic transients simulation*

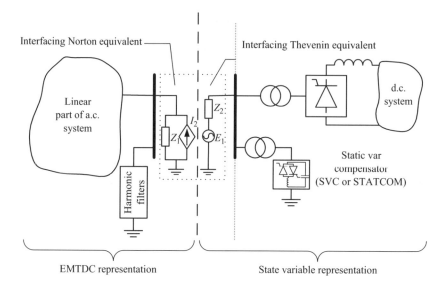

Figure 9.30   The divided HVDC system

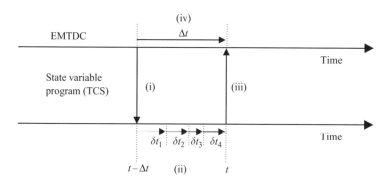

Figure 9.31   Timing synchronisation

The main EMTDC program controls the timing synchronisation, snapshot handling and operation of the state variable subprogram. The exchange of information between them takes place at the fixed time steps of the main program.

A Thevenin source equivalent is derived from the busbar voltages, and upon completion of a Δt step by the state variable subprogram, the resulting phase current is used as a Norton current injection at the converter busbar. Figure 9.31 illustrates the four steps involved in the interfacing process with reference to the case of Figure 9.30.

Step (i) :   The main program calls the state variable subprogram using the interface busbar voltages (and the converter firing angle orders, if the control system is represented in EMTDC, as mentioned in the following section, Figure 9.32) as inputs.

Power electronic systems

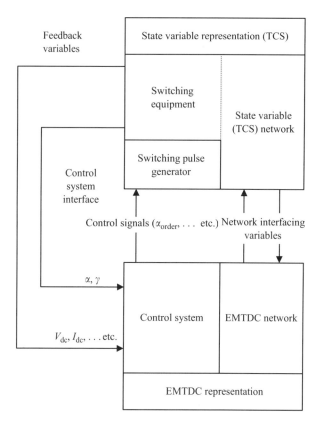

Figure 9.32   Control systems in EMTDC

Step (ii) : The state variable program is run with the new input voltages using variable time steps with an upper limit of Δt. The intermediate states of the interfacing three-phase source voltages are derived by the following phase-advancing technique:

$$V'_a = V_a \cos(\Delta t) + \frac{V_c - V_b}{\sqrt{3}} \sin(\Delta t) \qquad (9.16)$$

where, V_a, V_b, V_c are the phase voltages known at time t, and Δt is the required phase advance.

Step (iii) : At the end of each complete Δt run of step (ii) the interfacing Thevenin source currents are used to derive the Norton current sources to be injected into the system at the interface points.

Step (iv) : The rest of the system solution is obtained for a Δt interval, using these current injections.

A Δt value of 50 μs normally leads to stable solutions. The state variable multiple time steps vary from a fraction of a degree to the full Δt time, depending on the state of the system. As the system approaches steady state the number of intermediate steps is progressively reduced.

9.8.2 Control system representation

This section discusses the simulation of the control system specifically related to the non-linear components of the state variable (TCS) subsystem down to the level where the control order signals are derived (i.e. the firing signals to the converter and/or other non-linear components).

The converter controls can be modelled as part of the state variable program or included within the main (EMTDC) program. In each case the switching pulse generator includes the generation of signals required to trigger the switching (valve) elements and the EMTDC block represents the linear power network including the distributed transmission line models. When the control system is part of the TCS solution, the control system blocks are solved iteratively at every step of the state variable solution until convergence is reached. All the feedback variables are immediately available for further processing of the control system within the TCS program.

Instead, the control system can be represented within the EMTDC program, as shown in Figure 9.32. In this case the function library of EMTDC becomes available, allowing any generic or non-conventional control system to be built with the help of FORTRAN program statements. In this case the main program must be provided with all the feedback variables required to define the states of the switching equipment (e.g. the converter firing and extinction angles, d.c. voltage and current, commutation failure indicators, etc.). The control system is solved at every step of the main program sequentially; this is perfectly acceptable, as the inherent inaccuracy of the sequential function approach is rendered insignificant by the small calculation step needed to simulate the electric network and the usual delays and lags in power system controls [15].

9.9 Summary

The distinguishing feature of power electronic systems from other plant components is their frequent switching requirement. Accordingly, ways of accommodating frequent switching without greatly affecting the efficiency of the EMTP method have been discussed. The main issue in this respect is the use of interpolation techniques for the accurate placement of switching instants and subsequent resynchronisation with normal time grid.

Detailed consideration has also been given to the elimination of numerical oscillations, or chatter, that results from errors associated with the trapezoidal rule.

The EMTDC program, initially designed for HVDC systems, is well suited to the modelling of power electronic systems and has, therefore, been used as the main source of information. Thus the special characteristics of HVDC and FACTS devices have been described and typical systems simulated in PSCAD/EMTDC.

State variable analysis is better than numerical integrator substitution (NIS) for the modelling of power electronic equipment, but is inefficient to model the complete system. This has led to the development of hybrid programs that combine the two methods into one program. However, considerable advances have been made in NIS programs to handle frequent switching efficiently and thus the complex hybrid methods are less likely to be widely used.

9.10 References

1. DOMMEL, H. W.: 'Digital computer solution of electromagnetic transients in single- and multiphase networks', *IEEE Transactions on Power Apparatus and Systems*, 1969, **88** (2), pp. 734–41
2. TINNEY, W. F. and WALKER, J. W.: 'Direct solutions of sparse network equations by optimally ordered triangular factorization', *Proccedings of IEEE*, 1967, **55**, pp. 1801–9
3. GOLE, A. M. and SOOD, V. K.: 'A static compensator model for use with electromagnetic transients simulation programs', *IEEE Transactions on Power Delivery*, 1990, **5** (3), pp. 1398–1407
4. IRWIN, G. D., WOODFORD, D. A. and GOLE, A.: 'Precision simulation of PWM controllers', Proceedings of International Conference on *Power System Transients (IPST'2001)*, June 2001, pp. 161–5
5. LIN, J. and MARTI, J. R.: 'Implementation of the CDA procedure in EMTP', *IEEE Transactions on Power Systems*, 1990, **5** (2), pp. 394–402
6. MARTI, J. R. and LIN, J.: 'Suppression of numerical oscillations in the EMTP', *IEEE Transactions on Power Systems*, 1989, **4** (2), pp. 739–47
7. KRUGER, K. H. and LASSETER, R. H.: 'HVDC simulation using NETOMAC', Proceedings, IEEE Montec '86 Conference on *HVDC Power Transmission*, Sept/Oct 1986, pp. 47–50
8. KULICKE, B.: 'NETOMAC digital program for simulating electromechanical and electromagnetic transient phenomena in AC power systems', *Elektrizitätswirtschaft*, **1**, 1979, pp. 18–23
9. ARAUJO, A. E. A., DOMMEL, H. W. and MARTI, J. R.: 'Converter simulations with the EMTP: simultaneous solution and backtracking technique', IEEE/NTUA Athens Power Tech Conference: *Planning, Operation and Control of Today's Electric Power Systems*, Sept. 5–8, 1993, **2**, pp. 941–5
10. SZECHTMAN, M., WESS, T. and THIO, C. V.: 'First benchmark model for HVdc control studies', *ELECTRA*, 1991, **135**, pp. 55–75
11. HAMMAD, A. E.: 'Analysis of second harmonic instability for the Chateauguay HVdc/SVC scheme', *IEEE Transaction on Power Delivery*, 1992, **7** (1), pp. 410–15
12. WOODFORD, D. A.: 'Introduction to PSCAD/EMTDC V3', Manitoba HVdc Research Centre, Canada
13. ARRILLAGA, J., AL-KASHALI, H. J. and CAMPOS-BARROS, J. G.: 'General formulation for dynamic studies in power systems including static converters', *Proceedings of IEE*, 1977, **124** (11), pp. 1047–52
14. DAS, B. and GHOSH, A.: 'Generalised bridge converter model for electromagnetic transient analysis', *IEE Proc.-Gener. Transm. Distrib.*, 1998, **145** (4), pp. 423–9

15 ZAVAHIR, J. M., ARRILLAGA, J. and WATSON, N. R.: 'Hybrid electromagnetic transient simulation with the state variable representation of HVdc converter plant', *IEEE Transactions on Power Delivery*, 1993, **8** (3), pp. 1591–8
16 WOODFORD, D. A.: 'EMTDC users' manual', Manitoba HVdc Research Centre, Canada

Chapter 10

Frequency dependent network equivalents

10.1 Introduction

A detailed representation of the complete power system is not a practical proposition in terms of computation requirements. In general only a relatively small part of the system needs to be modelled in detail, with the rest of the system represented by an appropriate equivalent. However, the use of an equivalent circuit based on the fundamental frequency short-circuit level is inadequate for transient simulation, due to the presence of other frequency components.

The development of an effective frequency-dependent model is based on the relationship that exists between the time and frequency domains. In the time domain the system impulse response is convolved with the input excitation. In the frequency domain the convolution becomes a multiplication; if the frequency response is represented correctly, the time domain solution will be accurate.

An effective equivalent must represent the external network behaviour over a range of frequencies. The required frequency range depends on the phenomena under investigation, and, hence, the likely frequencies involved.

The use of frequency dependent network equivalents (FDNE) dates back to the late 1960s [1]–[3]. In these early models the external system was represented by an appropriate network of R, L, C components, their values chosen to ensure that the equivalent network had the same frequency response as the external system. These schemes can be implemented in existing transient programs with minimum change, but restrict the frequency response that can be represented. A more general equivalent, based on rational functions (in the s or z domains) is currently the preferred approach. The development of an FDNE involves the following processing stages:

- Derivation of the system response (either impedance or admittance) to be modelled by the equivalent.
- Fitting of model parameters (identification process).
- Implementation of the FDNE in the transient simulation program.

The FDNE cannot model non-linearities, therefore any component exhibiting significant non-linear behaviour must be removed from the processing. This will increase the number of ports in the equivalent, as every non-linear component will be connected to a new port.

Although the emphasis of this chapter is on frequency dependent network equivalents, the same identification techniques are applicable to the models of individual components. For example a frequency-dependent transmission line (or cable) equivalent can be obtained by fitting an appropriate model to the frequency response of its characteristic admittance and propagation constant (see section 6.3.1).

10.2 Position of FDNE

The main factors influencing the decision of how far back from the disturbance the equivalent should be placed are:

- the points in the system where the information is required
- the accuracy of the synthesised FDNE
- the accuracy of the frequency response of the model components in the transient simulation
- the power system topology
- the source of the disturbance

If approximations are made based on the assumption of a remote FDNE location, this will have to be several busbars away and include accurate models of the intervening components. In this respect, the better the FDNE the closer it can be to the source of the disturbance. The location of the FDNE will also depend on the characteristics of the transient simulation program.

The power system has two regions; the first is the area that must be modelled in detail, i.e. immediately surrounding the location of the source of the disturbance and areas of particular interest; the second is the region replaced by the FDNE.

10.3 Extent of system to be reduced

Ideally, the complete system should be included in the frequency scan of the reduction process, but this is not practical. The problem then is how to assess whether a sufficient system representation has been included. This requires judging how close the response of the system entered matches that of the complete system.

One possible way to decide is to perform a sensitivity study of the effect of adding more components on the frequency response and stop when the change they produce is sufficiently small. The effect of small loads fed via transmission lines can also be significant, as their combined harmonic impedances (i.e. line and load) can be small due to standing wave effects.

10.4 Frequency range

The range of the frequency scan and the FDNE synthesis will depend on the problem being studied. In all cases, however, the frequency scan range should extend beyond the maximum frequency of the phenomena under investigation. Moreover, the first resonance above the maximum frequency being considered should also be included in the scan range, because it will affect the frequency response in the upper part of the required frequency range.

Another important factor is the selection of the interval between frequency points, to ensure that all the peaks and troughs are accurately determined. Moreover this will impact on the number and position of the frequency points used for the calculation of the LSE (least square error) if optimisation techniques are applied. The system response at intermediate points can be found by interpolation; this is computationally more efficient than the direct determination of the response using smaller intervals. An interval of 5 Hz in conjunction with cubic spline interpolation yields practically the same system response derived at 1 Hz intervals, which is perfectly adequate for most applications. However cubic spline interpolation needs to be applied to both the real and imaginary parts of the system response.

10.5 System frequency response

The starting point in the development of the FDNE is the derivation of the external system driving point and transfer impedance (or admittance) matrices at the boundary busbar(s), over the frequency range of interest.

Whenever available, experimental data can be used for this purpose, but this is rarely the case, which leaves only time or frequency domain identification techniques. When using frequency domain identification, the required data to identify the model parameters can be obtained either from time or frequency domain simulation, as illustrated in Figure 10.1.

10.5.1 Frequency domain identification

The admittance or impedance seen from a terminal busbar can be calculated from current or voltage injections, as shown in Figures 10.2 and 10.3 respectively. The injections can be performed in the time domain, with multi-sine excitation, or in the frequency domain, where each frequency is considered independently. The frequency domain programs can generate any required frequency-dependent admittance as seen from the terminal busbars.

Because the admittance (and impedance) matrices are symmetrical, there are only six different responses to be fitted and these can be determined from three injection tests.

When using voltage injections the voltage source and series impedance need to be made sufficiently large so that the impedance does not adversely affect the main circuit. If made too small, the conductance term is large and may numerically swamp

254 *Power systems electromagnetic transients simulation*

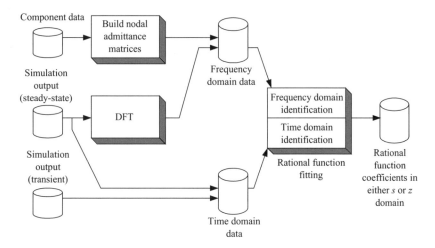

Figure 10.1 Curve-fitting options

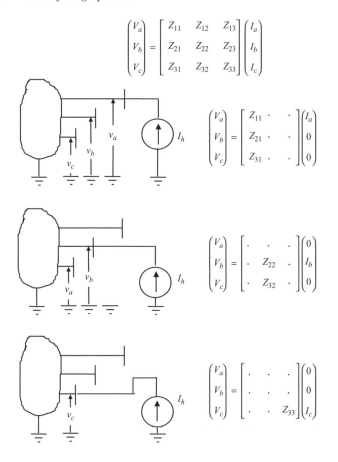

Figure 10.2 Current injection

Frequency dependent network equivalents

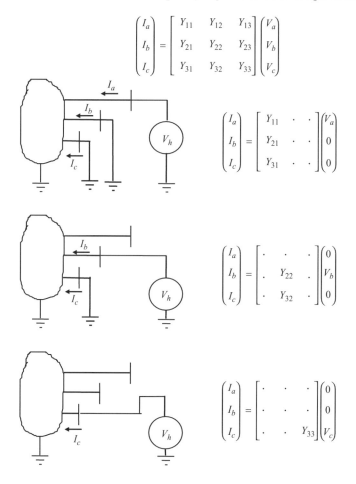

Figure 10.3   Voltage injection

out some of the circuit parameters that need to be identified. The use of current injections, shown in Figure 10.2, is simpler in this respect.

10.5.1.1 Time domain analysis

Figure 10.4 displays a schematic of a system drawn in DRAFT (PSCAD/EMTDC), where a multi-sine current injection is applied. In this case a range of sine waves is injected from 5 Hz up to 2500 Hz with 5 Hz spacing; all the magnitudes are 1.0 and the angles 0.0, hence the voltage is essentially the impedance. As the lowest frequency injected is 5 Hz all the sine waves add constructively every 0.2 seconds, resulting in a large peak. After the steady state is achieved, one 0.2 sec period is extracted from the time domain waveforms, as shown in Figure 10.5, and a DFT performed to obtain the required frequency response. This frequency response is shown in Figure 10.6. As has been shown in Figure 10.2 the current injection gives the impedances for the

256 *Power systems electromagnetic transients simulation*

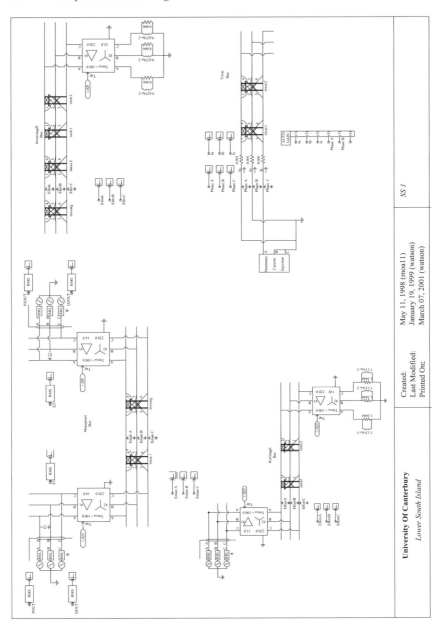

Figure 10.4   PSCAD/EMTDC schematic with current injection

submatrices. In the cases of a single port this is simply inverted; however in the more general multiport case the impedance matrix must be built and then a matrix inversion performed.

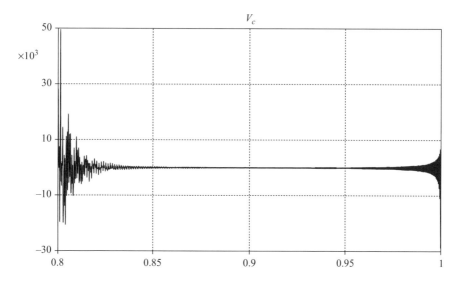

Figure 10.5 Voltage waveform from time domain simulation

10.5.1.2 Frequency domain analysis

Figure 10.7 depicts the process of generating the frequency response of an external network as seen from its ports. A complete nodal admittance matrix of the network to be equivalenced is formed with the connection ports ordered last, i.e.

$$[Y_f]\mathbf{V}_f = \mathbf{I}_f \tag{10.1}$$

where

[Y_f] is the admittance matrix at frequency f
$\mathbf{V}_f$ is the vector of nodal voltages at frequency f
$\mathbf{I}_f$ is the vector of nodal currents at frequency f.

The nodal admittance matrix is of the form:

$$[Y_f] = \begin{bmatrix} y_{11} & y_{12} & \cdots & y_{1i} & \cdots & y_{1k} & \cdots & y_{1N} \\ y_{21} & y_{22} & \cdots & y_{2i} & \cdots & y_{2k} & \cdots & y_{2N} \\ \vdots & \vdots & \ddots & \vdots & \ddots & \vdots & \ddots & \vdots \\ y_{i1} & y_{i2} & \cdots & y_{ii} & \cdots & y_{ik} & \cdots & y_{iN} \\ \vdots & \vdots & \ddots & \vdots & \ddots & \vdots & \ddots & \vdots \\ y_{k1} & y_{k2} & \cdots & y_{ki} & \cdots & y_{kk} & \cdots & y_{kN} \\ \vdots & \vdots & \ddots & \vdots & \ddots & \vdots & \ddots & \vdots \\ y_{N1} & y_{N2} & \cdots & y_{Ni} & \cdots & y_{Nk} & \cdots & y_{NN} \end{bmatrix} \tag{10.2}$$

where

y_{ki} is the mutual admittance between busbars k and i
y_{ii} is the self-admittance of busbar i.

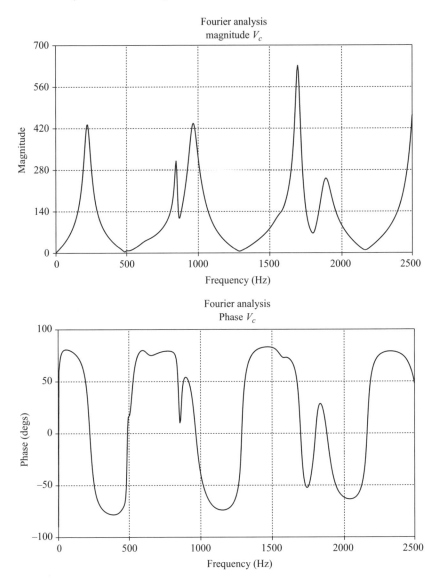

Figure 10.6 *Typical frequency response of a system*

Note that each element in the above matrix is a 3 × 3 matrix due to the three-phase nature of the power system, i.e.

$$y_{ki} = \begin{bmatrix} y_{aa} & y_{ab} & y_{ac} \\ y_{ba} & y_{bb} & y_{bc} \\ y_{ca} & y_{cb} & y_{cc} \end{bmatrix} \quad (10.3)$$

Frequency dependent network equivalents 259

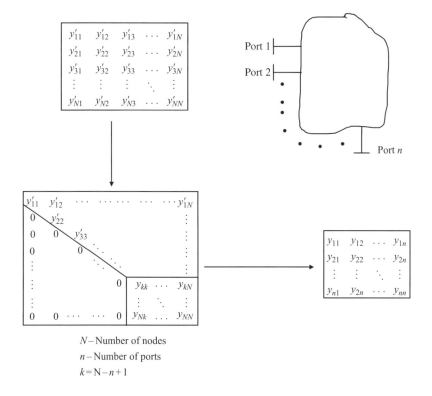

Figure 10.7 Reduction of admittance matrices

Gaussian elimination is performed on the matrix shown in 10.2, up to, but not including the connection ports i.e.

$$\begin{bmatrix} y'_{11} & y'_{12} & \cdots & & y'_{1k} & \cdots & y'_{1N} \\ 0 & y'_{22} & y'_{23} & \cdots & y'_{2k} & & y'_{2N} \\ 0 & 0 & y'_{33} & y'_{34} & y'_{3k} & & y'_{3N} \\ \vdots & \vdots & \ddots & \ddots & \vdots & & \vdots \\ \hline 0 & 0 & \cdots & 0 & y''_{kk} & \cdots & y''_{kN} \\ \vdots & \vdots & \cdots & 0 & \vdots & \ddots & \vdots \\ 0 & 0 & \cdots & 0 & y''_{Nk} & \cdots & y''_{NN} \end{bmatrix} \qquad (10.4)$$

The matrix equation based on the admittance matrix 10.4 is of the form:

$$\begin{bmatrix} [y_A] & [y_B] \\ 0 & [y_D] \end{bmatrix} \begin{bmatrix} V_{\text{internal}} \\ V_{\text{terminal}} \end{bmatrix} = \begin{bmatrix} 0 \\ I_{\text{terminal}} \end{bmatrix} \qquad (10.5)$$

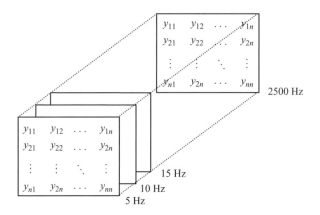

Figure 10.8   Multifrequency admittance matrix

The submatrix $[y_D]$ represents the network as seen from the terminal busbars. If there are n terminal busbars then renumbering to include only the terminal busbars gives:

$$\begin{bmatrix} y_{11} & \cdots & y_{1n} \\ \vdots & \ddots & \vdots \\ y_{n1} & \cdots & y_{nn} \end{bmatrix} \begin{pmatrix} V_1 \\ \vdots \\ V_n \end{pmatrix} = \begin{pmatrix} I_1 \\ \vdots \\ I_n \end{pmatrix} \quad (10.6)$$

This is performed for all the frequencies of interest, giving a set of submatrices as depicted in Figure 10.8.

The frequency response is then obtained by selecting the same element from each of the submatrices. The mutual terms are the negative of the off-diagonal terms of these reduced admittance matrices. The self-terms are the sum of all terms of a row (or column as the admittance matrix is symmetrical), i.e.

$$y_{\text{self } k} = \sum_{i=1}^{n} y_{ki} \quad (10.7)$$

The frequency response of the self and mutual elements, depicted in Figure 10.9, are matched and a FDNE such as in Figure 10.10 implemented. This is an admittance representation which is the most straightforward. An impedance based FDNE is achieved by inverting the submatrix of the reduced admittance matrices and matching each of the elements as functions of frequency. This implementation, shown in Figure 10.11 for three ports, is suitable for a state variable analysis, as an iterative procedure at each time point is required. Its advantages are that it is more intuitive, can overcome the topology restrictions of some programs and often results in more stable models. The frequency response is then fitted with a rational function or *RLC* network.

Transient analysis can also be performed on the system to obtain the FDNE by first using the steady-state time domain signals and then applying the discrete Fourier transform.

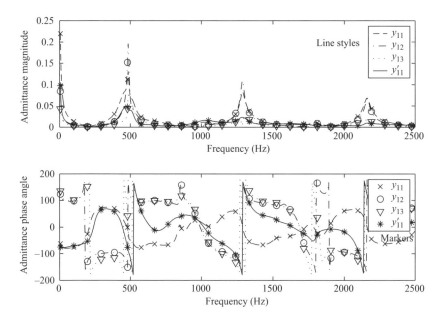

Figure 10.9 Frequency response

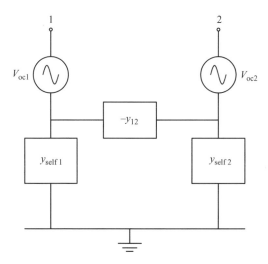

Figure 10.10 Two-port frequency dependent network equivalent (admittance implementation)

The advantage of forming the system nodal admittance matrix at each frequency is the simplicity by which the arbitrary frequency response of any given power system component can be represented. The transmission line is considered as the most frequency-dependent component and its dependence can be evaluated to great

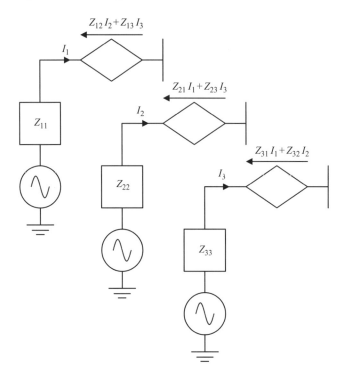

Figure 10.11 Three-phase frequency dependent network equivalent (impedance implementation)

accuracy. Other power system components are not modelled to the same accuracy at present due to lack of detailed data.

10.5.2 Time domain identification

Model identification can also be performed directly from time domain data. However, in order to identify the admittance or impedance at a particular frequency there must be a source of that frequency component. This source may be a steady-state type as in a multi-sine injection [4], or transient such as the ring down that occurs after a disturbance. Prony analysis (described in Appendix B) is the identification technique used for the ring down alternative.

10.6 Fitting of model parameters

10.6.1 RLC networks

The main reason for realising an *RLC* network is the simplicity of its implementation in existing transient analysis programs without requiring extensive modifications.

Frequency dependent network equivalents 263

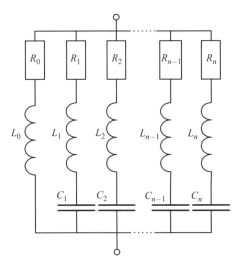

Figure 10.12 Ladder circuit of Hingorani and Burbery

The *RLC* network topology, however, influences the equations used for the fitting as well as the accuracy that can be achieved. The parallel form (Foster circuit) [1] represents reasonably well the transmission network response but cannot model an arbitrary frequency response. Although the synthesis of this circuit is direct, the method first ignores the losses to determine the *L* and *C* values for the required resonant frequencies and then determines the *R* values to match the response at minimum points. In practice an iterative optimisation procedure is necessary after this, to improve the fit [5]–[7].

Almost all proposed *RLC* equivalent networks are variations of the ladder circuit proposed by Hingorani and Burbery [1], as shown in Figure 10.12. Figure 10.13 shows the equivalent used by Morched and Brandwajn [6], which is the same except for the addition of an extra branch (C_∞ and R_∞) to shape the response at high frequencies. Do and Gavrilovic [8] used a series combination of parallel branches, which although looks different, is the dual of the ladder network.

The use of a limited number of *RLC* branches gives good matches at the selected frequencies, but their response at other frequencies is less accurate. For a fixed number of branches, the errors increase with a larger frequency range. Therefore the accuracy of an FDE can always be improved by increasing the number of branches, though at the cost of greater complexity.

The equivalent of multiphase circuits, with mutual coupling between the phases, requires the fitting of admittance matrices instead of scalar admittances.

10.6.2 *Rational function*

An alternative approach to *RLC* network fitting is to fit a rational function to a response and implement the rational function directly in the transient program. The fitting can

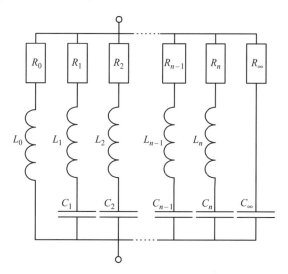

Figure 10.13   Ladder circuit of Morched and Brandwajn

be performed either in the *s*-domain

$$H(s) = e^{-s\tau} \frac{a_0 + a_1 \cdot s + a_2 \cdot s^2 + \cdots + a_N s^N}{1 + b_1 \cdot s + b_2 s^2 + \cdots + b_n \cdot s^n} \quad (10.8)$$

or in the *z*-domain

$$H(z) = e^{-l\Delta t} \frac{a_0 + a_1 z + a_2 z^2 + \cdots + a_n z^{-n}}{1 + b_1 z + b_2 z^2 + \cdots + b_n \cdot z^{-n}} \quad (10.9)$$

where $e^{-s\tau}$ or $e^{-l\Delta t}$ represent the transmission delay associated with the mutual coupling terms.

The *s*-domain has the advantage that the fitted parameters are independent of the time step; there is however a hidden error in its implementation. Moreover the fitting should be performed up to the Nyquist frequency for the smallest time step that is ever likely to be used. This results in poles being present at frequencies higher than the Nyquist frequency for normal simulation step size, which have no influence on the simulation results but add complexity.

The *z*-domain fitting gives Norton equivalents of simpler implementation and without introducing error. The fitting is performed only on frequencies up to the Nyquist frequency and, hence, all the poles are in the frequency range of interest. However the parameters are functions of the time step and hence the fitting must be performed again if the time step is altered.

Frequency dependent network equivalents 265

The two main classes of methods are:

1. Non-linear optimisation (e.g. vector-fitting and the Levenberg–Marquardt method), which are iterative methods.
2. Linearised least squares or weighted least squares (WLS). These are direct fast methods based on SVD or the normal equation approach for solving an over-determined linear system. To determine the coefficients the following equation is solved:

$$\begin{bmatrix} d_{11} & d_{12} & \cdots & d_{1,2m+1} \\ d_{21} & d_{22} & \cdots & d_{1,2m+1} \\ \vdots & \vdots & \ddots & \vdots \\ d_{k1} & d_{k2} & \cdots & d_{k,2m+1} \end{bmatrix} \cdot \begin{pmatrix} b_1 \\ b_2 \\ \vdots \\ b_m \\ a_0 \\ a_1 \\ \vdots \\ a_m \end{pmatrix} = \begin{pmatrix} -c(j\omega_1) \\ -c(j\omega_2) \\ \vdots \\ -c(j\omega_k) \\ -d(j\omega_1) \\ -d(j\omega_2) \\ \vdots \\ -d(j\omega_k) \end{pmatrix} \quad (10.10)$$

This equation is of the form $[D] \cdot \mathbf{x} = \mathbf{b}$ where
$\mathbf{b}$ is the vector of measurement points ($\mathbf{b}_i = H(j\omega_i) = c(j\omega_i) + jd(j\omega_i)$)
$[D]$ is the design matrix
$\mathbf{x}$ is the vector of coefficients to be determined.

When using the linearised least squares method the fitting can be carried out in the s or z-domain, using the frequency or time domain by simply changing the design matrix used. Details of this process are given in Appendix B and it should be noted that the design matrix represents an over-sampled system.

10.6.2.1 Error and figure of merit

The percentage error is not a useful index, as often the function to be fitted passes through zero. Instead, either the percentage of maximum value or the actual error can be used.

Some of the figures of merit (FOM) that have been used to rate the goodness of fit are:

$$\text{Error}_{\text{RMS}} = \frac{\sqrt{\sum_{i=1}^{n} \left(y_i^{\text{Fitted}} - y_i^{\text{Data}}\right)^2}}{n} \quad (10.11)$$

$$\text{Error}_{\text{Normalised}} = \frac{\sqrt{\sum_{i=1}^{n} \left(y_i^{\text{Fitted}} - y_i^{\text{Data}}\right)^2}}{\sqrt{\sum_{i=1}^{n} \left(y_i^{\text{Data}}\right)^2}} \quad (10.12)$$

$$\text{Error}_{\text{Max}} = MAX \left(y_i^{\text{Fitted}} - y_i^{\text{Data}}\right) \quad (10.13)$$

The fit must be stable for the simulation to be possible; of course the stability of the fit can be easily tested after performing the fit, the difficulty being the incorporation of stability criteria as part of the fitting process. Stability can be achieved by fitting only real poles in the left half plane (in the s-domain) but this greatly restricts the accuracy that can be achieved. Other approaches have been to mirror poles in the right half-plane into the left half-plane to ensure stability, or to remove them on the basis that the corresponding residual is small.

Since the left half s-plane maps to the unit circle in the z-plane, the stability criteria in this case is that the pole magnitude should be less than or equal to one. One way of determining this for both s and z-domains is to find the poles by calculating the roots of the characteristic equation (denominator), and checking that this criterion is met. Another method is to use the Jury table (z-domain) [9] or the s-domain equivalent of Routh–Hurwitz stability criteria [10]. The general rule is that as the order of the rational function is increased the fit is more accurate but less stable. So the task is to find the highest order stable fit.

In three-phase mutually coupled systems the admittance matrix, rather than a scalar admittance, must be fitted as function of frequency. Although the fitting of each element in the matrix may be stable, inaccuracies in the fit can result in the complete system having instabilities at some frequencies. Thus, rather than fitting each element independently, the answer is to ensure that the system of fitted terms is stable.

The least squares fitting process tends to smear the fitting error over the frequency range. Although this gives a good transient response, it results in a small but noticeable steady-state error. The ability to weight the fundamental frequency has also been incorporated in the formulation given in Appendix B. By giving a higher weighting to the fundamental frequency (typically 100) the steady-state error is removed, while the transient response is slightly worse due to higher errors at other frequencies.

10.7 Model implementation

Given a rational function in z, i.e.

$$H(z) = \frac{a_0 + a_1 z^{-1} + a_2 z^{-2} + \cdots + a_m z^{-m}}{1 + b_1 z^{-1} + b_2 z^{-2} + \cdots + b_m z^{-m}} = \frac{I(z)}{V(z)} \quad (10.14)$$

multiplying both sides by the denominators and rearranging gives:

$$I(z) = a_0 V(z) + \left(a_1 z^{-1} + a_2 z^{-2} + \cdots + a_m z^{-m} \right) V(z)$$
$$- \left(b_1 z^{-1} + b_2 z^{-2} + \cdots + a_m z^{-m} \right) I(z)$$
$$= G_{\text{equiv}} + I_{\text{History}} \quad (10.15)$$

Transforming back to discrete time:

$$i(n\Delta t) = a_0 v(n\Delta t) + a_1 v(n\Delta t - \Delta t) + a_2 \cdot v(n\Delta t - 2\Delta t)$$
$$+ \cdots + a_m \cdot v(n\Delta t - m\Delta t) - (b_1 i(n\Delta t - \Delta t) + b_2 i(n\Delta t - 2\Delta t)$$
$$+ \cdots + b_m i(n\Delta t - m\Delta t))$$
$$= G_{\text{equiv}} v(n\Delta t) + I_{\text{History}} \quad (10.16)$$

where

$$G_{\text{equiv}} = a_0$$
$$I_{\text{History}} = a_1 \cdot v(n\Delta t - \Delta t) + a_2 \cdot v(n\Delta t - 2\Delta t) + \cdots + a_m \cdot v(n\Delta t - m\Delta t)$$
$$- (b_1 i(n\Delta t - \Delta t) + b_2 \cdot i(n\Delta t - 2\Delta t) + \cdots + b_m \cdot i(n\Delta t - m\Delta t))$$

As mentioned in Chapter 2 this is often referred to as an ARMA (autoregressive moving average) model.

Hence any rational function in the z-domain is easily implemented without error, as it is simply a Norton equivalent with a conductance a_0 and a current source I_{History}, as depicted in Figure 2.3 (Chapter 2).

A rational function in s must be discretised in the same way as is done when solving the main circuit or a control function. Thus, with the help of the root-matching technique and partial fraction expansion, a high order rational function can be split into lower order rational functions (i.e. 1st or 2nd). Each 1st or 2nd term is turned into a Norton equivalent using the root-matching (or some other discretisation) technique and then the Norton current sources are added, as well as the conductances.

10.8 Examples

Figure 10.14 displays the frequency response of the following transfer function [11]:

$$f(s) = \frac{1}{s+5} + \frac{30 + j40}{s - (-100 - j500)} + \frac{30 - j40}{s - (-100 - j500)} + 0.5$$

The numerator and denominator coefficients are given in Table 10.1 while the poles and zeros are shown in Table 10.2. In practice the order of the response is not known and hence various orders are tried to determine the best.

Figure 10.15 shows a comparison of three different fitting methods, i.e. least squares fitting, vector fitting and non-linear optimisation. All gave acceptable fits with vector fitting performing the best followed by least squares fitting. The corresponding errors for the three methods are shown in Figure 10.16. The vector-fitting error is so close to zero that it makes the zero error grid line look thicker, while the dotted least squares fit is just above this.

Obtaining stable fits for 'well behaved' frequency responses is straightforward, whatever the method chosen. However the frequency response of transmission lines

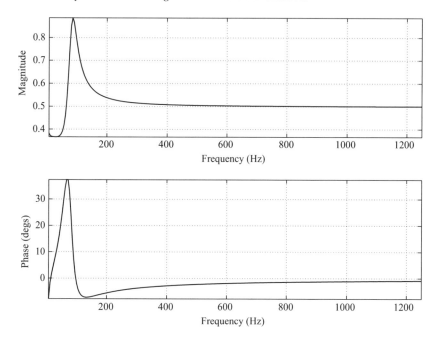

Figure 10.14   Magnitude and phase response of a rational function

Table 10.1   Numerator and denominator coefficients

	Numerator	Denominator
s^0	7.69230769e−001	1.00000000e+000
s^1	7.47692308e−002	2.00769231e−001
s^2	1.26538462e−004	1.57692308e−004
s^3	3.84615385e−007	7.69230769e−007

Table 10.2   Poles and zeros

Zero	Pole
−1.59266199e+002 + 4.07062658e+002 ∗ j	−1.00000000e+002 + 5.00000000e+002 ∗ j
−1.59266199e+002 − 4.07062658e+002 ∗ j	−1.00000000e+002 − 5.00000000e+002 ∗ j
−1.04676019e+001	−5.00000000e+000

Frequency dependent network equivalents 269

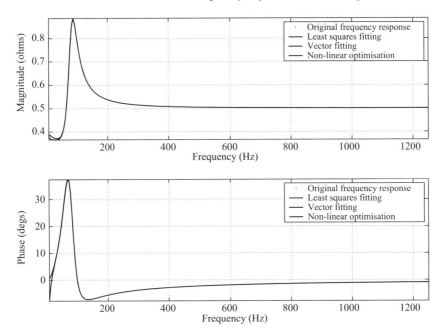

Figure 10.15   Comparison of methods for the fitting of a rational function

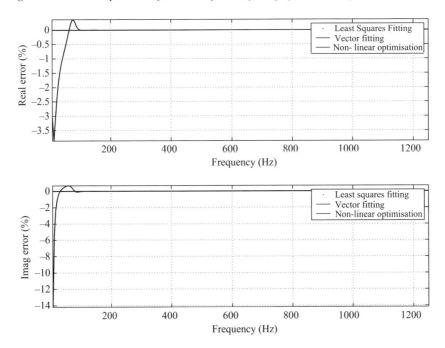

Figure 10.16   Error for various fitted methods

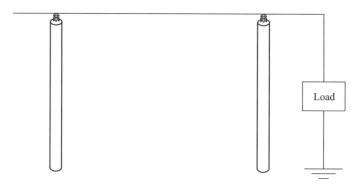

Figure 10.17 Small passive network

Table 10.3 Coefficients of z^{-1} (no weighting factors)

Term	Denominator	Numerator
z^{-0}	1	0.00187981208257
z^{-1}	−5.09271275503264	−0.00942678842550
z^{-2}	12.88062106081476	0.02312960416135
z^{-3}	−21.58018890110835	−0.03674152374824
z^{-4}	26.73613316059277	0.04159398757818
z^{-5}	−25.81247589702268	−0.03448198061263
z^{-6}	19.89428694917709	0.02039138329319
z^{-7}	−12.26856666212080	−0.00756861064417
z^{-8}	5.88983411589258	0.00077750907595
z^{-9}	−2.00963299687702	0.00074985289424
z^{-10}	0.36276901898885	−0.00029244729760

and cables complicates the fitting task, as their related hyperbolic function responses are difficult to fit. This is illustrated with reference to the simple system shown in Figure 10.17, consisting of a transmission line and a resistive load. A z-domain fit is performed with the parameters of Table 10.3 and the fit is shown in Figure 10.18. As is usually the case, the fit is good at higher frequencies but deteriorates at lower frequencies. As an error at the fundamental frequency is undesirable, a weighting factor must be applied to ensure a good fit at this frequency; however this is achieved at the expense of other frequencies. The coefficients obtained using the weighting factor are given in Table 10.4. Finally Figure 10.19 shows the comparison between the full system and FDNE for an energisation transient.

In order to use the same fitted network for an active FDNE, the same transmission line is used with a source impedance of 1 ohm. Figure 10.20 displays the test system,

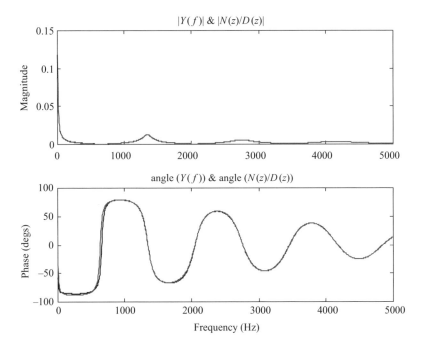

Figure 10.18 Magnitude and phase fit for the test system

Table 10.4   Coefficients of z^{-1} (weighting-factor)

Term	Denominator	Numerator
z^{-0}	1	1.8753222e−003
z^{-1}	−5.1223634e+000	−9.4562048e−003
z^{-2}	1.3002665e+001	2.3269772e−002
z^{-3}	−2.1840662e+001	−3.7014495e−002
z^{-4}	2.7116238e+001	4.1906856e−002
z^{-5}	−2.6233100e+001	−3.4689620e−002
z^{-6}	2.0264580e+001	2.0419347e−002
z^{-7}	−1.2531812e+001	−7.4643948e−003
z^{-8}	6.0380835e+000	6.4923773e−004
z^{-9}	−2.0707968e+000	8.2779560e−004
z^{-10}	3.7723943e−001	−3.1544461e−004

which involves energisation, fault inception and fault removal. The response using the FDNE with weighting factor is shown in Figure 10.21 and, as expected, no steady-state error can be observed. Using the fit without weighting factor gives a better representation during the transient but introduces a steady-state error. Figures 10.22

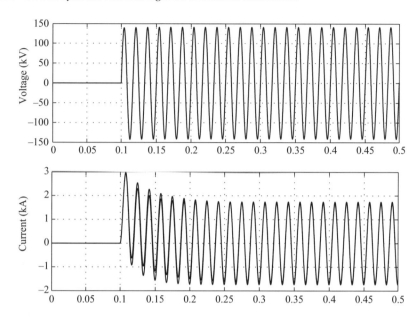

Figure 10.19 Comparison of full and a passive FDNE for an energisation transient

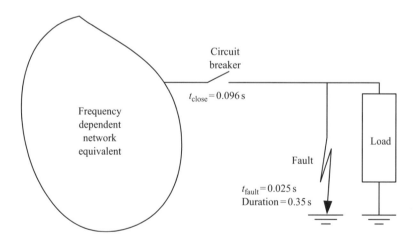

Figure 10.20 Active FDNE

and 10.23 show a detailed comparison for the latter case (i.e. without weighting factor). Slight differences are noticeable in the fault removal time, due to the requirement to remove the fault at current zero. Finally, when allowing current chopping the comparison in Figure 10.24 results.

Frequency dependent network equivalents 273

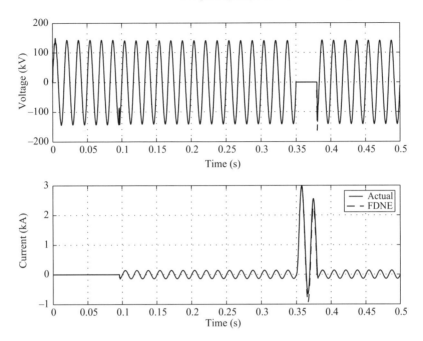

Figure 10.21   Comparison of active FDNE response

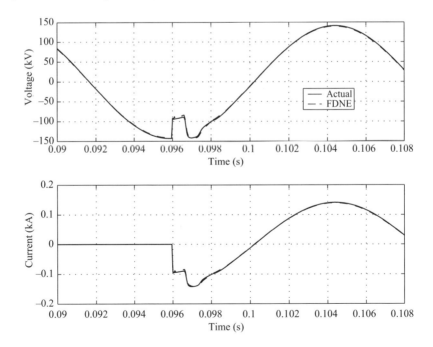

Figure 10.22   Energisation

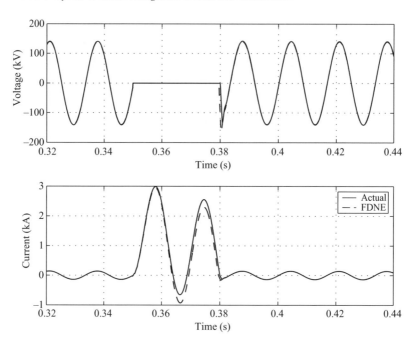

Figure 10.23 Fault inception and removal

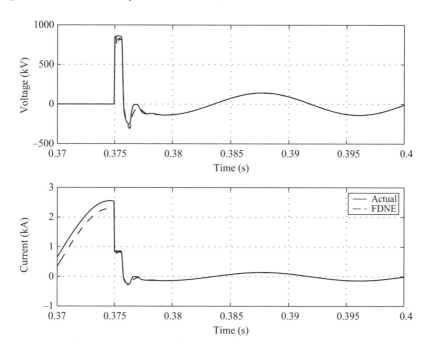

Figure 10.24 Fault inception and removal with current chopping

10.9 Summary

Frequency dependent network equivalents are important for modelling modern power systems due to their size and complexity. The first stage is to determine the response of the portion of the network to be replaced by an equivalent, as seen from its boundary busbar(s). This is most efficiently performed using frequency domain techniques to perform a frequency scan. Once determined, a rational function which is easily implemented can be fitted to match this response.

For simple responses, such as that of a single port, the techniques discussed give equally good fits. When there are multiple ports the responses are difficult to fit accurately with a stable rational function. The presence of transmission lines and cables providing a connection between the ports complicates the fitting task, as their related hyperbolic function responses are difficult to fit. This results in a time delay associated with the mutual coupling terms. Even if a stable fit for all the self and mutual terms is achieved, the overall FDNE can be unstable; this is caused by the matrices not being positive definite at some frequencies due to fitting errors at these frequencies.

Research work is still under way to find a computationally efficient technique to match the self and mutual terms, while ensuring a stable model. One approach is to match all self and mutual terms simultaneously by solving one large constrained optimisation problem.

The current techniques used for developing FDNE for use in transient studies have been reviewed. The fitting of a FDNE is still an art in that judgement must be exercised of the form and order of the rational function (or *RLC* circuit) to be used and the frequency range and sample points to be matched. The stability of each element in a multi port FDNE is essential and the combination of elements must be positive definite at each frequency.

10.10 References

1 HINGORANI, N. G. and BURBERY, M. F.: 'Simulation of AC system impedance in HVDC system studies', *IEEE Transactions on Power Apparatus and Systems*, 1970, **89** (5/6), pp. 820–8
2 BOWLES, J. P.: 'AC system and transformer representation for HV-DC transmission studies', *IEEE Transactions on Power Apparatus and Systems*, 1970, **89** (7), pp. 1603–9
3 CLERICI, A. and MARZIO, L.: 'Coordinated use of TNA and digital computer for switching surge studies', *IEEE Transactions on Power Apparatus and Systems*, 1970, **89**, pp. 1717–26
4 ABUR, A. and SINGH, H.: 'Time domain modeling of external systems for electromagnetic transients programs', *IEEE Transactions on Power Systems*, 1993, **8** (2), pp. 671–7
5 WATSON, N. R.: 'Frequency-dependent A.C. system equivalents for harmonic studies and transient convertor simulation' (Ph.D. thesis, University of Canterbury, New Zealand, 1987)

6 MORCHED, A. S. and BRANDWAJN, V.: 'Transmission network equivalents for electromagnetic transient studies', *IEEE Transactions on Power Apparatus and Systems*, 1983, **102** (9), pp. 2984–94
7 MORCHED, A. S., OTTEVANGERS, J. H. and MARTI, L.: 'Multi port frequency dependent network equivalents for the EMTP', *IEEE Transactions on Power Delivery*, Seattle, Washington, 1993, **8** (3), pp. 1402–12
8 DO, V. Q. and GAVRILOVIC, M. M.: 'An interactive pole-removal method for synthesis of power system equivalent networks', *IEEE Transactions on Power Apparatus and Systems*, 1984, **103** (8), pp. 2065–70
9 JURY, E. I.: 'Theory and application of the z-transform method' (John Wiley, New York, 1964)
10 OGATA, K.: 'Modern control engineering' (Prentice Hall International, Upper Saddle River, N. J., 3rd edition, 1997)
11 GUSTAVSEN, B. and SEMLYEN, A.: 'Rational approximation of frequency domain response by vector fitting', *IEEE Transaction on Power Delivery*, 1999, **14** (3), pp. 1052–61

Chapter 11
Steady state applications

11.1 Introduction

Knowledge of the initial conditions is critical to the solution of most power system transients. The electromagnetic transient packages usually include some type of frequency domain initialisation program [1]–[5] to try and simplify the user's task. These programs, however, are not part of the electromagnetic transient simulation discussed in this book. The starting point in the simulation of a system disturbance is the steady-state operating condition of the system prior to the disturbance.

The steady-state condition is often derived from a symmetrical (positive sequence) fundamental frequency power-flow program. If this information is read in to initialise the transient solution, the user must ensure that the model components used in the power-flow program represent adequately those of the electromagnetic transient program. In practice, component asymmetries and non-linearities will add imbalance and distortion to the steady-state waveforms.

Alternatively the steady-state solution can be achieved by the so-called *'brute force'* approach; the simulation is started without performing an initial calculation and is carried out long enough for the transient to settle down to a steady-state condition. Hence the electromagnetic transient programs themselves can be used to derive steady-state waveforms. It is, thus, an interesting matter to speculate whether the correct approach is to provide an 'exact' steady state initialisation for the EMTP method or to use the latter to derive the final steady-state waveforms. The latter alternative is discussed in this chapter with reference to power quality application.

A good introduction to the variety of topics considered under 'power quality' can be found in reference [6] and an in-depth description of the methods currently used for its assessment is given in reference [7].

An important part of power quality is steady state (and quasi-steady state) waveform distortion. The resulting information is sometimes presented in the time domain (e.g. notching) and more often in the frequency domain (e.g. harmonics and interharmonics).

Although their source of origin is a transient disturbance, i.e. a short-circuit, voltage sags are characterised by their (quasi) steady-state magnitude and duration, which, in general, will display three-phase imbalance; moreover, neither the voltage drop nor its recovery will take place instantaneously. Thus for specified fault conditions and locations the EMTP method provides an ideal tool to determine the voltage sag characteristics.

Randomly varying non-linear loads, such as arc furnaces, as well as substantial and varying harmonic (and interharmonic) content, cause voltage fluctuations that often produce flicker. The random nature of the load impedance variation with time prevents an accurate prediction of the phenomena. However the EMTP method can still help in the selection of compensating techniques, with arc models based on the experience of existing installations.

11.2 Initialisation

As already mentioned in the introduction, the electromagnetic transients program requires auxiliary facilities to initialise the steady-state condition, and only a three-phase harmonic power flow can provide a realistic start. However this is difficult and time consuming as it involves the preparation of another data set and transfer from one program to another, not to mention the difficulty in ensuring that both are modelling exactly the same system and to the same degree of accuracy.

Often a symmetrical fundamental frequency power-flow program is used due to familiarity with and availability of such programs. However failure to consider the imbalance and distortion can cause considerable oscillations, particularly if low frequency poorly damped resonant frequencies exist. For this reason PSCAD/EMTDC uses a 'black-start' approach whereby sources are ramped from zero up to their final value over a period of time, typically 0.05 s. This often results in reaching steady state quicker than initialising with power-flow results where the distortion and/or imbalance is ignored. Synchronous machines have long time constants and therefore special techniques are required for an efficient simulation. The rotor is normally locked to the system frequency and/or the resistance artificially changed to improve damping until the electrical transient has died away, then the rotor is released and the resistance reset to its correct value.

11.3 Harmonic assessment

Although the frequency domain provides accurate information of harmonic distortion in linear networks, conventional frequency domain algorithms are inadequate to represent the system non-linear components.

An early iterative method [8], referred to as IHA (for Iterative Harmonic Analysis), was developed to analyse the harmonic interaction of a.c.–d.c. power systems, whereby the converter response at each iteration was obtained from knowledge of the converter terminal voltage waveforms (which could be unbalanced and distorted).

Steady state applications 279

The resulting converter currents were then expressed in terms of harmonic current injections to be used in a new iteration of the a.c. system harmonic flow. This method, based on the fixed point iteration (or Gauss) concept, had convergence problems under weak a.c. system conditions. An alternative IHA based on Newton's method [9] provided higher reliability at the expense of greatly increased analytical complexity.

However the solution accuracy achieved with these early methods was very limited due to the oversimplified modelling of the converter (in particular the idealised representation of the converter switching instants).

An important step in solution accuracy was made with the appearance of the so-called harmonic domain [9], a full Newton solution that took into account the modulating effect of a.c. voltage and d.c. current distortion on the switching instants and converter control functions. This method performs a linearisation around the operating point that provides sufficient accuracy. In the present state of harmonic domain development the Jacobian matrix equation combines the system fundamental frequency three-phase load-flow and the system harmonic balance in the presence of multiple a.c.–d.c. converters. Although in principle any other type of non-linear component can be accommodated, the formulation of each new component requires considerable skill and effort. Accordingly a program for the calculation of the non-sinusoidal periodic steady state of the system may be of very high dimension and complexity.

11.4 Phase-dependent impedance of non-linear device

Using perturbations the transient programs can help to determine the phase-dependent impedance of a non-linear device. In the steady state any power system component can be represented by a voltage controlled current source: $I = F(V)$, where I and V are arrays of frequency phasors. The function F may be non-linear and non-analytic. If F is linear, it may include linear cross-coupling between frequencies, and may be non-analytic, i.e. frequency cross-coupling and phase dependence do not imply non-linearity in the frequency domain. The linearised response of F to a single applied frequency may be calculated by:

$$\begin{bmatrix} \Delta I_R \\ \Delta I_I \end{bmatrix} = \begin{bmatrix} \dfrac{\partial F_R}{\partial V_R} & \dfrac{\partial F_R}{\partial V_I} \\ \dfrac{\partial F_I}{\partial V_R} & \dfrac{\partial F_I}{\partial V_I} \end{bmatrix} \begin{bmatrix} \Delta V_R \\ \Delta V_I \end{bmatrix} \quad (11.1)$$

where F has been expanded into its component parts. If the Cauchy–Riemann conditions hold, then 11.1 can be written in complex form. In the periodic steady state, all passive components (e.g. *RLC* components) yield partial derivatives which satisfy the Cauchy–Riemann conditions. There is, additionally, no cross-coupling between harmonics for passive devices or circuits. With power electronic devices the Cauchy–Riemann conditions will not hold, and there will generally be cross-harmonic coupling as well.

In many cases it is desirable to ignore the phase dependence and obtain a complex impedance which is as near as possible to the average phase-dependent impedance.

Since the phase-dependent impedance describes a circle in the complex plane as a function of the phase angle of the applied voltage [10], the appropriate phase-independent impedance lies at the centre of the phase-dependent locus. Describing the phase-dependent impedance as

$$Z = \begin{bmatrix} z_{11} & z_{12} \\ z_{21} & z_{22} \end{bmatrix} \quad (11.2)$$

the phase independent component is given by:

$$Z = \begin{bmatrix} R & -X \\ X & R \end{bmatrix} \quad (11.3)$$

where

$$R = \tfrac{1}{2}(z_{11} + z_{22}) \quad (11.4)$$

$$X = \tfrac{1}{2}(z_{21} - z_{12}) \quad (11.5)$$

In complex form the impedance is then $\hat{Z} = R + jX$.

In most cases an accurate analytic description of a power electronic device is not available, so that the impedance must be obtained by perturbations of a steady-state model. Ideally, the model being perturbed should not be embedded in a larger system (e.g. a.c. or d.c. systems), and perturbations should be applied to control inputs as well as electrical terminals. The outcome from such an exhaustive study would be a harmonically cross-coupled admittance tensor completely describing the linearisation.

The simplest method for obtaining the impedance by perturbation is to sequentially apply perturbations in the system source, one frequency at a time, and calculate impedances from

$$Z_k = \frac{\Delta V_k}{\Delta I_k} \quad (11.6)$$

The Z_k obtained by this method includes the effect of coupling to the source impedance at frequencies coupled to k by the device, and the effect of phase dependency. This last means that for some k, Z_k will be located at some unknown position on the circumference of the phase-dependent impedance locus. The impedance at frequencies close to k will lie close to the centre of this locus, which can be obtained by applying two perturbations in quadrature. With the two perturbations of the quadrature method, enough information is available to resolve the impedance into two components; phase dependent and phase independent.

The quadrature method proceeds by first solving a base case at the frequency of interest to obtain the terminal voltage and total current: (V_{kb}, I_{kb}). Next, two perturbations are applied sequentially to obtain (V_{k1}, I_{k1}) and (V_{k2}, I_{k2}). If the source was initially something like

$$E_b = E \sin(\omega t) \quad (11.7)$$

then the two perturbations might be

$$E_1 = E\sin(\omega t) + \delta\sin(k\omega t) \tag{11.8}$$

$$E_2 = E\sin(\omega t) + \delta\sin(k\omega t + \pi/2) \tag{11.9}$$

where δ is small to avoid exciting any non-linearity. The impedance is obtained by first forming the differences in terminal voltage and injected current:

$$\Delta V_{k1} = V_{k1} - V_{kb} \tag{11.10}$$

$$\Delta V_{k2} = V_{k2} - V_{kb} \tag{11.11}$$

$$\Delta I_{k1} = I_{k1} - I_{kb} \tag{11.12}$$

$$\Delta I_{k2} = I_{k2} - I_{kb} \tag{11.13}$$

Taking real components, the linear model to be fitted states that

$$\begin{bmatrix} \Delta V_{k1R} \\ \Delta V_{k1I} \end{bmatrix} = \begin{bmatrix} z_{k11} & z_{k12} \\ z_{k21} & z_{k22} \end{bmatrix} \begin{bmatrix} \Delta I_{k1R} \\ \Delta I_{k1I} \end{bmatrix} \tag{11.14}$$

and

$$\begin{bmatrix} \Delta V_{k2R} \\ \Delta V_{k2I} \end{bmatrix} = \begin{bmatrix} z_{k11} & z_{k12} \\ z_{k21} & z_{k22} \end{bmatrix} \begin{bmatrix} \Delta I_{k2R} \\ \Delta I_{k2I} \end{bmatrix} \tag{11.15}$$

which permits a solution for the components z_{k11}, etc:

$$\begin{bmatrix} \Delta V_{k1R} \\ \Delta V_{k1I} \\ \Delta V_{k2R} \\ \Delta V_{k2I} \end{bmatrix} = \begin{bmatrix} \Delta I_{k1R} & \Delta I_{k1I} & 0 & 0 \\ 0 & 0 & \Delta I_{k1R} & \Delta I_{k1I} \\ \Delta I_{k2R} & \Delta I_{k2I} & 0 & 0 \\ 0 & 0 & \Delta I_{k2R} & \Delta I_{k2I} \end{bmatrix} \begin{bmatrix} z_{k11} \\ z_{k12} \\ z_{k21} \\ z_{k22} \end{bmatrix} \tag{11.16}$$

Finally the phase-independent impedance in complex form is:

$$\hat{Z}_k = \tfrac{1}{2}(z_{k11} + z_{k22}) + \tfrac{1}{2}j(z_{k21} - z_{k12}) \tag{11.17}$$

11.5 The time domain in an ancillary capacity

The next two sections review the increasing use of the time domain to try and find a simpler alternative to the harmonic solution. In this respect the flexibility of the EMTP method to represent complex non-linearities and control systems makes it an attractive alternative for the solution of harmonic problems. Two different modelling philosophies have been proposed. One, discussed in this section, is basically a frequency domain solution with periodic excursions into the time domain to update the contribution of the non-linear components. The alternative, discussed in section 11.6, is basically a time domain solution to the steady state followed by FFT processing of the resulting waveforms.

11.5.1 Iterative solution for time invariant non-linear components

In this method the time domain is used at every iteration of the frequency domain to derive a Norton equivalent for the non-linear component. The Norton admittance represents a linearisation, possibly approximate, of the component response to variations in the terminal voltage harmonics. For devices that can be described by a static (time invariant) voltage–current relationship,

$$i(t) = f(v(t)) \tag{11.18}$$

in the time domain, both the current injection and the Norton admittance can be calculated by an elegant procedure involving an excursion into the time domain. At each iteration, the applied voltage harmonics are inverse Fourier transformed to yield the voltage waveshape. The voltage waveshape is then applied point by point to the static voltage–current characteristic, to yield the current waveshape. By calculating the voltage and current waveshapes at $2n$ equispaced points, a FFT is readily applied to the current waveshape, to yield the total harmonic injection.

To derive the Norton admittance, the waveshape of the total derivative

$$\frac{dI}{dV} = \frac{di(t)}{dt} \frac{dt}{dv(t)} = \frac{di(t)/dt}{dv(t)/dt} \tag{11.19}$$

is calculated by dividing the point by point changes in the voltage and current waveshapes. Fourier transforming the total derivative yields columns of the Norton admittance matrix; in this matrix all the elements on any diagonal are equal, i.e. it has a Toeplitz structure. The Norton admittance calculated in this manner is actually the Jacobian for the source.

A typical non-linearity of this type is the transformer magnetising characteristic, for which the derivation of the Norton equivalent (shown in Figure 11.1) involves the following steps [11], illustrated in the flow diagram of Figure 11.2.

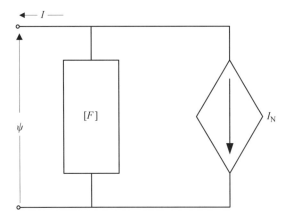

Figure 11.1 Norton equivalent circuit

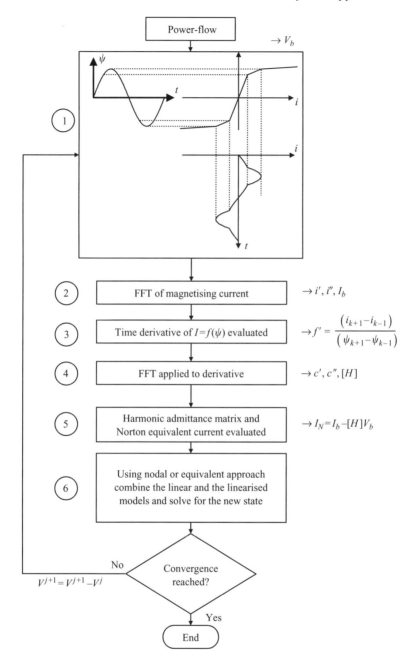

Figure 11.2   Description of the iterative algorithm

1 For each phase the voltage waveform is used to derive the corresponding flux wave and the latter is impressed, point by point, upon the experimental characteristic ϕ–I and the associated magnetising current is then determined in the time domain.
2 By means of an FFT the magnetising current is solved in the frequency domain and the Fourier coefficients i' and i'' are assembled into a base current vector I_b.
3 Using the magnetising current and flux as determined in step 1, the time derivative of the function $I = f(\phi)$ is evaluated.
4 The FFT is applied to the slope shape of step 3, and the Fourier coefficients c' and c'' obtained from this exercise are used to assemble the Toeplitz matrix $[H]$.
5 The Norton equivalent current source I_N, i.e. $I_N = I_b - [H]V_b$ is calculated.
6 The above linearised model is combined with the linear network as part of a Newton-type iterative solution as described in Figure 11.2, with the Jacobian defined by the matrix $[H]$.

11.5.2 Iterative solution for general non-linear components

Time-variant non-linear components, such as power electronic devices, do not fall into the category defined by equation 11.18. Instead their voltage–current relationships result from many interdependent factors, such as the phase and magnitude of each of the a.c. voltage and current harmonic components, control system functions, firing angle constraints, etc.

In these cases the converter Norton admittance matrix does not display the Toeplitz characteristic, and, in general, contains of the order of n^2 different elements, as opposed to the n elements obtained from the FFT. Thus, not only is the calculation of a harmonic Norton equivalent computationally difficult but, for accurate results, it has to be iteratively updated. The computational burden is thus further increased in direct proportion with the size of the system and the number of harmonics represented.

To extend the iterative algorithm to any type of non-linearity, a generally applicable time domain solution (such as the state variable or the EMTP methods) must be used to represent the behaviour of the non-linear components [12], [13].

As in the previous case, the system is divided into linear and non-linear parts. Again, the inputs to a component are the voltages at its terminal and the output, the terminal currents, and both of these will, in general, contain harmonics. The iterative solution proceeds in two stages.

In the first stage the periodic steady state of the individual components is initially derived from a load-flow program and then updated using voltage corrections from the second stage. The calculations are performed in the frequency domain where appropriate (e.g. in the case of transmission lines) and in the time domain otherwise.

The currents obtained in stage (i) are used in stage (ii) to derive the current mismatches Δi, expressed in the frequency domain. These become injections into a system-wide incremental harmonic admittance matrix Y, calculated in advance from such matrices for all the individual components. The equation $\Delta i = Y \Delta v$ is then solved for Δv to be used in stage (i) to update all bus voltages.

The first stage uses a modular approach, but in the second stage the voltage corrections are calculated globally, for the whole system. However, convergence is only achieved linearly, because of the approximations made on the accuracy of v. A separate iterative procedure is needed to model the controllers of active non-linear devices, such as a.c.–d.c. converters, and this procedure relies entirely on information from the previous iteration.

11.5.3 Acceleration techniques

Time domain simulation, whether performed by the EMTP, state variable or any other method, may require large computation times to reach steady state and thus the use of accelerating techniques [14], [15] is advocated to speed up the solution. These techniques take advantage of the two-point boundary value inherent in the steady-state condition. Thus a correction term is added to the initial state vector, calculated as a function of the residuum of the initial and final state vectors and the mapping derivative over the period. A concise version of the Poincaré method described in reference [14] is given here.

A non-linear system of state equations is expressed as:

$$\dot{\mathbf{x}} = g(\mathbf{x}, \mathbf{u}) \qquad x(t_0) = x_0 \qquad (11.20)$$

where $u = u(t)$ is the input and x_0 the vector of state variables at $t = t_0$ close to the periodic steady state.

This state is characterised by the condition

$$f(\mathbf{x}_0) = \mathbf{x}(t_0 + T) - \mathbf{x}(t_0) \qquad (11.21)$$

where $x(t_0 + T)$ is derived by numerical integration over the period t_0 to $t_0 + T$ of the state equations 11.20

Equation 11.21 represents a system of n non-linear algebraic equations with n unknown x_i and can thus be solved by the Newton–Raphson method.

The linearised form of equation 11.21 around an approximation $\mathbf{x}_0^{(k)}$ at step k of its solution is:

$$f(\mathbf{x}_0) \cong f\left(\mathbf{x}_0^{(k)}\right) + J^{(k)}\left(\mathbf{x}_0^{(k+1)} - \mathbf{x}_0^{(k)}\right) = 0 \qquad (11.22)$$

where $J^{(k)}$ is the Jacobian (the matrix of partial derivatives of $f(\mathbf{x}_0)$ with respect to $\mathbf{x}$, evaluated at $\mathbf{x}_0^{(k)}$). By approximating $J^{(k)}$ at each iteration k, using its definition, in addition to the mapping

$$\mathbf{x}_0^{(k)} \rightarrow f\left(\mathbf{x}_0^{(k)}\right) \qquad (11.23)$$

the mappings are

$$x_0^{(k)} + \varepsilon I_i \rightarrow f\left(x_0^{(k)} + \varepsilon I_i\right) \qquad i = 1, \ldots, n \qquad (11.24)$$

where I_i are the columns of the unit matrix and ε is a small scalar. $J^{(k)}$ is then assembled from the vectors

$$\frac{1}{\varepsilon}\left(f\left(\mathbf{x}_0^{(k)}+\varepsilon I_i\right)+f\left(\mathbf{x}_0^{(k)}\right)\right) \quad i=1,\ldots,n \quad (11.25)$$

obtained in equations 11.23 and 11.24.

Finally, using the above approximation $J^{(k)}$ of the Jacobian, the updated value $\mathbf{x}_0^{(k+1)}$ for $\mathbf{x}_0$ is obtained from equation 11.22.

The process described above is quasi-Newton but its convergence is close to quadratic. Therefore, as in a conventional Newton power-flow program, only three to five iterations are needed for convergence to a highly accurate solution, depending on the closeness of the initial state x_0 to the converged solution.

11.6 The time domain in the primary role

11.6.1 Basic time domain algorithm

Starting from standstill, the basic time domain uses a 'brute force' solution, i.e. the system equations are integrated until a reasonable steady state is reached. This is a very simple approach but can have very slow convergence when the network has components with light damping.

To alleviate this problem the use of acceleration techniques has been described in sections 11.5.2 and 11.5.3 with reference to the hybrid solution. However the number of periods to be processed in the time domain required by the acceleration technique is almost directly proportional to the number of state variables multiplied by the number of Newton iterations [14]. Therefore the solution efficiency reduces very rapidly as the a.c. system size increases. This is not a problem in the case of the hybrid algorithm, because the time domain solutions require no explicit representation of the a.c. network. On the other hand, when the solution is carried out entirely in the time domain, the a.c. system components are included in the formulation and thus the number of state variables is always large. Moreover, the time domain algorithm only requires a single transient simulation to steady state, and therefore the advantage of the acceleration technique is questionable in this case, considering its additional complexity.

On reaching the steady state within a specified tolerance, the voltage and current waveforms, represented by sets of discrete values at equally spaced intervals (corresponding with the integration steps), are subjected to FFT processing to derive the harmonic spectra.

11.6.2 Time step

The time step selection is critical to model accurately the resonant conditions when converters are involved. A resonant system modelled with 100 or 50 μs steps can miss a resonance, while the use of a 10 μs captures it. Moreover, the higher the resonant frequency the smaller the step should be. A possible way of checking the effectiveness of a given time step is to reduce the step and then compare the results with those obtained in the previous run. If there is a significant change around the resonant frequency, then the time step is too large.

The main reason for the small time-step requirement is the need to pin-point the commutation instants very accurately, as these have great influence on the positive feedback that appears to occur between the a.c. harmonic voltages and the corresponding driven converter currents.

11.6.3 DC system representation

It is essential to represent correctly the main components of the particular converter configuration. For instance, a voltage source converter should include the d.c. capacitor explicitly, while a current source converter should instead include the series inductor.

The inverter end representation, although less critical, may still have some effect. An ideal d.c. current source or a series $R-L$ load representation are the simplest solutions; in the latter case the R is based on the d.c. load-flow operating point and the inductance should be roughly twice the inverter a.c. inductance (including transformer leakage plus any a.c. system inductance). A pure resistance is not advised as this will produce an overdamped d.c. system, which may lead to inaccurate results.

11.6.4 AC system representation

The main advantage claimed by the hybrid frequency/time domain methods, described in section 11.5, over conventional time domain solutions is their ability to model accurately the frequency dependence of the a.c. system components (particularly the transmission lines). Thus, if the time domain is going to be favoured in future harmonic simulations, the accuracy of its frequency dependent components needs to be greatly improved.

The use of a frequency dependent equivalent avoids the need to model any significant part of the a.c. system in detail, yet can still provide an accurate matching of the system impedance across the harmonic frequency spectra [16]. The derivation of frequency dependent equivalents is described in Chapter 10.

On completion of the time domain simulation, the FFT-derived harmonic current spectrum at the converter terminals needs to be injected into the full a.c. system to determine the harmonic flows throughout the actual system components.

By way of example, the test system of Figure 11.3 includes part of the primary transmission system connected to the rectifier end of the New Zealand HVDC link [17]. Though not shown in the diagram, the converter terminal also contains a set of filters as per the CIGRE benchmark model [18].

The corresponding frequency dependent equivalent circuit is shown in Figure 11.4 and its component values in Table 11.1. A graph of the impedance magnitude of the actual rectifier a.c. system based on its modelled parameters, and the frequency dependent equivalent, is given in Figure 11.5. It can be seen that this equivalent provides a very good match for the impedance of the actual system up to about the

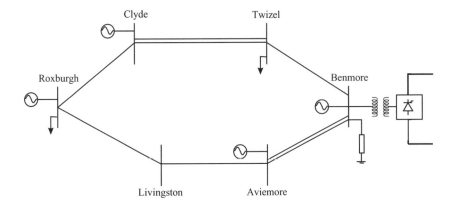

Figure 11.3 Test system at the rectifier end of a d.c. link

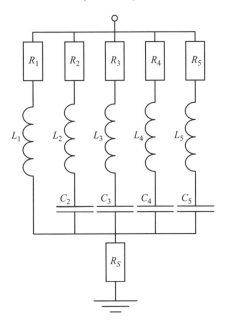

Figure 11.4 Frequency dependent network equivalent of the test system

17th harmonic. Of course the use of extra parallel branches in the equivalent circuit will extend the range of frequency matching further.

11.7 Voltage sags

Considering the financial implications of industrial plant disruptions resulting from voltage sags, their mitigation by means of active power electronic devices is on

Table 11.1 Frequency dependent equivalent circuit parameters

	$R(\Omega)$	$L(H)$	$C(\mu F)$
Arm 1	17.0	0.092674	–
Arm 2	0.50	0.079359	1.8988
Arm 3	25.1	0.388620	0.1369
Arm 4	6.02	0.048338	0.7987
Arm 5	13.6	0.030883	0.3031
Series R	1.2	–	–

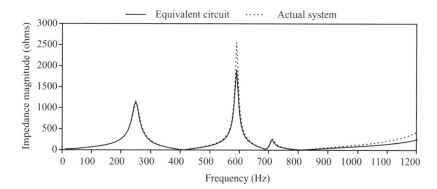

Figure 11.5 Impedance/frequency of the frequency dependent equivalent

the increase. Cost-effective solutions require a good deal of computer simulation of the power system, including its protection and control, to minimise the mitigation requirements.

For a given type of fault and location the characteristics of voltage sags are mainly influenced by the dynamic components of the power system, namely the synchronous generators and the induction motors. The modelling of these components must therefore include all the elements influencing their subtransient and transient responses to the short-circuit, and, in the case of the synchronous generator, the automatic voltage regulator.

Present regulations only specify sags by their fundamental frequency magnitude and duration and, therefore, the representation of the system passive components is less critical, e.g. a lumped impedance is sufficient to model the transmission lines.

When the system contains large converter plant, the fundamental frequency simplification is inadequate to represent the behaviour of the converter plant during system faults. The converter normal operation is then disrupted, undergoing uncontrollable switching and commutation failures and the result is an extremely distorted voltage at the converter terminals. Under these conditions it is important to model the frequency dependence of the transmission system, as described in Chapter 10.

The present state of electromagnetic transient simulation programs is perfectly adequate to represent all the conditions discussed above. The models of synchronous and induction machines described in Chapter 7 meet all the requirements for accurate voltage sag simulation. In particular the flexible representation of power electronic devices and their controllers, especially in the PSCAD/EMTDC package, provides sufficient detail of voltage waveform distortion to model realistically the behaviour of the non-linear devices following system short-circuits.

The use of a real-time digital simulator permits, via digital to analogue conversion and amplification, the inclusion of actual physical components such as protective relays and controls. It also permits testing the ability of power electronic equipment to operate during simulated voltage sag conditions.

11.7.1 Examples

First the EMTP program is used to illustrate the effect of induction motors on the characteristics of voltage sags following fault conditions.

The fault condition is a three-phase short-circuit of 206 ms duration, placed at a feeder connected to the same busbar as the induction motor plant [19].

Figure 11.6 shows the voltage variation at the common busbar. A deep sag is observed during the fault, which in the absence of the motor would have established itself immediately at the final level of 35 per cent. However the reduction in electromagnetic torque that follows the voltage drop causes a speed reduction and the motor

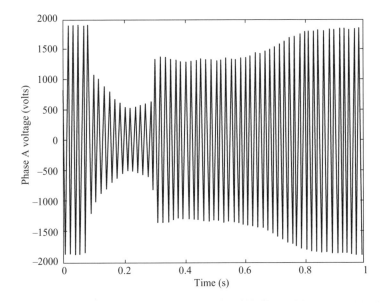

Figure 11.6   Voltage sag at a plant bus due to a three-phase fault

Steady state applications 291

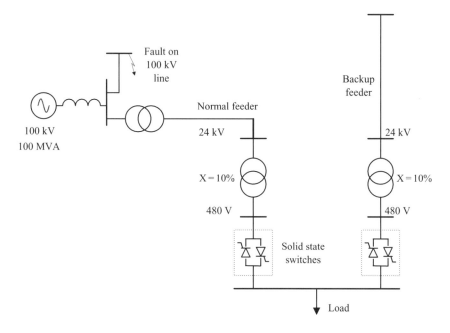

Figure 11.7 *Test circuit for transfer switch*

goes temporarily into a generating mode, thus contributing to the fault current; as a result, the presence of the motor increases the terminal voltage for a short decaying period.

The motor reacceleration following fault clearance requires extra reactive current, which slows the voltage recovery. Thus the figure displays a second sag of 75 per cent magnitude and 500 ms duration.

Of course the characteristics of these two sags are very dependent on the protection system. The EMTP program is therefore an ideal tool to perform sensitivity studies to assess the effect of different fault locations and protection philosophies. The 5th order induction motor model used by the EMTP program is perfectly adequate for this purpose.

The second example involves the use of a fast solid state transfer switch (SSTS) [20], as shown in Figure 11.7, to protect the load from voltage sags. The need for such a rapid transfer is dictated by the proliferation of sensitive equipment such as computers, programmable drives and consumer electronics.

Each phase of the SSTS is a parallel back to back thyristor arrangement. The switch which is on, has the thyristors pulsed continuously. On detection of a sag, these firing pulses are stopped, its thyristors are now subjected to a high reverse voltage from the other feeder and are thus turned off immediately. Current interruption is thus achieved at subcycle intervals.

The sag detection is achieved by continuous comparison of the voltage waveform with an ideal sinusoid in phase with it and of a magnitude equal to the pre-sag value.

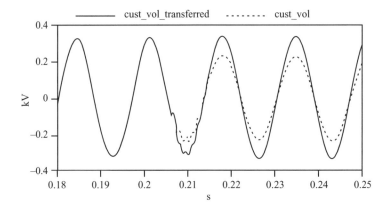

Figure 11.8   Transfer for a 30 per cent sag at 0.8 power factor with a 3325 kVA load

The latter is constructed using the fundamental frequency component of the FFT of the voltage waveform from the previous cycle.

The voltage waveforms derived from PSCAD/EMTDC simulation, following a disturbance in the circuit of Figure 11.7, are shown in Figure 11.8. The continuous trace shows that the feeder transfer is achieved within quarter of a cycle and with minimal transients. The dotted line shows the voltage that would have appeared at the load in the absence of a transfer switch.

11.8   Voltage fluctuations

Low frequency voltage fluctuations give rise to the flicker effect, defined as the variation in electric lamp luminosity which affects human vision. The problem frequencies causing flicker are in the region of 0.5–25 Hz, the most critical value being 8.3 Hz, for which even a 0.3 per cent voltage amplitude variation can reach the perceptibility threshold.

The main cause of voltage fluctuation is the electric arc furnace (EAF), due to the continuous non-linear variation of the arc resistance, particularly during the melting cycle. A physical analysis of the arc length variation is impractical due to the varying metal scrap shapes, the erratic electromagnetic forces and the arc-electrode positions. Instead, the EAF is normally represented by simplified deterministic or stochastic models, with the purpose of determining the effect of possible compensation techniques.

By way of example, Figure 11.9 shows a single line diagram of an 80 MVA arc furnace system fed from a 138 kV bus with a 2500 MVA short-circuit capacity [21]. The EAF transformer secondary voltage is 15 kV and the EAF operates at 900 V. The EAF behaviour is simulated in the PSCAD/EMTDC program by a chaotic arc model and the power delivered to the EAF is kept constant at 80 MVA by adjusting the tap changers on the EAF transformers.

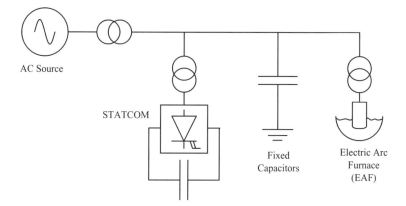

Figure 11.9   EAF system single line diagram

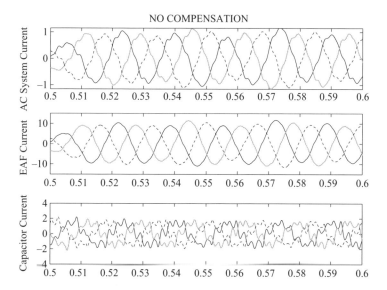

Figure 11.10   EAF without compensation

The results shown in Figure 11.10, corresponding to the initial case without any compensation, illustrate a totally unacceptable distortion in the supplied current. Figure 11.11 shows that the addition of a 64 MVAr static VAR compensator (SVC) to the 15 kV busbar, improves considerably the supply current waveform. Finally, the effect of installing a ±32 MVAr static compensator (STATCOM) in the 15 kV bus is illustrated in Figure 11.12. The STATCOM is able to dynamically eliminate the harmonics and the current fluctuations on the source side by injecting the precise currents needed. It is these current fluctuations which result in voltage flicker. These results further demonstrate the role of electromagnetic transient simulation in the solution of power quality problems.

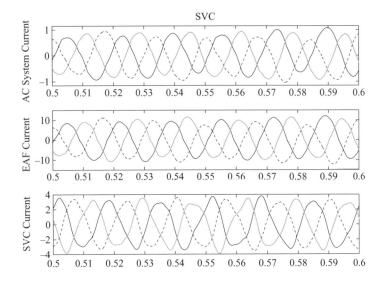

Figure 11.11   EAF with SVC compensation

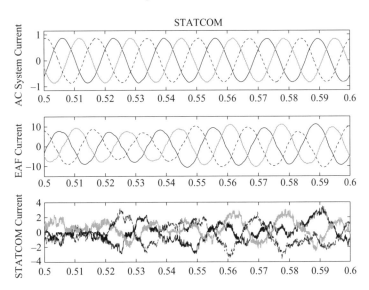

Figure 11.12   EAF with STATCOM compensation

11.8.1   Modelling of flicker penetration

The simple circuit of Figure 11.9 is typical of the test systems used to simulate arc furnaces and flicker levels, i.e. a radial feeder connected to a source specified by the MVA fault level, i.e. the voltage fluctuations are only available at the arc furnace terminals and there is practically no information on flicker penetration.

To illustrate the use of the PSCAD/EMTDC package to simulate flicker penetration [22] a PSCAD user component has been implemented that models the digital

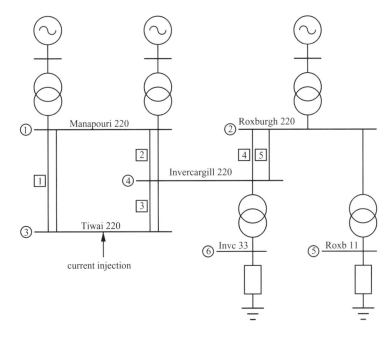

Figure 11.13 Test system for flicker penetration (the circles indicate busbars and the squares transmission lines)

version of the IEC flickermeter. The unit receives the input signal from the time domain simulation and produces the instantaneous flicker level (IFL) as well as the short-term and long-term flicker severity indices (P_{st}, P_{lt}). Moreover, a number of these components are needed to study the propagation of flicker levels throughout the power system.

However, the observation time for the P_{st} index is 10 minutes, resulting in very long runs. For example to complete ten minutes of simulation of the nine-bus system shown in Figure 11.13 requires about twelve hours of running time in an UltraSPARC computer (300 MHz). In Figure 11.13 the flicker injection, at the Tiwai busbar, consists of three sinusoidally amplitude modulated current sources that operate at $50 \pm f$ Hz.

The voltages at the load and transmission system buses are monitored by 18 identical flicker meters. To reduce the simulation burden the observation time for the P_{st} evaluation was set to 10 seconds instead of 10 minutes. A control block allows stepping automatically through the list of specified frequencies (1–35 Hz) during the simulation run and also to selectively record the output channels.

Figure 11.14 is an example of the flicker severities monitored at the various points provided with the virtual flicker meters used with the EMTDC program. For comparison the figure also includes the flicker levels derived from steady-state frequency domain analysis. The flicker severity is highest at the Tiwai bus, which has to be expected since it is the point of injection. Comparing Figures 11.14(b) and (e) and Figures 11.14(d) and (f) respectively it can be seen that, for a positive sequence injection, flicker propagates almost without attenuation from the transmission to the load busbars.

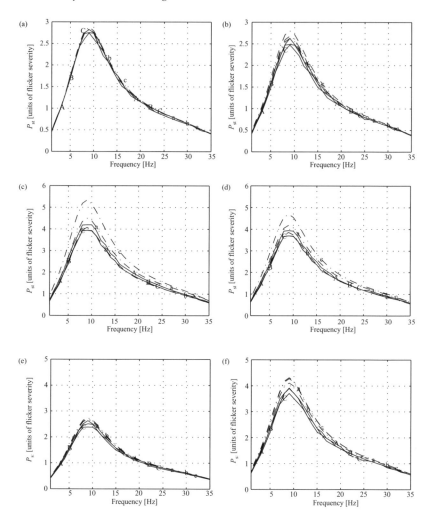

Figure 11.14 Comparison of P_{st} indices resulting from a positive sequence current injection at Tiwai. PSCAD/EMTDC results are shown as solid lines (phases A, B, C), frequency domain results as dash-dotted lines (phases a, b, c).

11.9 Voltage notching

Voltage notches are caused by the brief phase to phase short-circuits that occur during the commutation process in line-commutated current sourced a.c.–d.c. converters. For a specified firing angle, the notch duration is directly proportional to the source inductance and the d.c. current; its depth reduces as the observation point separates from the converter terminals, i.e. with increasing inductance between them.

Steady state applications 297

In distribution systems with low short-circuit levels, voltage notches can excite the natural frequency created by the capacitance of lines and other shunt capacitances in parallel with the source inductance, thus causing significant voltage waveform distortion.

The EMTP simulation can be used to calculate the voltage distortion at various points of the distribution system and to evaluate possible solutions to the problem.

11.9.1 Example

Figure 11.15 shows a 25 kV distribution system supplied from a 10 MVA transformer connected to a 144 kV transmission system [23]. The feeder on the right includes a six-pulse converter adjustable speed drive (ASD) controlling a 6000 HP induction motor. The ASD is connected to the 4.16 kV bus of a 7.5 MVA transformer and a set of filters tuned to the 5, 7 and 11 harmonics is also connected to that bus.

The second feeder, on the left of the circuit diagram, supplies a motor load of 800 HP at 4.16 kV in parallel with a capacitor for surge protection. This feeder also supplies other smaller motor loads at 480 V which include power factor correction capacitors. Under certain operating conditions the voltage notches produced by the ASD excited a parallel resonance between the line capacitance and the system source inductance and thus produced significant oscillations on the 25 kV bus. Furthermore, the oscillations were magnified at the 4.16 kV busbar by the surge capacitor of the 800 HP motor, which failed as a result.

A preliminary study carried out to find the system frequency response produced the impedance versus frequency plot of Figure 11.16, which shows a parallel resonance at a frequency just above the 60th harmonic. The EMTP program was then used to model the circuit of Figure 11.15 under different operating conditions. The results of the simulation are shown in Figures 11.17 and 11.18 for the voltage waveforms at the 25 kV and 4.16 kV buses respectively. Figure 11.17 clearly shows the notch related oscillations at the resonant frequency and Figure 11.18 the amplification caused by the surge capacitor at the terminals of the 800 HP motor. On the other hand the simulation showed no problem at the 480 V bus. Possible solutions are the use of a capacitor bank at the 25 kV bus or additional filters (of the bandpass type) at the ASD terminals. However, solutions based on added passive components may themselves excite lower-order resonances. For instance, in the present example, the use of a 1200 kVAr capacitor bank caused unacceptable 13th harmonic distortion, whereas a 2400 kVAr reduced the total voltage harmonic distortion to an insignificant level.

11.10 Discussion

Three different approaches are possible for the simulation of power system harmonics. These are the harmonic domain, the time domain and a hybrid combination of the conventional frequency and time domains.

The harmonic domain includes a linearised representation of the non-linear components around the operating point in a full Newton solution. The fundamental

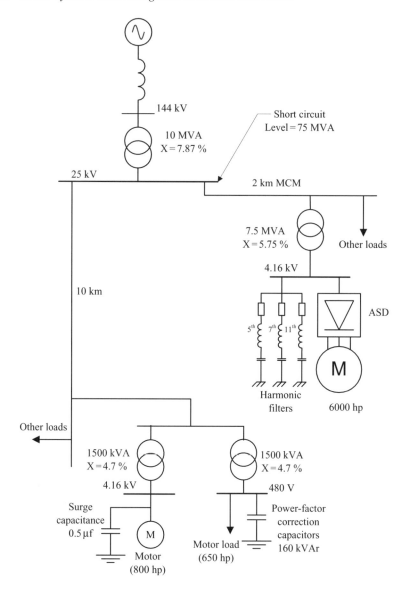

Figure 11.15 Test system for the simulation of voltage notching

frequency load-flow is also incorporated in the Newton solution and thus provides the ideal tool for general steady-state assessment. However the complexity of the formulation to derive the system Jacobian may well prevent its final acceptability.

The hybrid proposal takes advantage of the characteristics of the frequency and time domains for the linear and non-linear components respectively. The hybrid algorithm is conceptually simpler and more flexible than the harmonic domain but it is not a full Newton solution and therefore not as reliable under weak system conditions.

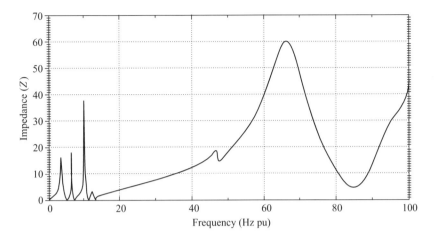

Figure 11.16 Impedance/frequency spectrum at the 25 kV bus

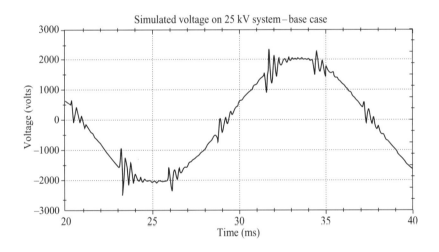

Figure 11.17 Simulated 25 kV system voltage with drive in operation

A direct time domain solution, particularly with the EMTP method, is the simplest and most reliable, but the least accurate due to the approximate modelling of the linear network components at harmonic frequencies. The latter can be overcome with the use of frequency dependent equivalents. A preliminary study of the linear part of the network provides a reduced equivalent circuit to any required matching accuracy. Then all that is needed is a single 'brute force' transient to steady state run followed by FFT processing of the resulting waveforms.

While there is still work to be done on the subject of frequency dependent equivalents, it can be confidently predicted that its final incorporation will place the

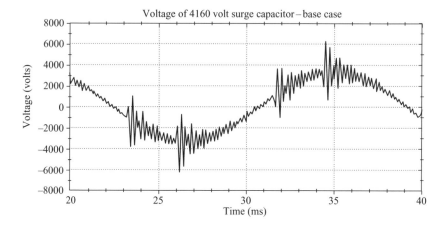

Figure 11.18 Simulated waveform at the 4.16 kV bus (surge capacitor location)

electromagnetic transient alternative in the driving seat for the assessment of power system harmonics.

Modelling of voltage sags and voltage interruptions requires accurate representation of the dynamic characteristics of the main system components, particularly the synchronous generators and induction motors, power electronic equipment and their protection and control. The EMT programs meet all these requirements adequately and can thus be used with confidence in the simulation of sag characteristics, their effects and the role of sag compensation devices.

Subject to the unpredictability of the arc furnace characteristics, EMT simulation with either deterministic or stochastic models of the arc behaviour can be used to investigate possible mitigation techniques. Flicker penetration can also be predicted with these programs, although the derivation of the IEC short and long-term flicker indices is currently computationally prohibitive. However, real-time digital simulators should make this task easier.

11.11 References

1 LOMBARD, X., MASHEREDJIAN, J., LEFEVRE, S. and KIENY, C.: 'Implementation of a new harmonic initialisation method in EMTP', *IEEE Transactions on Power Delivery*, 1995, **10** (3), pp. 1343–42
2 PERKINS, B. K., MARTI, J. R. and DOMMEL, H. W.: 'Nonlinear elements in the EMTP: steady state intialisation', *IEEE Transactions on Power Apparatus and Systems*, 1995, **10** (2), pp. 593–601
3 WANG, X., WOODFORD, D. A., KUFFEL, R. and WIERCKX, R.: 'A real-time transmission line model for a digital TNA', *IEEE Transactions on Power Delivery*, 1996, **11** (2), pp. 1092–7

4 MURERE, G., LEFEVRE, S. and DO, X. D.: 'A generalised harmonic balanced method for EMTP initialisation', *IEEE Transactions on Power Delivery*, 1995, **10** (3), pp. 1353–9
5 XU, W., MARTI, J. R. and DOMMEL, H. W.: 'A multi-phase harmonic load-flow solution technique', *IEEE Transactions on Power Systems*, 1991, **6** (1), pp. 174–82
6 HEYDT, G. T.: 'Electric power quality' (Stars in a Circle Publication, West LaFayette, 1991)
7 ARRILLAGA, J., WATSON, N. R. and CHEN, S.: 'Power system quality assessment' (John Wiley, Chichester, 2000)
8 YACAMINI, R. and DE OLIVEIRA, J. C.: 'Harmonics in multiple converter systems: a generalised approach', *Proceedings of IEE on Generation, Transmission and Distribution (Part C)*, 1980, **127** (2), pp. 96–106
9 SMITH, B. C., ARRILLAGA, J., WOOD, A. R. and WATSON, N. R.: 'A review of iterative harmonic analysis for AC-DC power systems', Proceedings of International Conference on *Harmonics and Quality of Power (ICHQP)*, Las Vegas, 1996, pp. 314–19
10 SMITH, B. C.: 'A harmonic domain model for the interaction of the HVdc convertor with ac and dc systems' (Ph.D. thesis, University of Canterbury, New Zealand, Private Bag 4800, Christchurch, New Zealand, 1996)
11 SEMLYEN, A., ACHA, A. and ARRILLAGA, J.: 'Newton-type algorithms for the harmonic phasor analysis of non-linear power circuits in periodical steady state with special reference to magnetic non-linearities', *IEEE Transactions on Power Delivery*, 1992, **7** (3), pp. 1090–9
12 SEMLYEN, A. and MEDINA, A.: 'Computation of the periodic steady state in system with non-linear components using a hybrid time and frequency domain methodology', *IEEE Transactions on Power Systems*, 1995, **10** (3), pp. 1498–1504
13 USAOLA, J. and MAYORDOMO, J. G.: 'Multifrequency analysis with time domain simulation', *ETEP*, 1996, **6** (1), pp. 53–9
14 SEMLYEN, A. and SHLASH, M.: 'Principles of modular harmonic power flow methodology', *Proceedings of IEE on Generation, Transmission and Distribution (Part C)*, 2000, **147** (1), pp. 1–6
15 USAOLA, J. and MAYORDOMO, J. G.: 'Fast steady state technique for harmonic analysis', Proceedings of International Conference on *Harmonics and Quality of Power (ICHQP IV)*, 1990, Budapest, pp. 336–42
16 WATSON, N. R. and IRWIN, G. D.: 'Electromagnetic transient simulation of power systems using root-matching techniques', *Proceedings IEE, Part C*, 1998, **145** (5), pp. 481–6
17 ANDERSON, G. W. J., ARNOLD, C. P., WATSON, N. R. and ARRILLAGA, J.: 'A new hybrid ac-dc transient stability program', International Conference on *Power Systems Transients (IPST'95)*, September 1995, pp. 535–40
18 SZECHTMAN, M., WESS, T. and THIO, C. V.: 'First benchmark model for HVdc control studies', *ELECTRA*, 1991, **135**, pp. 55–75
19 BOLLEN, M. H. J., YALCINKAYA, G. and HAZZA, G.: 'The use of electromagnetic transient programs for voltage sag analysis', Proceedings of 10th

International Conference on *Harmonics and Quality of Power (ICHQP'98)*, Athens, October 14–16, 1998, pp. 598–603
20 GOLE, A. M. and PALAV, L.: 'Modelling of custom power devices in PSCAD/EMTDC', *Manitoba HVdc Research Centre Journal*, 1998, **11** (1)
21 WOODFORD, D. A.: 'Flicker reduction in electric arc furnaces', *Manitoba HVdc Research Centre Journal*, 2001, **11** (7)
22 KEPPLER, T.: 'Flicker measurement and propagation in power systems' (Ph.D. thesis, University of Canterbury, New Zealand, Private Bag 4800, Christchurch, New Zealand, 1996)
23 TANG, L., McGRANAGHAN, M., FERRARO, R., MORGANSON, S. and HUNT, B.: 'Voltage notching interaction caused by large adjustable speed drives on distribution systems with low short-circuit capacities', *IEEE Transactions on Power Delivery*, 1996, **11** (3), pp. 1444–53

Chapter 12
Mixed time-frame simulation

12.1 Introduction

The use of a single time frame throughout the simulation is inefficient for studies involving widely varying time constants. A typical example is multimachine transient stability assessment when the system contains HVDC converters. In such cases the stability levels are affected by both the long time constant of the electromechanical response of the generators and the short time constant of the converter's power electronic control.

It is, of course, possible to include the equations of motion of the generators in the electromagnetic transient programs to represent the electromechanical behaviour of multimachine power systems. However, considering the different time constants influencing the electromechanical and electromagnetic behaviour, such approach would be extremely inefficient. Electromagnetic transient simulations use steps of (typically) 50 μs, whereas the stability programs use steps at least 200 times larger.

To reduce the computational requirements the NETOMAC package [1] has two separate modes. An instantaneous mode is used to model components in three-phase detail with small time steps in a similar way to the EMTP/EMTDC programs [2]. The alternative is a stability mode and uses r.m.s. quantities at fundamental frequency only, with increased time-step lengths. The program can switch between the two modes as required while running. The HVDC converter is either modelled elementally by resistive, inductive and capacitive components, or by quasi-steady-state equations, depending on the simulation mode. In either mode, however, the entire system must be modelled in the same way. When it is necessary to run in the instantaneous mode, a system of any substantial size would still be very computationally intensive.

A more efficient alternative is the use of a hybrid algorithm [3], [4] that takes advantage of the computationally inexpensive dynamic representation of the a.c. system in a stability program, and the accurate dynamic modelling of the power electronic components.

The slow dynamics of the a.c. system are sufficiently represented by the stability program while, at the same time, the fast dynamic response of the power electronic

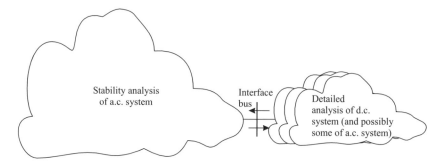

Figure 12.1 The hybrid concept

plant is accurately represented by electromagnetic simulation. A hybrid approach is particularly useful to study the impact of a.c. system dynamics, particularly weak a.c. systems, on the transient performance of HVDC converters. Disturbance response studies, control assessment and temporary overvoltage consequences are all typical examples for which a hybrid package is suited.

The basic concept, shown in Figure 12.1, is not restricted to a.c./d.c. applications only. A particular part of an a.c. system may sometimes require detailed three-phase modelling and this same hybrid approach can then be used. Applications include the detailed analysis of synchronous or static compensators, FACTS devices, or the frequency dependent effects of transmission lines.

Detailed modelling can also be applied to more than one independent part of the complete system. For example, if an a.c. system contains two HVDC links, then both links can be modelled independently in detail and their behaviour included in one overall a.c. electromechanical stability program.

12.2 Description of the hybrid algorithm

The proposed hybrid algorithm utilises electromechanical simulation as the steering program while the electromagnetic transients program is called as a subroutine. The interfacing code is written in separate routines to minimise the number of modifications and thus make it easily applicable to any stability and dynamic simulation programs. To make the description more concise, the component programs are referred to as TS (for transient stability) and EMTDC (for electromagnetic transient simulation). The combined hybrid algorithm is called TSE.

With reference to Figure 12.2(a), initially the TSE hybrid reads in the data files, and runs the entire network in the stability program, until electromechanical steady-state equilibrium is reached. The quasi-steady-state representation of the converter is perfectly adequate as no fault or disturbance has yet been applied. At a selectable point in time prior to a network disturbance occurring, the TS network is split up into the two independent and isolated systems, system 1 and system 2.

Mixed time-frame simulation 305

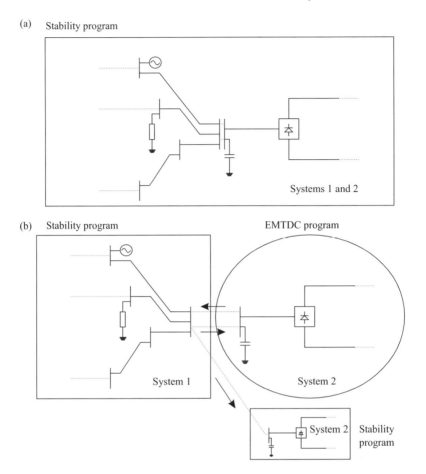

Figure 12.2   Example of interfacing procedure

For the sake of clarity system 1 is classified as the a.c. part of the system modelled by the stability program TS, while system 2 is the part of the system modelled in detail by EMTDC.

The snapshot data file is now used to initialise the EMTDC program used, instead of the TS representation of system 2. The two programs are then interfaced and the network disturbance can be applied. The system 2 representation in TS is isolated but kept up to date during the interfacing at each TS time step to allow tracking between programs. The a.c. network of system 1 modelled in the stability program also supplies interface data to this system 2 network in TS as shown in Figure 12.2(b).

While the disturbance effects abate, the quasi-steady-state representation of system 2 in TS and the EMTDC representation of system 2 are tracked. If both of these system 2 models produce the same results within a predefined tolerance and over a set period, the complete system can then be reconnected and used by TS, and the EMTDC

306 *Power systems electromagnetic transients simulation*

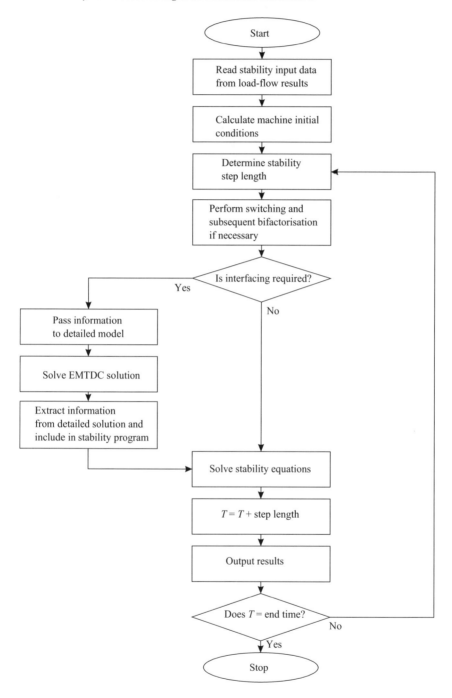

Figure 12.3 Modified TS steering routine

representation terminated. This allows better computational efficiency, particularly for long simulation runs.

12.2.1 Individual program modifications

To enable EMTDC to be called as a subroutine from TS requires a small number of changes to its structure. The EMTDC algorithm is split into three distinct segments, an initialising segment, the main time loop, and a termination segment. This allows TS to call the main time loop for discrete periods as required when interfacing. The EMTDC options, which are normally available when beginning a simulation run, are moved to the interface data file and read from there. The equivalent circuit source values, which TS updates periodically, are located in the user accessible DSDYN file of EMTDC (described in Appendix A).

A TS program, such as the one described in reference [5], requires only minor modifications. The first is a call of the interfacing routine during the TS main time loop as shown in Figure 12.3. The complete TS network is also split into system 1 and system 2 and isolated at the interface points, but this is performed in separate code to TS. The only other direct modification inside TS is the inclusion of the interface current injections at each TS network solution.

12.2.2 Data flow

Data for the detailed EMTDC model is entered in the program database via the PSCAD graphics. Equivalent circuits are used at each interface point to represent the rest of the system not included in the detailed model. This system is then run until steady state is reached and a 'snapshot' taken. The snapshot holds all the relevant data for the components at that point in time and can be used as the starting point when interfacing the detailed model with the stability program.

The stability program is initialised conventionally through power flow results via a data file. An interface data file is also read by the TSE hybrid and contains information such as the number and location of interface buses, analysis options, and timing information.

12.3 TS/EMTDC interface

Hybrid simulation requires exchange of information between the two separate programs. The information that must be transferred from one program to the other must be sufficient to determine the power flow in or out of the interface. Possible parameters to be used are the real power P, the reactive power Q, the voltage $\tilde{V}$ and the current $\tilde{I}$ at the interface (Figure 12.4). Phase angle information is also required if separate phase frames of reference are to be maintained. An equivalent circuit representing the network modelled in the stability program is used in EMTDC and vice versa. The equivalent circuits are as shown in Figure 12.5, where $\tilde{E}_1$ and $\tilde{Z}_1$ represent the equivalent circuit of system 1 and $\tilde{I}_c$ and $\tilde{Z}_2$ the equivalent circuit of system 2.

308 *Power systems electromagnetic transients simulation*

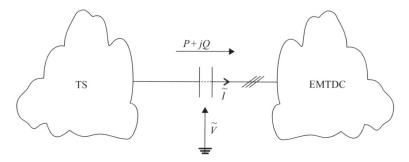

Figure 12.4 Hybrid interface

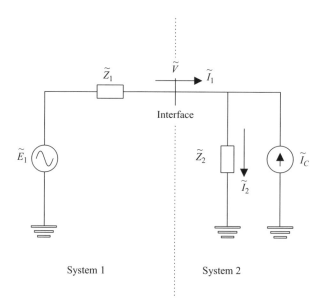

Figure 12.5 Representative circuit

12.3.1 Equivalent impedances

The complexity of the equivalent impedance representation varies considerably between the two programs.

In the TS program, $\tilde{I}_c$ and $\tilde{Z}_2$ represent the detailed part of the system modelled by EMTDC. TS, being positive-sequence and fundamental-frequency based, is concerned only with the fundamental real and reactive power in or out-flow through the interface. The equivalent impedance $\tilde{Z}_2$ is then arbitrary, since the current source $\tilde{I}_c$ can be varied to provide the correct power flow.

To avoid any possible numerical instability, a constant value of $\tilde{Z}_2$, estimated from the initial power flow results, is used for the duration of the simulation.

The EMTDC program represents system 1 by a Thevenin equivalent ($\tilde{E}_1$ and $\tilde{Z}_1$) as shown in Figure 12.5. The simplest $\tilde{Z}_1$ is an R–L series impedance, representing the fundamental frequency equivalent of system 1. It can be derived from the results of a power flow and a fault analysis at the interface bus.

The power flow provides an initial current through the interface bus and the initial interface bus voltage. A fault analysis can easily determine the fault current through the interface for a short-circuit fault to ground. If the network requiring conversion to an equivalent circuit is represented by a Thevenin source E_1 and Thevenin impedance $\tilde{Z}_1$, as shown in Figure 12.6, these values can thus be found as follows.

From the power flow circuit:

$$\tilde{E}_1 = \tilde{I}_n \tilde{Z}_1 + \tilde{V} \tag{12.1}$$

and from the fault circuit:

$$\tilde{E}_1 = \tilde{I}_F \tilde{Z}_1 \tag{12.2}$$

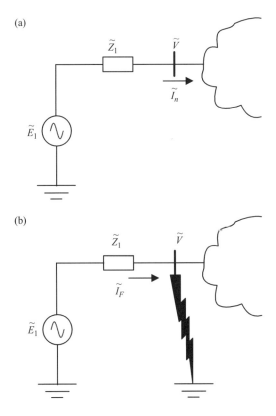

Figure 12.6 Derivation of Thevenin equivalent circuit: (a) power-flow circuit (b) fault circuit

310 *Power systems electromagnetic transients simulation*

Combining these two equations:

$$\tilde{Z}_1 = \frac{\tilde{V}}{\tilde{I}_F - \tilde{I}_n} \qquad (12.3)$$

$\tilde{E}_1$ can then be found from either equation 12.1 or 12.2.

During a transient, the impedance of the synchronous machines in system 1 can change. The net effect on the fundamental power in or out of the equivalent circuit, however, can be represented by varying the source $\tilde{E}_1$ and keeping $\tilde{Z}_1$ constant.

EMTDC is a 'point on wave' type program, and consequently involves frequencies other than the fundamental. A more advanced equivalent impedance capable of representing different frequencies is used in section 12.6.

12.3.2 Equivalent sources

Information from the EMTDC model representing system 2 (in Figure 12.5) is used to modify the source of the equivalent circuit of system 2 in the stability program. Similarly, data from TS is used to modify the source of the equivalent circuit of system 1 in EMTDC. These equivalent sources are normally updated at each TS step length (refer to section 12.5). From Figure 12.5, if both $\tilde{Z}_1$ and $\tilde{Z}_2$ are known, additional information is still necessary to determine update values for the sources $\tilde{I}_c$ and $\tilde{E}_1$. This information can be selected from the interface parameters of voltage $\tilde{V}$, current $\tilde{I}_1$, real power P, reactive power Q and power factor angle ϕ.

The interface voltage and current, along with the phase angle between them, are used to interchange information between programs.

12.3.3 Phase and sequence data conversions

An efficient recursive curve fitting algorithm is described in section 12.4 to extract fundamental frequency information from the discrete point oriented waveforms produced by detailed programs such as EMTDC.

Analysis of the discrete data from EMTDC is performed over a fundamental period interval, but staggered to produce results at intervals less than a fundamental period. This allows the greatest accuracy in deriving fundamental results from distorted waveforms.

The stability program requires only positive sequence data, so data from the three a.c. phases at the interface(s) is analysed and converted to a positive sequence by conventional means. The positive sequence voltage, for example, can be derived as follows:

$$\tilde{V}_{\text{ps}} = \tfrac{1}{3} \left(\tilde{V}_a + \tilde{a}\tilde{V}_b + \tilde{a}^2\tilde{V}_c \right) \qquad (12.4)$$

where

$\tilde{V}_{\text{ps}}$ = positive sequence voltage
$\tilde{V}_a, \tilde{V}_b, \tilde{V}_c$ = phase voltages
$\tilde{a}$ = 120 degree forward rotation vector (i.e. $a = 1\angle 120°$).

Positive sequence data from the stability program is converted to three-phase through simple multiplication of the rotation vector, i.e. for the voltage:

$$\tilde{V}_a = \tilde{V}_{ps} \tag{12.5}$$

$$\tilde{V}_b = \tilde{a}^2 \tilde{V}_{ps} \tag{12.6}$$

$$\tilde{V}_c = \tilde{a} \tilde{V}_{ps} \tag{12.7}$$

12.3.4 Interface variables derivation

In Figure 12.5, $\tilde{E}_1$ and $\tilde{Z}_1$ represent the equivalent circuit of system 1 modelled in EMTDC, while $\tilde{Z}_2$ and $\tilde{I}_c$ represent the equivalent circuit of system 2 modelled in the stability program. $\tilde{V}$ is the interface voltage and $\tilde{I}_1$ the current through the interface which is assumed to be in the direction shown.

From the detailed EMTDC simulation, the magnitude of the interface voltage and current are measured, along with the phase angle between them. This information is used to modify the equivalent circuit source ($\tilde{I}_c$) of system 2 in TS. The updated $\tilde{I}_c$ value can be derived as follows:

From Figure 12.5

$$\tilde{E}_1 = \tilde{I}_1 \tilde{Z}_1 + \tilde{V} \tag{12.8}$$

$$\tilde{V} = \tilde{I}_2 \tilde{Z}_2 \tag{12.9}$$

$$\tilde{I}_2 = \tilde{I}_1 + \tilde{I}_c \tag{12.10}$$

From equations 12.9 and 12.10

$$\tilde{V} = \tilde{I}_1 \tilde{Z}_2 + \tilde{I}_c \tilde{Z}_2 \tag{12.11}$$

From equation 12.8

$$\tilde{E}_1 = I_1 Z_1 \angle(\theta_{I_1} + \theta_{Z_1}) + V \angle \theta_V$$
$$= I_1 Z_1 \cos(\theta_{I_1} + \theta_{Z_1}) + j I_1 Z_1 \sin(\theta_{I_1} + \theta_{Z_1})$$
$$+ V \cos(\theta_V) + j V \sin(\theta_V) \tag{12.12}$$

and

$$\theta_{I_1} = \theta_V - \phi \tag{12.13}$$

where ϕ is the displacement angle between the voltage and the current. Thus, equation 12.12 can be written as

$$\tilde{E}_1 = I_1 Z_1 \cos(\theta_V + \beta) + j I_1 Z_1 \sin(\theta_V + \beta) + V \cos \theta_V + j \sin \theta_V$$
$$= I_1 Z_1 (\cos \theta_V \cos \beta - \sin \theta_V \sin \beta) + V \cos \theta_V$$
$$+ j[I_1 Z_1 (\sin \theta_V \cos \beta + \cos \theta_V \sin \beta) + V \sin \theta_V] \tag{12.14}$$

where $\beta = \theta_{Z_1} - \phi$.

If $\tilde{E}_1 = E_{1r} + jE_{1i}$ then equating real terms only

$$E_{1r} = (I_1 Z_1 \cos(\beta) + V)\cos(\theta_V) + (-I_1 Z_1 \sin(\beta))\sin(\theta_V) \quad (12.15)$$

where $\tilde{Z}_1$ is known and constant throughout the simulation.

From the EMTDC results, the values of V, I, and the phase difference ϕ are also known and hence so is β. $\tilde{E}_1$ can be determined in the TS phase reference frame from knowledge of $\tilde{Z}_1$ and the previous values of interface current and voltage from TS, through the use of equation 12.8.

From equation 12.15, making

$$A = I_1 Z_1 \cos(\beta) + V \quad (12.16)$$

$$B = -I_1 Z_1 \sin(\beta) \quad (12.17)$$

and remembering that

$$A\cos(\theta_V) + B\sin(\theta_V) = \sqrt{A^2 + B^2}\cos((\theta_V) \pm \psi) \quad (12.18)$$

where

$$\left(\psi = \tan^{-1}\left[\frac{-B}{A}\right]\right) \quad (12.19)$$

the voltage angle θ_V in the TS phase reference frame can be calculated, i.e.

$$\theta_V = \cos^{-1}\left[\frac{E_{1r}}{\sqrt{A^2 + B^2}}\right] - \psi \quad (12.20)$$

The equivalent current source $\tilde{I}_c$ can be calculated by rearranging equation 12.11:

$$\tilde{I}_c = \frac{V}{Z_2}\angle(\theta_V - \theta_{Z_2}) - I_1 \angle \theta_{I_1} \quad (12.21)$$

where θ_{I_1} is obtained from equation 12.13.

In a similar way, data from the transient stability program simulation can be used to calculate a new Thevenin source voltage magnitude for the equivalent circuit of system 1 in the EMTDC program. Knowing the voltage and current magnitude at the TS program interface and the phase difference between them, by a similar analysis the voltage angle in the EMTDC phase reference frame is:

$$\theta_V = \cos^{-1}\left[\frac{I_{cr}}{\sqrt{C^2 + D^2}}\right] - \psi \quad (12.22)$$

where I_{cr} is the real part of $\tilde{I}_c$ and

$$C = \frac{V}{Z_2}\cos\theta_{Z_2} - I_1\cos\phi \qquad (12.23)$$

$$D = \frac{V}{Z_2}\sin\theta_{Z_2} - I_1\sin\phi \qquad (12.24)$$

$$\phi = \theta_V - \theta_{I_1} \qquad (12.25)$$

$$\psi = \tan^{-1}\left[\frac{-D}{C}\right] \qquad (12.26)$$

Knowing the EMTDC voltage angle θ_V allows calculation of the EMTDC current angle θ_{I_1} from equation 12.25. The magnitude value of E_1 can then be derived from equation 12.8.

12.4 EMTDC to TS data transfer

A significant difference between TS and EMTDC is that in TS, sinusoidal waveforms are assumed. However, during faults the EMTDC waveforms are very distorted.

The total r.m.s. power is not always equivalent to either the fundamental frequency power nor the fundamental frequency positive sequence power. A comparison of these three powers following a single-phase fault at the inverter end of a d.c. link is shown in Figure 12.7. The difference between the total r.m.s. power and the positive sequence power can be seen to be highly significant during the fault.

The most appropriate power to transfer from EMTDC to TS is then the fundamental frequency positive sequence power. This, however, requires knowledge of both the fundamental frequency positive sequence voltage and the fundamental frequency positive sequence current. These two variables contain all the relevant information and, hence, the use of any other power variable to transfer information becomes unnecessary.

12.4.1 Data extraction from converter waveforms

At each step of the transient stability program, power transfer information needs to be derived from the distorted converter waveforms. This can be achieved using the FFT, which provides accurate information for the whole frequency spectrum. However, only the fundamental frequency is used in the stability program and a simpler recursive least squares curve fitting algorithm (CFA) (described in Appendix B.5 [4]), provides sufficient accuracy.

12.5 Interaction protocol

The data from each program must be interchanged at appropriate points during the hybrid simulation run. The timing of this data interchange between the TS and

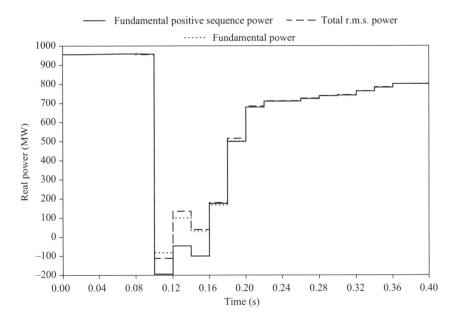

Figure 12.7   Comparison of total r.m.s. power, fundamental frequency power and fundamental frequency positive sequence power

EMTDC programs is important, particularly around discontinuities caused by fault application and removal.

The interfacing philosophy for TS step lengths which are less than a fundamental period is shown in Figure 12.8. A portion of the figure is sequentially numbered to show the order of occurrence of the variable interchange. In the example, the stability step length is exactly one half of a fundamental period.

Following the sequential numbering on Figure 12.8, at a particular point in time, the EMTDC and TS programs are concurrent and the TS information from system 1 is passed to update the system 1 equivalent in EMTDC. This is shown by the arrow marked 1. EMTDC is then called for a length of half a fundamental period (arrow 2) and the curve fitted results over the last full fundamental period processed and passed back to update the system 2 equivalent in TS (arrow 3). The information over this period is passed back to TS at the mid-point of the EMTDC analysis window which is half a period behind the current EMTDC time. TS is then run to catch up to EMTDC (arrow 4), and the new information over this simulation run used to again update the system 1 equivalent in EMTDC (arrow 5). This protocol continues until any discontinuity in the network occurs.

When a network change such as a fault application or removal occurs, the interaction protocol is modified to that shown in Figure 12.9. The curve fitting analysis process is also modified to avoid applying an analysis window over any point of discontinuity.

Mixed time-frame simulation 315

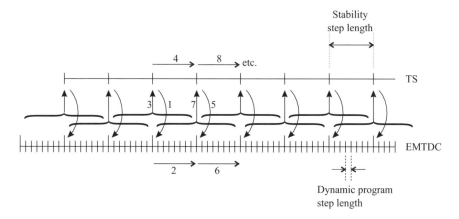

Figure 12.8 Normal interaction protocol

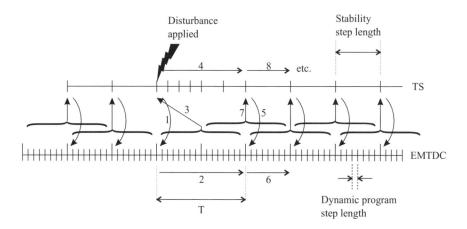

Figure 12.9 Interaction protocol around a disturbance

The sequential numbering in Figure 12.9 explains the flow of events. At the fault time, the interface variables are passed from TS to the system 1 equivalent in EMTDC in the usual manner, as shown by the arrow marked 1. Neither system 1 nor system 2 have yet been solved with the network change. The fault is now applied in EMTDC, which is then run for a full fundamental period length past the fault application (arrow 2), and the information obtained over this period passed back to TS (arrow 3). The fault is now also applied to the TS program which is then solved for a period until it has again reached EMTDC's position in time (arrow 4). The normal interaction protocol is then followed until any other discontinuity is reached.

A full period analysis after the fault is applied is necessary to accurately extract the fundamental frequency component of the interface variables. The mechanically controlled nature of the a.c. system implies a dynamically slow response to any disturbance and so, for this reason, it is considered acceptable to run EMTDC for a full period without updating the system 1 equivalent circuit during this time.

12.6 Interface location

The original intention of the initial hybrid algorithm [6] was to model the a.c. and d.c. solutions separately. The point of interface location was consequently the converter bus terminal. The detailed d.c. link model included all equipment connected to the converter bus, such as the a.c. filters, and every other a.c. component was modelled within the stability analysis. A fundamental frequency Thevenin's equivalent was used to represent the stability program in the detailed solution and vice versa.

An alternative approach has been proposed [7] where the interface location is extended out from the converter bus into the a.c. system. This approach maintains that, particularly for weak a.c. systems, a fundamental frequency equivalent representing the a.c. system is not sufficiently adequate at the converter terminals. In this case, the extent of the a.c. system to be included in the d.c. system depends on phase imbalance and waveform distortion.

Although the above concept has some advantages, it also suffers from many disadvantages. The concept is proposed, in particular, for weak a.c. systems. A weak a.c. system, however, is likely to have any major generation capability far removed from the converter terminal bus as local generation serves to enhance system strength. If the generation is, indeed, far removed out into the a.c. system, then the distance required for an interface location to achieve considerably less phase imbalance and waveform distortion is also likely to be significant.

The primary advantage of a hybrid solution is in accurately providing the d.c. dynamic response to a transient stability program, and in efficiently representing the dynamic response of a considerably sized a.c. system to the d.c. solution. Extending the interface some distance into the a.c. system, where the effects of a system disturbance are almost negligible, diminishes the hybrid advantage. If a sizeable portion of the a.c. system requires modelling in detail before an interface to a transient stability program can occur, then one might question the use of a hybrid solution at all and instead use a more conventional approach of a detailed solution with a.c. equivalent circuits at the system cut-off points.

Another significant disadvantage in an extended interface is that a.c. systems may well be heavily interconnected. The further into the system that an interface is moved, the greater the number of interface locations required. The hybrid interfacing complexity is thus increased and the computational efficiency of the hybrid solution decreased. The requirement for a detailed representation of a significant portion of the a.c. system serves to decrease this efficiency, as does the increased amount of processing required for variable extraction at each interface location.

The advantages of using the converter bus as the interface point are:

- The detailed system is kept to a minimum.
- Interfacing complexity is low.

- Converter terminal equipment, such as filters, synchronous condensers, SVCs, etc. can still be modelled in detail.

The major drawback of the detailed solution is in not seeing a true picture of the a.c. system, since the equivalent circuit is fundamental-frequency based. Waveform distortion and imbalance also make it difficult to extract the fundamental frequency information necessary to transfer to the stability program.

The problem of waveform distortion for transfer of data from EMTDC to TS is dependent on the accuracy of the technique for extraction of interfacing variable information. If fundamental-frequency quantities can be accurately measured under distorted conditions, then the problem is solved. Section 12.4 has described an efficient way to extract the fundamental frequency quantities from distorted waveforms.

Moreover, a simple fundamental frequency equivalent circuit is insufficient to represent the correct impedance of the a.c. system at each frequency. Instead, this can be achieved by using a fully frequency dependent equivalent circuit of the a.c. system [8] at the converter terminal instead of just a fundamental frequency equivalent. A frequency dependent equivalent avoids the need for modelling any significant portion of the a.c. system in detail, yet still provides an accurate picture of the system impedance across its frequency spectra.

12.7 Test system and results

The test system shown in Figure 11.3 is also used here. As explained in section 12.6, the high levels of current distortion produced by the converter during the disturbance require a frequency dependent model of the a.c. system.

A three-phase short-circuit is applied to the rectifier terminals of the link at $t = 1.7$ s and cleared three cycles later.

The rectifier d.c. currents, displayed for the three solutions in Figure 12.10, show a very similar variation for the TSE and EMTDC solutions, except for the region between $t = 2.03$ s and $t = 2.14$ s but the difference with the TS only solution is very large.

Figure 12.11 compares the fundamental positive sequence real and resistive powers across the converter interface for the TS and TSE solutions.

The main differences in real power occur during the link power ramp. The difference is almost a direct relation to the d.c. current difference between TS and TSE shown in Figure 12.10. The oscillation in d.c. voltage and current as the rectifier terminal is de-blocked is also evident.

As for the reactive power Q, prior to the fault, a small amount is flowing into the system due to surplus MVArs at the converter terminal. The fault reduces this power flow to zero. When the fault is removed and the a.c. voltage overshoots in TSE, the reactive MVArs also overshoot in TSE and since the d.c. link is shut down, a considerable amount of reactive power flows into the system.

Finally, the machine angle swings with respect to the Clyde generator, shown in Figure 12.12, indicating that the system is transiently stable.

318 *Power systems electromagnetic transients simulation*

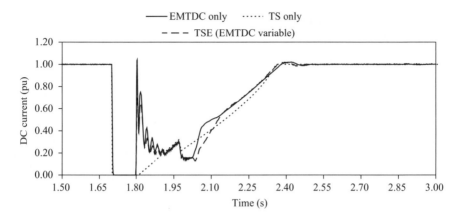

Figure 12.10 Rectifier terminal d.c. current comparisons

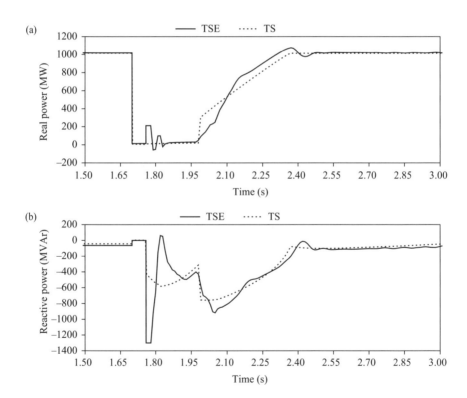

Figure 12.11 Real and reactive power across interface

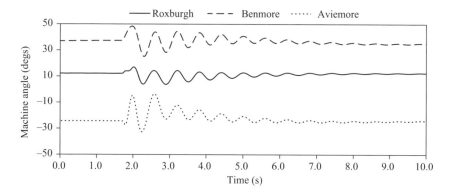

Figure 12.12   Machine variables – TSE (TS variables)

12.8   Discussion

It has been shown that mixed time-frame simulation requires elaborate interfaces between the component programs. Therefore, considering the increased cheap computer power availability, it would be difficult to justify its use purely in terms of computation efficiency.

The EMTP method has already proved its value in practically all types of power systems transient. Its effectiveness has also been extended to problems involving electromechanical oscillations, like in the case of subsynchronous resonance.

The only possible area of application for the mixed time-frame solution is in multimachine transient stability when the system contains HVDC transmission. The criterion used to decide on the prospective use of a mixed time-frame solution in this case is the inability of the power electronic controller to take the specified control action within the integration steps of the stability program. It has been shown in the chapter that the inverter behaviour of a conventional HVDC link is unpredictable during a.c. system faults. This is caused by the occurrence of commutation failures and by the filter's response. Thus the effectiveness of the mixed time-frame alternative has been clearly demonstrated for this application.

However the criterion is not met by other power electronic devices, such as FACTS. These do not suffer from commutation failures, either because they do not use inverters (e.g. thyristor controlled series capacitors, static VAR compensators, etc.) or they use turn-off switching (e.g. STATCOM and unified power flow controller). In all these cases the use of a quasi-steady-state model will be perfectly adequate for the needs of the stability study.

12.9   References

1   KULICKE, B.: 'NETOMAC digital program for simulating electromechanical and electromagnetic transient phenomena in AC power systems', *Elektrizitätswirtschaft*, **1**, 1979, pp. 18–23

2 WOODFORD, D. A., INO, T., MATHUR, R. M., GOLE, A. M. and WIERCKX, R.: 'Validation of digital simulation of HVdc transients by field tests', IEE Conference Publication on AC and DC power transmission, 1985, **255**, pp. 377–81

3 ANDERSON, G. W. J., ARNOLD, C. P., WATSON, N. R. and ARRILLAGA, J.: 'A new hybrid ac-dc transient stability program', International Conference on *Power Systems Transients (IPST '95)*, September 1995, pp. 535–40

4 ANDERSON, G. W. J.: 'Hybrid simulation of ac-dc power systems' (Ph.D. thesis, University of Canterbury, New Zealand, Private Bag 4800, Christchurch, New Zealand, 1995)

5 ARRILLAGA, J. and WATSON, N. R.: 'Computer modelling of electrical power systems' (John Wiley, Chichester, 2nd edition, 2001)

6 HEFFERNAN, M. D., TURNER, K. S. and ARRILLAGA, J.: 'Computation of AC-DC system disturbances, parts I, II and III', *IEEE Transactions on Power Apparatus and Systems*, 1981, **100** (11), pp. 4341–63

7 REEVE, J. and ADAPA, R.: 'A new approach to dynamic analysis of AC networks incorporating detailed modelling of DC systems, part I and II', *IEEE Transactions on Power Delivery*, 1988, **3** (4), pp. 2005–19

8 WATSON, N. R.: 'Frequency-dependent A.C. system equivalents for harmonic studies and transient convertor simulation' (Ph.D. thesis, University of Canterbury, New Zealand, 1987)

Chapter 13
Transient simulation in real time

13.1 Introduction

Traditionally the simulation of transient phenomena in real time has been carried out on analogue simulators. However their modelling limitations and costly maintenance, coupled with the availability of cheap computing power, has restricted their continued use and further development. Instead, all the recent development effort has gone into digital transient network analysers (DTNA) [1], [2].

Computing speed by itself would not justify the use of real-time simulation, as there is no possibility of human interaction with information derived in real time. The purpose of their existence is two-fold, i.e. the need to test control [3], [4] and protection [5]–[8] equipment in the power network environment and the simulation of system performance taking into account the dynamics of such equipment.

In the 'normal' close-loop testing mode, the real-time digital simulator must perform continuously all the necessary calculations in a time step less than that of actual time. This allows closed-loop testing involving the actual hardware, which in turn influences the simulation model, as indicated in Figure 13.1. Typical examples of signals that can be fed back are the relay contacts controlling the circuit breaker in simulation and the controller modifying the firing angle of a converter model.

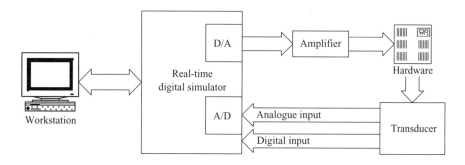

Figure 13.1 *Schematic of real-time digital simulator*

322 *Power systems electromagnetic transients simulation*

The processing power required to solve the system equations in real time is immense and the key to achieving it with present computer technology is the use of parallel processing. Chapter 4 has shown that the presence of transmission lines (with a travelling wave time of one time step or more) results in a block diagonal structure of the conductance matrix, with each block being a subsystem. The propagation of a disturbance from one end of the line to the other is delayed by the line travelling time. Therefore, the voltages and currents in a subsystem can be calculated at time t without information about the voltages and currents in the other subsystems at this time step. Thus by splitting the system into subsystems the calculations can be performed in parallel without loss of accuracy.

Although in the present state of development DTNAs are limited in the size and detail of system representation, they are already a considerable improvement on the conventional TNAs in this respect. The main advantages of digital over analogue simulators are:

- Cost
- Better representation of components, particularly high-frequency phenomena
- Faster and easier preparation for tests
- Ease and flexibility for entering new models
- Better consistency (repeatability) in simulation results.

Some applications use dedicated architectures to perform the parallel processing. For instance the RTDS uses DSPs (digital signal processors) to perform the calculations. However, the ever increasing processing power of computers is encouraging the development of real-time systems that will run on standard parallel computers. Eventually this is likely to result in lower cost as well as provide portability of software and simplify future upgrading as computer systems advance [9].

Regardless of the type of DTNA hardware, real-time simulation requires interfacing with 'physical' equipment. The main interface components are digital to analogue converters (DACs), amplifiers and analogue to digital converters (ADCs).

13.2 Simulation with dedicated architectures

The first commercial real-time digital simulator was released by RTDS Technologies in 1991; an early prototype is shown in Figure 13.2. The RTDS (in the middle of the picture) was interfaced to the controller of an HVDC converter (shown on the left) to assess its performance; the amplifiers needed to interface the digital and analogue parts are shown on the right of the picture.

However, recent developments have made it possible to achieve real-time large-scale simulation of power systems using fully digital techniques. The latter provide more capability, accuracy and flexibility at a much lower cost. The new HVDC control equipment, include the phase-locked oscillator with phase limits and frequency correction, various inner control loops (I_{dc}, V_{dc} AC overvoltage, and γ limit control), control loop selection, voltage dependent current order limits (VDCOL),

Transient simulation in real time 323

Figure 13.2 Prototype real-time digital simulator

and balancing, and power trim control. An improved firing algorithm (IFA) has also been added to overcome the jitter effect that results when firing pulses arrive asynchronously during a time-step, as double interpolation is not used.

The largest RTDS delivered so far (to the Korean KEPCO network) simulates in real-time (with 50 μs time step) a system of 160 buses, 41 generators, 131 transmission lines, 78 transformers and 60 dynamic loads.

The RTDS can be operated with or without user interaction (i.e. on batch mode), whereby the equipment can be subjected to thousands of tests without supervision. In that mode the simulator provides detail reports on the equipment's response to each test case.

The main hardware and software components of the present RTDS design are discussed next.

13.2.1 Hardware

The RTDS architecture consists of one or more racks installed in a cubicle that also house the auxiliary components (power supplies, cooling fans, ... etc.). A rack, illustrated in Figure 13.3, contains up to 18 processor cards and two communication cards. Currently two types of processor cards are available, i.e. the tandem processor card (TPC) and the triple processor card (3PC). Two types of communication card are also required to perform the simulations, i.e. the workstation interface card (WIC) and the inter-rack communications card (IRC). The functions of the various cards are as

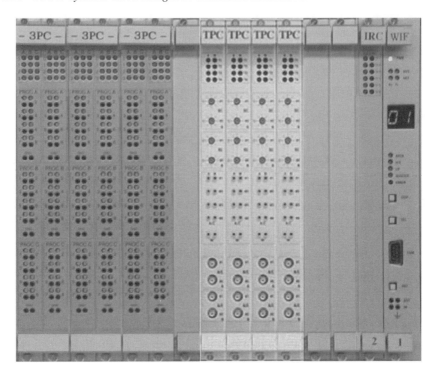

Figure 13.3 Basic RTDS rack

follows:

Tandem processor card
The TPC is used to perform the computations required to model the power system. One TPC contains two independent digital signal processors (DSPs) and its hardware is not dedicated to a particular system component. Therefore, it may participate in the modelling of a transformer in one case, while being used to model a synchronous machine or a transmission line in another case.

Triple processor card
The 3PC is used to model complex components, such as FACTS devices, which cannot be modelled by a TPC. The 3PC is also used to model components which require an excessive number of TPC processors. Each 3PC contains three analogue devices (ADSP21062), based on the SHARC (Super Harvard ARchitecture) chip; these enable the board to perform approximately six times as many instructions as a TPC in any given period.

Similarly to the TPC, the function of a given processor is not component dedicated.

Inter-rack communication card
The IRC card permits direct communications between the rack in which it is installed and up to six other racks. In a multirack simulation, the equations representing

different parts of the power system can be solved in parallel on the individual racks and the required data exchanged between them via the IRC communication channels.

Thus a multirack RTDS is able to simulate large power systems and still maintain real-time operation. The IRC communication channels are dedicated and different from the Ethernet communications between the host workstation and the simulator.

Workstation interface card

The WIC is an M68020-based card, whose primary function is to handle the communications requests between the RTDS simulator and the host workstation. Each card contains an Ethernet transceiver and is assigned its own Ethernet address, thus allowing the connection of the RTDS racks to any standard Ethernet-based local area network.

All the low level communication requests between the simulator and the host workstation are handled by the high level software running on the host workstation and the multitasking operating system being run by the WIC's M68020 processor.

RTDS simulation uses two basic software tools, a Library of Models and Compilers and PSCAD, a Graphical User Interface.

PSCAD allows the user to select a pictorial representation of the power system or control system components from the library in order to build the desired circuit. The structure of PSCAD is described in Appendix A with reference to the EMTDC program. Although initially the RTDS PSCAD was the original EMTDC version, due to the RTDS special requirements, it has now developed into a different product. The latter also provides a script language to help the user to describe a sequence of commands to be used for either simulation, output processing or circuit modification. This facility, coupled with the multi-run feature, allows many runs to be performed quickly under a variety of operating conditions.

Once the system has been drawn and the parameters entered, the appropriate compiler automatically generates the low level code necessary to perform the simulation using the RTDS. Therefore this software determines the function of each processor card for each simulation. In addition, the compiler automatically assigns the role that each DSP will play during the simulation, based on the required circuit layout and the available RTDS hardware. It also produces a user readable file to direct the user to I/O points which may be required for interfacing of physical measurement, protection or control equipment. Finally, subsystems of tightly coupled components can be identified and assigned to different RTDS racks in order to reduce the computational burden on processors.

The control system software allows customisation of control system modules. It also provides greater flexibility for the development of sequences of events for the simulations.

13.2.2 RTDS applications

Protective relay testing

Combined with appropriate voltage and current amplification, the RTDS can be used to perform closed-loop relay tests, ranging from the application of simple voltage and current waveforms through to complicated sequencing within a complex power

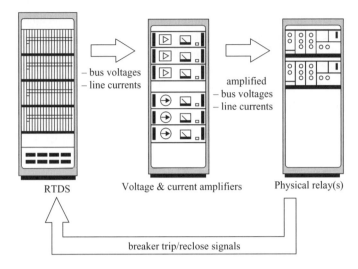

Figure 13.4 RTDS relay set-up

system model. The availability of an extensive library, which includes measurement transducers, permits testing the relays under realistic system conditions. The relay is normally connected via analogue output channels to voltage and current amplifiers. Auxiliary contacts of the output relay are, in turn, connected back to circuit breaker models using the RTDS digital input ports. A sketch of the relay testing facility is shown in Figure 13.4.

By way of example, Figure 13.5 shows a typical set of voltages and currents at the location of a distance protection relay [5]. The fault condition was a line-to-line short on the high voltage side of a generator step-up transformer connected to a transmission line. The diagrams indicate the position of the relay trip signal, the circuit breakers opening (at current zero crossings) and the reclosing of the circuit breaker after fault removal.

Control system testing

Similarly to the concept described above for protection relay testing, the RTDS can be applied to the evaluation and testing of control equipment. The signals required by the control system (analogue and/or digital) are produced during the power system simulation, while the controller outputs are connected to input points on the particular power system component under simulation. This process closes the loop and permits the evaluation of the effect of the control system on the system under test.

Figure 13.6 illustrates a typical configuration for HVDC control system tests, where analogue voltage and current signals are passed to the control equipment, which in turn issues firing pulses to the HVDC converter valves in the power system model [9].

Figure 13.7 shows typical captured d.c. voltage and current waveforms that occur following a three-phase line to ground fault at the inverter end a.c. system.

Transient simulation in real time 327

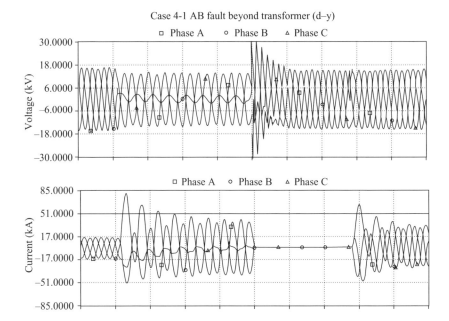

Figure 13.5   Phase distance relay results

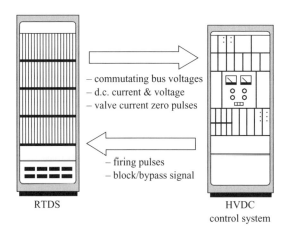

Figure 13.6   HVDC control system testing

13.3 Real-time implementation on standard computers

This section describes a DTNA that can perform real-time tests on a standard multipurpose parallel computer. The interaction between the real equipment under test and the simulated power system is carried out at every time step. A program based on the parallel processing architecture is used to reduce the solution time [10], [11].

328 *Power systems electromagnetic transients simulation*

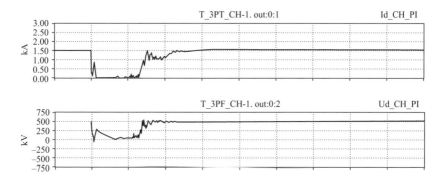

Figure 13.7 *Typical output waveforms from an HVDC control study*

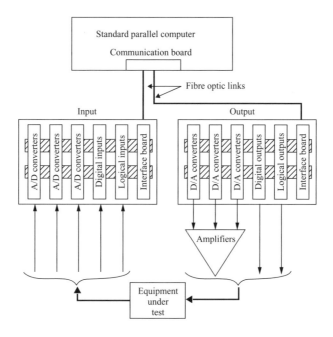

Figure 13.8 *General structure of the DTNA system*

The general structure of the DTNA system is shown in Figure 13.8. A standard HP-CONVEX computer is used, with an internal architecture based on a crossbar that permits complete intercommunication between the different processors. This increases the computing power linearly with the number of processors, unlike most computers, which soon reach their limit due to bus congestion.

The basic unit input/output (I/O) design uses two VME racks (for up to 32 analogue channels) and allows the testing of three relays simultaneously. Additional VME racks and I/O boards can be used to increase the number of test components. The only special-purpose device to be added to the standard computer is a communication board, needed to interface the computer and the I/O systems.

Each board provides four independent 16-bit ADC and DAC converters, allowing the simultaneous sampling of four analogue inputs. Moreover, all the boards are synchronised to ensure that all the signals are sampled at exactly the same time.

Each of the digital and logical I/O units provides up to 96 logical channels or 12 digital channels. Most standard buses are able to handle large quantities of data but require relatively long times to initialise each transmission. In this application, however, the data sent at each time step is small but the transmission speed must be fast; thus, the VME based architecture must meet such requirements. Like other EMTP based algorithms, the ARENE's version uses a linear interpolation to detect the switching instants, i.e. when a switching occurs at t_x (in the time step between t and $t + \Delta t$) then the solution is interpolated back to t_x. However, as some of the equipment (e.g. the D/A converters and amplifiers) need equal spacing between data points, the new values at t_x are used as $t + \Delta t$ values. Then, in the next step an extrapolation is performed to get back on to the $t + 2\Delta t$ step [12]–[15].

Finally the characteristics and power rating of the amplifiers depend on the equipment to be tested.

13.3.1 Example of real-time test

The test system shown in Figure 13.9 consists of three lines, each 120 km long and a distance relay (under test). The relay is the only real piece of equipment, the rest of the system being represented in the digital simulator and the solution step used is 100 μs. The simulated currents and voltages monitored by the current and voltage

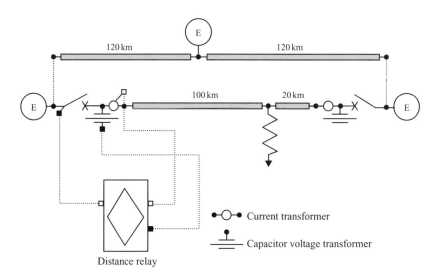

Figure 13.9   Test system

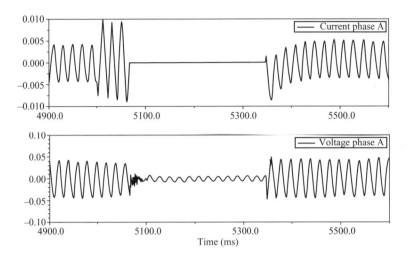

Figure 13.10 Current and voltage waveforms following a single-phase short-circuit

transformers are sent to the I/O converters and to the amplifiers. The relays are directly connected to these amplifiers.

The test conditions are as follows: initially a 5 s run is carried out to achieve the steady state. Then a single-phase fault is applied to one of the lines 100 km away from the relay location.

Some of the results from the real-time simulation are illustrated in Figure 13.10. The top graph shows the current in the faulty phase, monitored on the secondary of the simulated current transformer. The lower graph shows the voltage of the faulty phase, monitored on the secondary of the simulated capacitive voltage transformer.

Important information derived from these graphs is the presence of some residual voltage in the faulty phase, due to capacitive coupling to other phases (even though the line is opened at both ends). The self-extinguishing fault disappears after 100 ms. The relay recloser sends a closing order to the breakers after 330 ms. Then after a transient period the current returns to the steady-state condition.

13.4 Summary

Advances in digital parallel processing, combined with the ability of power systems to be processed by means of subsystems, provides the basis for real-time transient simulation.

Simulation in real-time permits realistic testing of the behaviour of control and protection systems. This requires the addition of digital to analogue and analogue to digital converters, as well as analogue signal amplifiers.

The original, and at present still the main application in the market, is a simulator based on dedicated architecture called RTDS (real-time digital simulator). This unit practically replaced all the scale-down physical simulators and can potentially represent any size system,

The development of multipurpose parallel computing is now providing the basis for real-time simulation using standard computers instead of dedicated architectures, and should eventually provide a more economical solution.

13.5 References

1 KUFFEL, P., GIESBRECHT, J., MAGUIRE, T., WIERCKX, R. P. and McLAREN, P.: 'RTDS – a fully digital power system simulator operating in real-time', Proceedings of the ICDS Conference, College Station, Texas, USA, April 1995, pp. 19–24
2 WIERCKX, R. P.: 'Fully digital real time electromagnetic transient simulator', *IERE Workshop on New Issues in Power System Simulation*, 1992, **VII**, pp. 128–228
3 BRANDT, D., WACHAL, R., VALIQUETTE, R. and WIERCKX, R. P.: 'Closed loop testing of a joint VAr controller using a digital real-time simulator for HVdc system and control studies', *IEEE Transactions on Power Systems*, 1991, **6** (3), pp. 1140–6.
4 WIERCKX, R. P., GIESBRECHT W. J., KUFFEL, R. *et al*.: 'Validation of a fully digital real time electromagnetic transient simulator for HVdc system and control studies', Proceedings of the Athens Power Tech. Conference, September 1993, pp. 751–9
5 McLAREN, P. G., KUFFEL, R., GIESBRECHT, W. J., WIERCKX, R. P. and ARENDT, L.: 'A real time digital simulator for testing relays', *IEEE Transactions on Power Delivery*, January 1992, **7** (1), pp. 207–13
6 KUFFEL, R., McLAREN, P., YALLA, M. and WANG, X.: 'Testing of the Beckwith electric M-0430 multifunction protection relay using a real-time digital simulator (RTDS)', Proceedings of International Conference on *Digital Power System Simulators (ICDS)*, College Station, Texas, USA, April 1995, pp. 49–54.
7 McLAREN, P., DIRKS, R. P., JAYASINGHE, R. P., SWIFT, G. W. and ZHANG, Z.: 'Using a real time digital simulator to develop an accurate model of a digital relay', Proceedings of International Conference on *Digital Power System Simulators, ICDS'95*, April 1995, p. 173
8 McLAREN, P., SWIFT, G. W., DIRKS, R. P. *et al*.: 'Comparisons of relay transient test results using various testing technologies', Proceedings of Second International Conference on *Digital Power System Simulators, ICDS'97*, May 1997, pp. 57–62
9 DUCHEN, H., LAGERKVIST, M., KUFFEL, R. and WIERCKX, R.: 'HVDC simulation and control system testing using a real-time digital simulator (RTDS)', Proceedings of the ICDS Conference, College Station, Texas, USA, April 1995, p. 213

10 STRUNZ, K. and MULLER, S.: 'New trends in protective relay testing', Proceedings of Fifth International Power Engineering Conference (IPEC), May 2001, **1**, pp. 456–60
11 STRUNZ, K., MARTINOLE, P., MULLER, S. and HUET, O.: 'Control system testing in electricity market places', Proceedings of Fifth International Power Engineering Conference (IPEC), May 2001
12 STRUNZ, K., LOMBARD, X., HUET, O., MARTI, J. R., LINARES, L. and DOMMEL, H. W.: 'Real time nodal analysis-based solution techniques for simulations of electromagnetic transients in power electronic systems', Procccdings of Thirteenth Power System Computation Conference (PSCC), June 1999, Trondheim, Norway, pp. 1047–53
13 STRUNZ, K. and FROMONT, H.: 'Exact modelling of interaction between gate pulse generators and power electronic switches for digital real time simulators', Proceedings of Fifth Brazilian Power Electronics Conference (COBEP), September 1999, pp. 203–8
14 STRUNZ, K., LINARES, L., MARTI, J. R., HUET, O. and LOMBARD, X.: 'Efficient and accurate representation of asynchronous network structure changing phenomena in digital real time simulators', *IEEE Transactions on Power Systems*, 2000, **15** (2), pp. 586–92
15 STRUNZ, K.: 'Real time high speed precision simulators of HDC extinction advance angle', Proceedings of International Conference on *Power Systems Technology (PowerCon2000)*, December 2000, pp. 1065–70

Appendix A
Structure of the PSCAD/EMTDC program

PSCAD/EMTDC version 2 consists of a set of programs which enable the efficient simulation of a wide variety of power system networks. EMTDC (Electromagnetic Transient and DC) [1], [2], although based on the EMTP method, introduced a number of modifications so that switching discontinuities could be accommodated accurately and quickly [3], the primary motivation being the simulation of HVDC systems. PSCAD (Power Systems Computer Aided Design) is a graphical Unix-based user interface for the EMTDC program. PSCAD consists of software enabling the user to enter a circuit graphically, create new custom components, solve transmission line and cable parameters, interact with an EMTDC simulation while in progress and to process the results of a simulation [4].

The programs comprising PSCAD version 2 are interfaced by a large number of data files which are managed by a program called FILEMANAGER. This program also provides an environment within which to call the other five programs and to perform housekeeping tasks associated with the Unix system, as illustrated in Figure A.1. The starting point for any study with EMTDC is to create a graphical sketch of the circuit to be solved using the DRAFT program. DRAFT provides the user with a canvas area and a selection of component libraries (shown in Figure A.2).

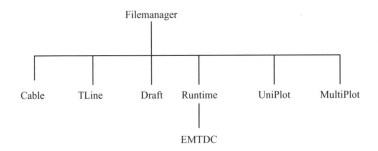

Figure A.1   The PSCAD/EMTDC Version 2 suite

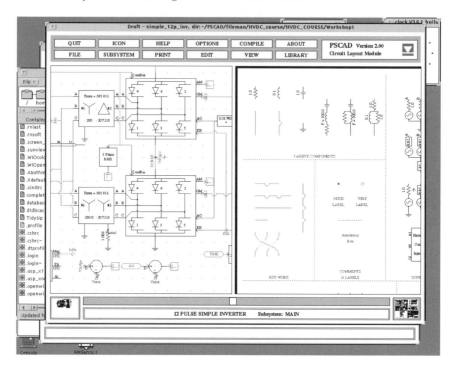

Figure A.2   DRAFT program

A library is a set of component icons, any of which can be dragged to the canvas area and connected to other components by bus-work icons. Associated with each component icon is a form into which component parameters can be entered. The user can create component icons, the forms to go with them and FORTRAN code to describe how the component acts dynamically in a circuit. Typical components are multi-winding transformers, six-pulse groups, control blocks, filters, synchronous machines, circuit-breakers, timing logic, etc.

The output from DRAFT is a set of files which are used by EMTDC. EMTDC is called from the PSCAD RUNTIME program, which permits interactions with the simulation while it is in progress. Figure A.3 shows RUNTIME plotting the output variables as EMTDC simulates. RUNTIME enables the user to create buttons, slides, dials and plots connected to variables used as input or output to the simulation (shown in Figure A.4). At the end of simulation, RUNTIME copies the time evolution of specified variables into data files. The complete state of the system at the end of simulation can also be copied into a snapshot file, which can then be used as the starting point for future simulations. The output data files from EMTDC can be plotted and manipulated by the plotting programs UNIPLOT or MULTIPLOT. MULTIPLOT allows multiple pages to be laid out, with multiple plots per page and the results from different runs shown together. Figure A.5 shows a MULTIPLOT display of the results from two different simulations. A calculator function and off-line DFT function are

Structure of the PSCAD/EMTDC program 335

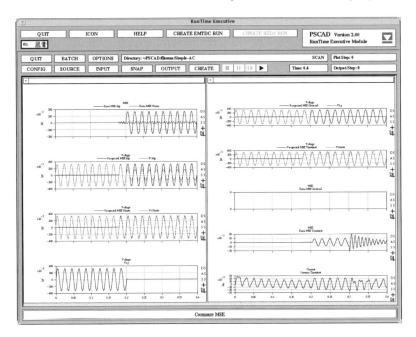

Figure A.3   RUNTIME program

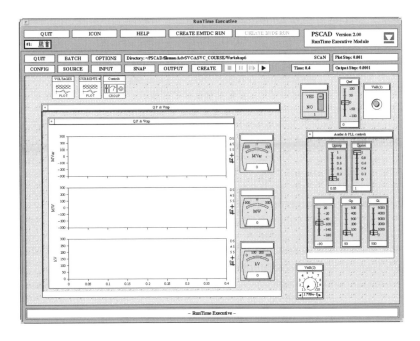

Figure A.4   RUNTIME program showing controls and metering available

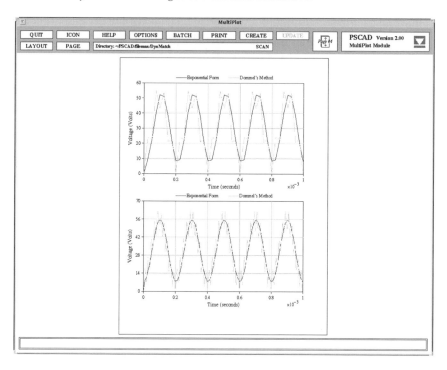

Figure A.5 MULTIPLOT program

also very useful features. The output files can also be processed by other packages, such as MATLAB, or user-written programs, if desired. Ensure % is the first character in the title so that the files do not need to be manually inserted after each simulation run if MATLAB is to be used for post-processing.

All the intermediate files associated with the PSCAD suite are in text format and can be inspected and edited. As well as compiling a circuit schematic to input files required by EMTDC, DRAFT also saves a text-file description of the schematic, which can be readily distributed to other PSCAD users. A simplified description of the PSCAD/EMTDC suite is illustrated in Figure A.6. Not shown are many batch files, operating system interface files, set-up files, etc.

EMTDC consists of a main program primarily responsible for finding the network solution at every time step, input and output, and supporting user-defined component models. The user must supply two FORTRAN source-code subroutines to EMTDC – DSDYN.F and DSOUT.F. Usually these subroutines are automatically generated by DRAFT but they can be completely written or edited by hand. At the start of simulation these subroutines are compiled and linked with the main EMTDC object code.

DSDYN is called each time step before the network is solved and provides an opportunity for user-defined models to access node voltages, branch currents or internal variables. The versatility of this approach to user-defined component modules means that EMTDC has enjoyed wide success as a research tool. A flowchart for the

Structure of the PSCAD/EMTDC program 337

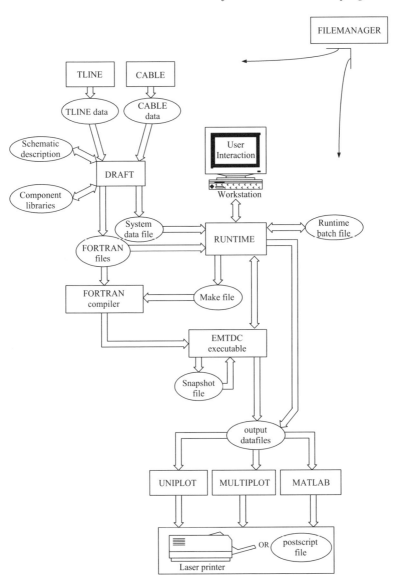

Figure A.6   Interaction in PSCAD/EMTDC Version 2

EMTDC program, illustrated in Figure A.7, indicates that the DSOUT subroutine is called after the network solution. The purpose of the subroutine is to process variables prior to being written to an output file. Again, the user has responsibility for supplying this FORTRAN code, usually automatically from DRAFT. The external multiple-run loop in Figure A.7 permits automatic optimisation of system parameters for some specified goal, or the determination of the effect of variation in system parameters.

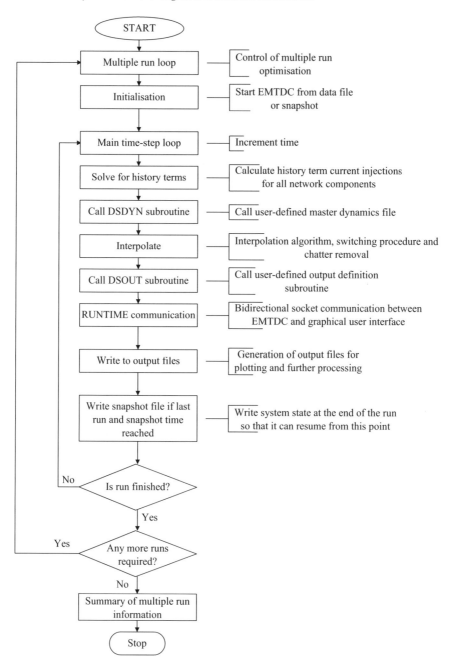

Figure A.7   PSCAD/EMTDC flow chart

Structure of the PSCAD/EMTDC program 339

The main component models used in EMTDC, i.e. transmission lines, synchronous generators and transformers, as well as control and switching modelling techniques, have already been discussed in previous chapters.

Due to the popularity of the WINDOWS operating system on personal computers, a complete rewrite of the successful UNIX version was performed, resulting in PSCAD version 3. New features include:

- The function of DRAFT and RUNTIME has been combined so that plots are put on the circuit schematic (as shown in Figure A.8).
- The new graphical user interface also supports: hierarchical design of circuit pages and localised data generation only for modified pages, single-line diagram data entry, direct plotting of all simulation voltages, currents and control signals, without writing to output files and more flexible multiple-run control.
- A MATLAB to PSCAD/EMTDC interface has been developed. The interface enables controls or devices to be developed in MATLAB, and then connected in any sequence to EMTDC components. Full access to the MATLAB toolboxes will be supported, as well as the full range of MATLAB 2D and 3D plotting commands.
- EMTDC V3 includes ideal switches with zero resistance, ideal voltage sources, improved storage methods and faster switching operations. Fortran 90/95 will be given greater support.
- A new solution algorithm (the root-matching technique) is implemented for control circuits which eliminates the errors due to trapezoidal integration but which is still numerically stable.

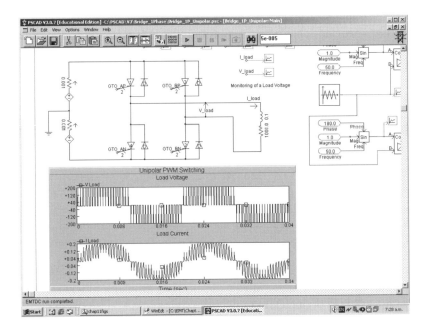

Figure A.8 PSCAD Version 3 interface

- New transmission-line and cable models using the phase domain (as opposed to modal domain) techniques coupled with more efficient curve fitting algorithms have been implemented, although the old models are available for compatibility purposes.

To date an equivalent for the very powerful MULTIPLOT post-processing program is not available, necessitating exporting to MATLAB for processing and plotting.

PSCAD version 2 had many branch quantities that were accessed using the node numbers of its terminals (e.g. CDC, EDC, GDC, CCDC, etc.). These have been replaced by arrays (GEQ, CBR, EBR, CCBR, etc.) that are indexed by branch numbers. For example CBR(10,2) is the 10th branch in subsystem 2. This allows an infinite number of branches in parallel whereas version 2 only allowed three switched branches in parallel. Version 2 had a time delay in the plotting of current through individual parallel switches (only in plotting but not in calculations). This was because the main algorithm only computed the current through all the switches in parallel, and the allocation of current in individual switches was calculated from a subroutine called from DSDYN. Old version 2 code can still run on version 3, as interface functions have been developed that scan through all branches until a branch with the correct sending and receiving nodes is located. Version 2 code that modifies the conductance matrix GDC directly needs to be manually changed to GEQ.

Version 4 of PSCAD/EMTDC is at present being developed. In version 3 a circuit can be split into subpages using page components on the main page. If there are ten page components on the main page connected by transmission lines or cables, then there will be ten subsystems regardless of the number of subpages branching off other pages. Version 4 has a new single line diagram capability as well as a new transmission line and cable interface consisting of one object, instead of the three currently used (sending end, receiving end and line constants information page). The main page will show multiple pages with transmission lines directly connected to electrical connections on the subpage components. PSCAD will optimally determine the subsystem splitting and will form subsystems wherever possible.

A.1 References

1. WOODFORD, D. A., INO, T., MATHUR, R. M., GOLE, A. M. and WIERCKX, R.: 'Validation of digital simulation of HVdc transients by field tests', *IEE Conference Publication on AC and DC power transmission*, 1985, **255**, pp. 377–81
2. WOODFORD, D. A., GOLE, A. M. and MENZIES, R. W.: 'Digital simulation of DC links and AC machines', *IEEE Transactions on Power Apparatus and Systems*, 1983, **102** (6), pp. 1616–23
3. KUFFEL, P., KENT, K. and IRWIN, G. D.: 'The implementation and effectiveness of linear interpolation within digital simulation', *Electrical Power and Energy Systems*, 1997, **19** (4), pp. 221–4
4. Manitoba HVdc Research Centre: 'PSCAD/EMTDC power systems simulation software tutorial manual', 1994

Appendix B
System identification techniques

B.1 s-domain identification (frequency domain)

The rational function in s to be fitted to the frequency domain data is:

$$H(s) = \frac{a_0 + a_1 s^1 + a_2 s^2 + \cdots + a_N s^N}{1 + b_1 s^1 + b_2 s^2 + \cdots + b_n s^n} \quad \text{(B.1)}$$

where $N \leq n$.

The frequency response of equation B.1 is:

$$H(j\omega) = \frac{\sum_{k=0}^{N} (a_k (j\omega)^k)}{\sum_{k=0}^{n} (b_k (j\omega)^k)} \quad \text{(B.2)}$$

where $b_0 = 1$.

Letting the sample data be $c(j\omega) + jd(j\omega)$, and equating it to equation B.2 yields

$$c(j\omega) + jd(j\omega) = \frac{(a_0 - a_2\omega_k^2 + a_4\omega_k^4 - \cdots) + j(a_1\omega_k - a_3\omega_k^3 - a_5\omega_k^5 - \cdots)}{(1 - b_2\omega_k^2 + b_4\omega_k^4 - \cdots) + j(b_1\omega_k - b_3\omega_k^3 - b_5\omega_k^5 - \cdots)} \quad \text{(B.3)}$$

or

$$(c(j\omega) + jd(j\omega))\left((1 - b_2\omega_k^2 + b_4\omega_k^4 - \cdots) + j(b_1\omega_k - b_3\omega_k^3 - b_5\omega_k^5 - \cdots)\right)$$
$$= (a_0 - a_2\omega_k^2 + a_4\omega_k^4 - \cdots) + j(a_1\omega_k - a_3\omega_k^3 - a_5\omega_k^5 - \cdots) \quad \text{(B.4)}$$

Splitting into real and imaginary parts yields:

$$-d_k(j\omega) \cdot (b_1\omega_k - b_3\omega_k^3 - b_5\omega_k^5 - \cdots) + c(j\omega)(b_2\omega_k^2 + b_4\omega_k^4 - \cdots)$$
$$- (a_0 - a_2\omega_k^2 + a_4\omega_k^4 - \cdots) = -c(j\omega)$$

$$d_k(j\omega) \cdot (-b_2\omega_k^2 + b_4\omega_k^4 - \cdots) + c(j\omega) \cdot (b_1\omega_k - b_3\omega_k^3 - b_5\omega_k^5 - \cdots)$$
$$- (a_1\omega_k - a_3\omega_k^3 - a_5\omega_k^5 - \cdots) = -d(j\omega)$$

This must hold for each sample point and therefore assembling into a matrix equation gives

$$
\begin{bmatrix}
-d(j\omega_1)\omega_1 & -c(j\omega_1)\omega_1^2 & d(j\omega_1)\omega_1^3 & \cdots & t_1 & -1 & 0 & \omega_1^2 & 0 & \omega_1^4 & \cdots & t_2 \\
-d(j\omega_2)\omega_2 & -c(j\omega_2)\omega_2^2 & d(j\omega_2)\omega_2^3 & \cdots & t_1 & -1 & 0 & \omega_2^2 & 0 & \omega_2^4 & \cdots & t_2 \\
\vdots & \vdots & \vdots & \ddots & \vdots & \vdots & \vdots & \vdots & \vdots & \vdots & \ddots & \vdots \\
-d(j\omega_k)\omega_k & -c(j\omega_k)\omega_k^2 & d(j\omega_k)\omega_k^3 & \cdots & t_1 & -1 & 0 & \omega_k^2 & 0 & \omega_k^4 & \cdots & t_2 \\
c(j\omega_1)\omega_1 & d(j\omega_1)\omega_1^2 & c(j\omega_1)\omega_1^3 & \cdots & t_3 & 0 & -\omega_1 & 0 & \omega_1^3 & 0 & \cdots & t_4 \\
c(j\omega_2)\omega_2 & d(j\omega_2)\omega_2^2 & c(j\omega_2)\omega_2^3 & \cdots & t_3 & 0 & -\omega_2 & 0 & \omega_2^3 & 0 & \cdots & t_4 \\
\vdots & \vdots & \vdots & \ddots & \vdots & \vdots & \vdots & \vdots & \vdots & \vdots & \ddots & \vdots \\
c(j\omega_k)\omega_k & d(j\omega_k)\omega_k^2 & c(j\omega_k)\omega_k^3 & \cdots & t_3 & 0 & -\omega_k & 0 & \omega_k^3 & 0 & \cdots & t_4
\end{bmatrix}
\begin{pmatrix} b_1 \\ b_2 \\ \vdots \\ b_n \\ a_0 \\ a_1 \\ \vdots \\ a_n \end{pmatrix}
$$

$$
=
\begin{pmatrix}
-c(j\omega_1) \\
-c(j\omega_2) \\
\vdots \\
-c(j\omega_k) \\
-d(j\omega_1) \\
-d(j\omega_2) \\
\vdots \\
-d(j\omega_k)
\end{pmatrix}
\tag{B.5}
$$

where the terms t_1, t_2, t_3, and t_4 are

$$t_1 = \left(\sin\left(\frac{-l\pi}{2}\right)\omega_k^l d(j\omega_k) + \cos\left(\frac{l\pi}{2}\right)\omega_k^l c(j\omega_k)\right)$$

$$t_2 = \cos\left(\frac{-l\pi}{2}\right)\omega_k^l$$

$$t_3 = \left(\cos\left(\frac{l\pi}{2}\right)\omega_k^l d(j\omega_k) + \sin\left(\frac{l\pi}{2}\right)\omega_k^l c(j\omega_k)\right)$$

$$t_4 = \sin\left(\frac{-l\pi}{2}\right)\omega_k^l$$

l = column number

k = row or sample number.

Equation B.5 is of the form:

$$
\begin{bmatrix} [A_{11}] & [A_{12}] \\ [A_{21}] & [A_{22}] \end{bmatrix} \begin{pmatrix} \mathbf{a} \\ \mathbf{b} \end{pmatrix} = \begin{pmatrix} \mathbf{C} \\ \mathbf{D} \end{pmatrix}
\tag{B.6}
$$

where

$\mathbf{a}^T = (a_0, a_1, a_2, a_3, \ldots, a_n)$
$\mathbf{b}^T = (b_1, b_2, b_3, \ldots, b_n)$
$\mathbf{C}^T = (-c(j\omega_1), -c(j\omega_2), -c(j\omega_3), \ldots, -c(j\omega_k))$
$\mathbf{D}^T = (-d(j\omega_1), -d(j\omega_2), -d(j\omega_3), \ldots, -d(j\omega_k))$

Equation B.6 is solved for the required coefficients (a's and b's).

B.2 z-domain identification (frequency domain)

The rational function in the z-domain to be fitted is:

$$H(z) = \frac{a_0 + a_1 z^{-1} + a_2 z^{-2} + \cdots + a_n z^{-n}}{1 + b_1 z^{-1} + b_2 z^{-2} + \cdots + b_n z^{-n}} \quad (B.7)$$

Evaluating the frequency response of the rational function in the z-domain and equating it to the sample data $(c(j\omega) + jd(j\omega))$ yields

$$c(j\omega) + jd(j\omega) = \frac{\sum_{k=0}^{n} \left(a_k e^{-kj\omega \Delta t}\right)}{1 + \sum_{k=1}^{n} \left(b_k e^{-kj\omega \Delta t}\right)} \quad (B.8)$$

Multiplying both sides by the denominator and rearranging gives:

$$-c(j\omega) - jd(j\omega) = -\sum_{k=1}^{n} \left((b_k(c(j\omega) + jd(j\omega)) - a_k)e^{-kj\omega \Delta t}\right) + a_0$$

Splitting into real and imaginary components gives:

$$\sum_{k=1}^{n} \left((b_k c(j\omega) - a_k) \cos(k\omega \Delta t) - b_k d(j\omega) \sin(k\omega \Delta t)\right) - a_0 = -c(j\omega) \quad (B.9)$$

for the real part and

$$\sum_{k=1}^{n} \left(-(b_k c(j\omega) - a_k) \sin(k\omega \Delta t) + b_k d(j\omega) \cos(k\omega \Delta t)\right) = -d(j\omega) \quad (B.10)$$

for the imaginary part.

Grouping in terms of the coefficients that are to be solved for (a_k and b_k) yields:

$$\sum_{k=1}^{n} \left(b_k (c(j\omega) \cos(k\omega \Delta t) + d(j\omega) \sin(k\omega \Delta t)) - a_k \cos(k\omega \Delta t)\right) - a_0 = -c(j\omega)$$

$$(B.11)$$

$$\sum_{k=1}^{n} \left(b_k (d(j\omega) \cos(k\omega \Delta t) - c(j\omega) \sin(k\omega \Delta t)) + a_k \sin(k\omega \Delta t)\right) = -d(j\omega)$$

$$(B.12)$$

This must hold for all sample points. Combining these equations for each sample point in matrix form gives the following matrix equation to be solved:

$$\begin{bmatrix} [A_{11}] & [A_{12}] \\ [A_{21}] & [A_{22}] \end{bmatrix} \begin{pmatrix} \mathbf{a} \\ \mathbf{b} \end{pmatrix} = \begin{pmatrix} \mathbf{C} \\ \mathbf{D} \end{pmatrix} \quad (B.13)$$

where
$$\mathbf{a}^T = (a_0, a_1, a_2, a_3, \ldots, a_n)$$
$$\mathbf{b}^T = (b_1, b_2, b_3, \ldots, b_n)$$
$$\mathbf{C}^T = (-c(j\omega_1), -c(j\omega_2), -c(j\omega_3), \ldots, -c(j\omega_m))$$
$$\mathbf{D}^T = (-d(j\omega_1), -d(j\omega_2), -d(j\omega_3), \ldots, -d(j\omega_m))$$

This is the same equation as for the s-domain fitting, except that the matrix $[A]$ (called the design matrix) comprises four different submatrices $[A_{11}]$, $[A_{12}]$, $[A_{21}]$, and $[A_{22}]$, i.e.

$$A_{11} = \begin{pmatrix} -1 & -\cos(\omega_1 \Delta t) & \cdots & -\cos(n\omega_1 \Delta t) \\ -1 & -\cos(\omega_2 \Delta t) & \cdots & -\cos(n\omega_2 \Delta t) \\ \vdots & \vdots & \ddots & \vdots \\ -1 & -\cos(\omega_m \Delta t) & \cdots & -\cos(n\omega_m \Delta t) \end{pmatrix}$$

$$A_{21} = \begin{pmatrix} 0 & \sin(\omega_1 \Delta t) & \cdots & \sin(n\omega_1 \Delta t) \\ 0 & \sin(\omega_2 \Delta t) & \cdots & \sin(n\omega_2 \Delta t) \\ \vdots & \vdots & \ddots & \vdots \\ 0 & \sin(\omega_m \Delta t) & \cdots & \sin(n\omega_m \Delta t) \end{pmatrix}$$

$$A_{12} = \begin{pmatrix} R_{11} & R_{12} & \cdots & R_{1n} \\ R_{21} & R_{22} & \cdots & R_{2n} \\ \vdots & \vdots & \ddots & \vdots \\ R_{m1} & R_{m2} & \cdots & R_{mn} \end{pmatrix}$$

$$A_{22} = \begin{pmatrix} S_{11} & S_{12} & \cdots & S_{1n} \\ S_{21} & S_{22} & \cdots & S_{2n} \\ \vdots & \vdots & \ddots & \vdots \\ S_{m1} & S_{m2} & \cdots & S_{mn} \end{pmatrix}$$

where
$R_{ik} = c(k\omega_i) \cdot \cos(k\omega_i \Delta t) + d(k\omega_i) \cdot \sin(k\omega_i \Delta t)$
$S_{ik} = d(k\omega_i) \cdot \cos(k\omega_i \Delta t) - c(k\omega_i) \cdot \sin(k\omega_i \Delta t)$
n = order of the rational function
m = number of frequency sample points.

As the number of sample points exceeds the number of unknown coefficients singular value decomposition is used to solve equation B.13.

The least squares approach 'smears out' the fitting error over the frequency spectrum. This is undesirable as it is important to obtain accurately the steady-state

System identification techniques 345

condition. Adding weighting factors allows this to be achieved. The power frequency is typically given a weighting of 100 (the other frequencies are weighted 1.0).
Adding weighting factors results in equations B.11 and B.12 becoming:

$$-c(j\omega)w(j\omega) = \sum_{k=1}^{n}(w(j\omega)b_k(c(j\omega)\cos(k\omega\Delta t) + d(j\omega)\sin(k\omega\Delta t))$$
$$- a_k w(j\omega)\sin(k\omega\Delta t)) - w(j\omega)a_0 \qquad (B.14)$$

$$-d(j\omega)w(j\omega) = \sum_{k=1}^{n}(w(j\omega)b_k(d(j\omega)\cos(k\omega\Delta t) - c(j\omega)\sin(k\omega\Delta t))$$
$$- a_k w(j\omega)\sin(k\omega\Delta t)) \qquad (B.15)$$

B.3 z-domain identification (time domain)

When the sampled data consists of samples in time rather than frequency, a rational function in the z-domain can be identified, provided the system has been excited by a waveform that contains the frequency components at which the matching is required. This is achieved by a multi-sine injection.

Given a rational function of the form of equation B.7, if admittance is being fitted then

$$I(z)\left(1 + b_1 z^{-1} + b_2 z^{-2} + \cdots + b_n z^{-n}\right)$$
$$= \left(a_0 + a_1 z^{-1} + a_2 z^{-2} + \cdots + a_n z^{-n}\right)V(z) \qquad (B.16)$$

or

$$I(z) = -I(z)\left(b_1 z^{-1} + b_2 z^{-2} + \cdots + b_n z^{-n}\right)$$
$$+ \left(a_0 + a_1 z^{-1} + a_2 z^{-2} + \cdots + a_n z^{-n}\right)V(z) \qquad (B.17)$$

Taking the inverse z-transform gives:

$$i(t) = -\sum_{k=1}^{n} b_k i(t - k\Delta t) + \sum_{k=0}^{n} a_k v(t - k\Delta t) \qquad (B.18)$$

and in matrix form

$$\begin{bmatrix} v(t_1) & v(t_1-\Delta t) & v(t_1-2\Delta t) & \cdots & v(t_1-n\Delta t) & -i(t_1-\Delta t) & -i(t_1-2\Delta t) & \cdots & i(t_1-n\Delta t) \\ v(t_2) & v(t_2-\Delta t) & v(t_2-2\Delta t) & \cdots & v(t_2-n\Delta t) & -i(t_2-\Delta t) & -i(t_2-2\Delta t) & \cdots & i(t_2-n\Delta t) \\ \vdots & \vdots & \vdots & \vdots & \vdots & \vdots & & \vdots \\ v(t_k) & v(t_k-\Delta t) & v(t_k-2\Delta t) & \cdots & v(t_k-n\Delta t) & -i(t_k-\Delta t) & -i(t_k-2\Delta t) & \cdots & i(t_k-n\Delta t) \end{bmatrix}$$

$$\times \begin{pmatrix} a_0 \\ a_1 \\ \vdots \\ a_n \\ b_1 \\ b_2 \\ \vdots \\ b_n \end{pmatrix} = \begin{pmatrix} i(t_1) \\ i(t_2) \\ \vdots \\ i(t_k) \end{pmatrix} \quad (B.19)$$

where

$k =$ time sample number
$n =$ order of the rational function ($k > p$, i.e. over-sampled).

The time step must be chosen sufficiently small to avoid aliasing, i.e. $\Delta t = 1/(K_1 f_{\max})$, where $K_1 > 2$ (Nyquist criteria) and $f_{\max}$ is the highest frequency injected. For instance if $K_1 = 10$ and $\Delta t = 50\,\mu s$ there will be 4000 samples points per cycle (20 ms for 50 Hz). This equivalent is easily extended to multi-port equivalents [1]. For an m-port equivalent there will be $m(m+1)/2$ rational functions to be fitted.

B.4 Prony analysis

Prony analysis identifies a rational function that will have a prescribed time-domain response [2]. Given the rational function:

$$H(z) = \frac{Y(z)}{U(z)} = \frac{a_0 + a_1 z^{-1} + a_2 z^{-2} + \cdots + a_n z^{-N}}{1 + b_1 z^{-1} + b_2 z^{-2} + \cdots + b_d z^{-n}} \quad (B.20)$$

the impulse response of $h(k)$ is related to $H(z)$ by the z-transform, i.e.

$$H(z) = \sum_{k=0}^{\infty} h(k) z^{-1} \quad (B.21)$$

which can be written as

$$Y(z)\left(1 + b_1 z^{-1} + b_2 z^{-2} + \cdots + b_d z^{-n}\right)$$
$$= U(z)\left(a_0 + a_1 z^{-1} + a_2 z^{-2} + \cdots + a_n z^{-N}\right) \quad (B.22)$$

This is the z-domain equivalent of a convolution in the time domain. Using the first L terms of the impulse response the convolution can be expressed as:

$$\begin{bmatrix} h_0 & 0 & 0 & \cdots & 0 \\ h_1 & h_0 & 0 & \cdots & 0 \\ h_2 & h_1 & h_0 & \cdots & 0 \\ \vdots & \vdots & \vdots & \ddots & \vdots \\ h_n & h_{n-1} & h_{n-2} & \cdots & h_0 \\ h_{n+1} & h_n & h_{n-1} & \cdots & h_1 \\ \vdots & \vdots & \vdots & \ddots & \vdots \\ h_L & h_{L-1} & h_{L-2} & \cdots & h_{L-n} \end{bmatrix} \begin{pmatrix} 1 \\ b_1 \\ b_2 \\ \vdots \\ b_n \end{pmatrix} = \begin{pmatrix} -a_0 \\ -a_1 \\ \vdots \\ -a_N \\ 0 \\ 0 \\ \vdots \\ 0 \end{pmatrix} \quad \text{(B.23)}$$

Partitioning gives:

$$\begin{pmatrix} \mathbf{a} \\ \mathbf{0} \end{pmatrix} = \left[\begin{array}{c|c} [H_1] & \\ \hline [h_1] & [H_2] \end{array} \right] \begin{pmatrix} 1 \\ \mathbf{b} \end{pmatrix} \quad \text{(B.24)}$$

The dimensions of the vectors and matrices are:
- $\mathbf{a}$ $(N+1)$ vector
- $\mathbf{b}$ $(n+1)$ vector
- $[H_1]$ $(N+1) \times (n+1)$ matrix
- $[h_1]$ vector of last $(L-N)$ terms of impulse response
- $[H_2]$ $(L-N) \times (n)$ matrix.

The b coefficients are determined by using the sample points more than n time steps after the input has been removed. When this occurs the output is no longer a function of the input (equation B.22) but only depends on the b coefficients and previous output values (lower partition of equation B.24), i.e.

$$\mathbf{0} = [h_1] + [H_2]\,\mathbf{b}$$

or

$$[h_1] = -[H_2]\mathbf{b} \quad \text{(B.25)}$$

Once the b coefficients are determined the a coefficients are obtained from the upper partition of equation B.24, i.e.

$$\mathbf{b} = [H_1]\mathbf{a}$$

When $L = N + n$ then H_2 is square and, if non-singular, a unique solution for b is obtained. If H_2 is singular many solutions exist, in which case $h(k)$ can be generated by a lower order system.

When $m > n + N$ the system is over-determined and the task is to find a and b coefficients that give the best fit (minimise the error). This can be obtained solving equation B.25 using the SVD or normal equation approach, i.e.

$$[H_2]^T [h_1] = -[H_2]^T [H_2]\mathbf{b}$$

B.5 Recursive least-squares curve-fitting algorithm

A least-squares curve fitting algorithm is described here to extract the fundamental frequency data based on a least squared error technique. We assume a sinusoidal signal with a frequency of ω radians/sec and a phase shift of ψ relative to some arbitrary time T_0, i.e.

$$y(t) = A \sin(\omega t - \psi) \tag{B.26}$$

where $\psi = \omega T_0$.

This can be rewritten as

$$y(t) = A \sin(\omega t) \cos(\omega T_0) - A \cos(\omega t) \sin(\omega T_0) \tag{B.27}$$

Letting $C_1 = A \cos(\omega T_0)$ and $C_2 = A \sin(\omega T_0)$ and representing $\sin(\omega t)$ and $\cos(\omega t)$ by functions $F_1(t)$ and $F_2(t)$ respectively, then:

$$y(t) = C_1 F_1(t) + C_2 F_2(t) \tag{B.28}$$

$F_1(t)$ and $F_2(t)$ are known if the fundamental frequency ω is known. However, the amplitude and phase of the fundamental frequency need to be found, so equation B.28 has to be solved for C_1 and C_2. If the signal $y(t)$ is distorted, then its deviation from a sinusoid can be described by an error function E, i.e.

$$x(t) = y(t) + E \tag{B.29}$$

For a least squares method of curve fitting, the size of the error function is measured by the sum of the individual residual squared values, such that:

$$E = \sum_{i=1}^{n} \{x_i - y_i\}^2 \tag{B.30}$$

where $x_i = x(t_0 + i \Delta t)$ and $y_i = y(t_0 + i \Delta t)$. From equation B.28

$$E = \sum_{i=1}^{n} \{x_i - C_1 F_1(t_i) - C_2 F_2(t_i)\}^2 \tag{B.31}$$

where the residual value r at each discrete step is defined as:

$$r_i = x_i - C_1 F_1(t_i) - C_2 F_2(t_i) \tag{B.32}$$

In matrix form:

$$\begin{bmatrix} r_1 \\ r_2 \\ \vdots \\ r_n \end{bmatrix} = \begin{bmatrix} x_1 \\ x_2 \\ \vdots \\ x_n \end{bmatrix} - \begin{bmatrix} F_1(t_1) & F_2(t_1) \\ F_1(t_2) & F_2(t_2) \\ \vdots & \vdots \\ F_1(t_n) & F_2(t_n) \end{bmatrix} \begin{bmatrix} C_1 \\ C_2 \end{bmatrix} \tag{B.33}$$

or
$$[r] = [X] - [F][C] \tag{B.34}$$

The error component can be described in terms of the residual matrix as follows:

$$\begin{aligned}
E = [r]^T[r] &= r_1^2 + r_2^2 + \cdots + r_n^2 \\
&= [[X] - [F][C]]^T [[X] - [F][C]] \\
&= [X]^T[X] - [C]^T[F]^T[X] - [X]^T[F][C] + [C]^T[F]^T[F][C]
\end{aligned} \tag{B.35}$$

This error then needs to be minimised, i.e.

$$\frac{\partial E}{\partial C} = -2[F]^T[X] + 2[F]^T[F][C] = 0$$

$$[F]^T[F][C] = [F]^T[X] \tag{B.36}$$

$$[C] = [[F]^T[F]]^{-1}[F]^T[X]$$

If $[A] = [F]^T[F]$ and $[B] = [F]^T[X]$ then:

$$[C] = [A]^{-1}[B] \tag{B.37}$$

Therefore

$$[A] = \begin{bmatrix} F_1 \\ F_2 \end{bmatrix} \begin{bmatrix} F_1 & F_2 \end{bmatrix} = \begin{bmatrix} F_1 F_1(t_i) & F_1 F_2(t_i) \\ F_2 F_1(t_i) & F_2 F_2(t_i) \end{bmatrix} = \begin{bmatrix} a_{11} & a_{12} \\ a_{21} & a_{22} \end{bmatrix}$$

Elements of matrix $[A]$ can then be derived as follows:

$$\begin{aligned}
a_{11_n} &= \begin{bmatrix} F_1(t_1) \\ \vdots \\ F_1(t_n) \end{bmatrix}^T \begin{bmatrix} F_1(t_1) \\ \vdots \\ F_1(t_n) \end{bmatrix} \\
&= \sum_{i=1}^{n-1} F_1^2(t_i) + F_1^2(t_n) \\
&= a_{11_{n-1}} + F_1^2(t_n)
\end{aligned} \tag{B.38}$$

etc.
Similarly:

$$[B] = \begin{bmatrix} F_1(t_i)x(t_i) \\ F_2(t_i)x(t_i) \end{bmatrix} = \begin{bmatrix} b_1 \\ b_2 \end{bmatrix}$$

and

$$b_{1_n} = b_{1_{n-1}} + F_1(t_n)x(t_n) \tag{B.39}$$

$$b_{2_n} = b_{2_{n-1}} + F_2(t_n)x(t_n) \tag{B.40}$$

From these matrix element equations, C_1 and C_2 can be calculated recursively using sequential data.

B.6 References

1 ABUR, A. and SINGH, H.: 'Time domain modeling of external systems for electromagnetic transients programs', *IEEE Transactions on Power Systems*, 1993, **8** (2), pp. 671–77
2 PARK, T. W. and BURRUS, C. S.: 'Digital filter design' (John Wiley Interscience, New York, 1987)

Appendix C
Numerical integration

C.1 Review of classical methods

Numerical integration is needed to calculate the solution $x(t + \Delta t)$ at time $t + \Delta t$ from knowledge of previous time points. The local truncation error (LTE) is the error introduced by the solution at $x(t + \Delta t)$ assuming that the previous time points are exact. Thus the total error in the solution $x(t + \Delta t)$ is determined by LTE and the build-up of error at previous time points (i.e. its propagation through the solution). The stability characteristics of the integration algorithm are a function of how this error propagates.

A numerical integration algorithm is either **explicit** or **implicit**. In an explicit integration algorithm the integral of a function f, from t to $t + \Delta t$, is obtained without using $f(t + \Delta t)$. An example of explicit integration is the forward Euler method:

$$x(t + \Delta t) = x(t) + \Delta t \, f(x(t), t) \tag{C.1}$$

In an implicit integration algorithm $f(x(t + \Delta t), t + \Delta t)$ is required to calculate the solution at $x(t + \Delta t)$. Examples are, the backward Euler method, i.e.

$$x(t + \Delta t) = x(t) + \Delta t \, f(x(t + \Delta t), t + \Delta t) \tag{C.2}$$

and the trapezoidal rule, i.e.

$$x(t + \Delta t) = x(t) + \frac{\Delta t}{2} [f(x(t), t) + f(x(t + \Delta t), t + \Delta t)] \tag{C.3}$$

There are various ways of developing numerical integration algorithms, such as manipulation of Taylor series expansions or the use of numerical solution by polynomial approximation. Among the wealth of material from the literature, only a few of the classical numerical integration algorithms have been selected for presentation here [1]–[3].

Runge–Kutta (fourth-order):

$$x(t + \Delta t) = x(t) + \Delta t \left[\tfrac{1}{6}k_1 + \tfrac{1}{3}k_2 + \tfrac{1}{3}k_3 + \tfrac{1}{6}k_4\right] \quad (C.4)$$

$$k_1 = f(x(t), t)$$
$$k_2 = f\left(x(t) + \frac{\Delta t}{2}k_1, x(t) + \frac{\Delta t}{2}\right)$$
$$k_3 = f\left(x(t) + \frac{\Delta t}{2}k_2, x(t) + \frac{\Delta t}{2}\right) \quad (C.5)$$
$$k_4 = f(x(t) + \Delta t\, k_3, x(t) + \Delta t)$$

Adams–Bashforth (third-order):

$$x(t + \Delta t) = x(t) + \Delta t \left[\tfrac{23}{12} f(x(t), t) - \tfrac{16}{12} f(x(t - \Delta t), t - \Delta t) \right.$$
$$\left. + \tfrac{5}{12} f(x(t - 2\Delta t), t - 2\Delta t)\right] \quad (C.6)$$

Adams–Moulton (fourth-order):

$$x(t + \Delta t) = x(t) + \Delta t \left[\tfrac{9}{24} f(x(t + \Delta t), t + \Delta t) + \tfrac{19}{24} f(x(t), t) \right.$$
$$\left. - \tfrac{5}{24} f(x(t - \Delta t), t - \Delta t) + \tfrac{1}{24} f(x(t - 2\Delta t), t - 2\Delta t)\right] \quad (C.7)$$

The method proposed by Gear is based on the equation:

$$\dot{x}(t + \Delta t) = \sum_{i=0}^{k} \alpha_i x(t + (1 - i)\Delta t) \quad (C.8)$$

This method was modified by Shichman for circuit simulation using a variable time step. The Gear second-order method is:

$$x(t + \Delta t) = \frac{4}{3}x(t) - \frac{1}{3}x(t - \Delta t) + \frac{2\Delta t}{3} f(x(t + \Delta t), t + \Delta t) \quad (C.9)$$

Numerical integration can be considered a sampled approximation of continuous integration, as depicted in Figure C.1. The properties of the sample-and-hold (reconstruction) determine the characteristics of the numerical integration formula. Due to the phase and magnitude errors in the process, compensation can be applied to generate a new integration formula. Consideration of numerical integration from a sample data viewpoint leads to the following tunable integration formula [4]:

$$y_{n+1} = y_n + \lambda \Delta t \left(\gamma f_{n+1} + (1 - \gamma) f_n\right) \quad (C.10)$$

where λ is the gain parameter, γ is the phase parameter and y_{n+1} represents the y value at $t + \Delta t$.

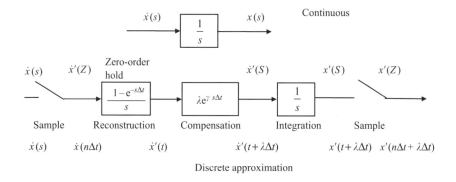

Figure C.1   Numerical integration from the sampled data viewpoint

Table C.1   Classical integration formulae as special cases of the tunable integrator

γ	Integration Rule	Formula
0	Forward Euler	$y_{n+1} = y_n + \Delta t f_n$
$\frac{1}{2}$	Trapezoidal	$y_{n+1} = y_n + \frac{\Delta t}{2}(f_{n+1} + f_n)$
1	Backward Euler	$y_{n+1} = y_n + \Delta t f_{n+1}$
$\frac{3}{2}$	Adams–Bashforth 2nd order	$y_{n+1} = y_n + \frac{\Delta t}{2}(3f_{n+1} - f_n)$
	Tunable	$y_{n+1} = y_n + \lambda \Delta t (\gamma f_{n+1} + (1-\gamma) f_n)$

If $\lambda = 1$ and $\gamma = (1 + \alpha)/2$ then the trapezoidal rule with damping is obtained [5]. The selection of integer multiples of half for the phase parameter produces the classical integration formulae shown in Table C.1. These formulae are actually the same integrator, differing only in the amount of phase shift of the integrand.

With respect to the differential equation:

$$\dot{y}_{n+1} = f(y,t) \qquad (C.11)$$

Table C.2 shows the various integration rules in the form of an integrator and differentiator.

Using numerical integration substitution for the differential equations of an inductor and a capacitor gives the Norton equivalent values shown in Tables C.3 and C.4 respectively.

Table C.2 Integrator formulae

Integration rule	Integrator	Differentiator
Trapezoidal rule	$y_{n+1} = y_n + \frac{\Delta t}{2}(f_{n+1} + f_n)$	$f_{n+1} \approx -f_n + \frac{2}{\Delta t}(y_{n+1} - y_n)$
Forward Euler	$y_{n+1} = y_n + \Delta t f_n$	$f_{n+1} \approx \frac{1}{\Delta t}(y_{n+2} - y_{n+1})$
Backward Euler	$y_{n+1} = y_n + \Delta t f_{n+1}$	$f_{n+1} \approx \frac{1}{\Delta t}(y_{n+1} - y_n)$
Gear 2nd order	$y_{n+1} = \frac{4}{3}y_n - \frac{1}{3}y_{n-1} + \frac{2\Delta t}{3}f_{n+1}$	$f_{n+1} \approx \frac{1}{\Delta t}\left(\frac{3}{2}y_{n+1} - 2y_n + \frac{1}{2}y_{n-1}\right)$
Tunable	$y_{n+1} = y_n + \lambda \Delta t (\gamma f_{n+1} + (1-\gamma)f_n)$	$f_{n+1} \approx \frac{-(1-\gamma)}{\gamma}f_n + \frac{1}{\gamma \lambda \Delta t}(y_{n+1} - y_n)$

Table C.3 Linear inductor

Integration Rule	G_{eff}	I_{History}
Trapezoidal	$\frac{\Delta t}{2L}$	$i_n + \frac{\Delta t}{2L}v_n$
Backward Euler	$\frac{\Delta t}{L}$	i_n
Forward Euler	$-$[1]	$i_n + \frac{\Delta t}{L}v_n$
Gear 2nd order	$\frac{2\Delta t}{3L}$	$\frac{4}{3}i_n - \frac{1}{3}i_{n-1}$
Tunable	$\frac{\lambda \Delta t \gamma}{L}$	$i_n + \frac{\lambda \Delta t}{L}(1-\gamma)v_n$

C.2 Truncation error of integration formulae

The exact expression of y_{n+1} in a Taylor series is:

$$y_{n+1} = y_n + \Delta t \frac{dy_n}{dt} + \frac{\Delta t^2}{2}\frac{d^2 y_n}{dt^2} + \frac{\Delta t^3}{3!}\frac{d^3 y_n}{dt^3} + O(\Delta t^4) \tag{C.12}$$

---

[1] Forward Euler does not contain a derivative term at $t_k + \Delta t$ hence difficult to apply

Table C.4 Linear capacitor

Integration rule	G_{eff}	I_{History}
Trapezoidal	$\dfrac{2C}{\Delta t}$	$-\dfrac{2C}{\Delta t}v_n - i_n$
Backward Euler	$\dfrac{C}{\Delta t}$	$-\dfrac{C}{\Delta t}v_n$
Gear 2nd order	$\dfrac{3C}{2\Delta t}$	$-\dfrac{2C}{\Delta t}v_n + \dfrac{C}{2\Delta t}v_{n-1}$
Tunable	$\dfrac{C}{\lambda\Delta t\gamma}$	$\dfrac{-(1-y)}{y}i_n - \dfrac{C}{\lambda\Delta+\gamma}v_n$

where $O(\Delta t^4)$ represents fourth and higher order terms. The derivative at $n+1$ can also be expressed as a Taylor series, i.e.

$$\frac{dy_{n+1}}{dt} = \frac{dy_n}{dt} + \Delta t \frac{d^2 y_n}{dt^2} + \frac{\Delta t^2}{2}\frac{d^3 y_n}{dt^3} + O(\Delta t^4) \tag{C.13}$$

If equation C.12 is used in the trapezoidal rule then the trapezoidal estimate is:

$$\hat{y}_{n+1} = y_n + \frac{\Delta t}{2}\left(\frac{dy_n}{dt} + \frac{dy_{n+1}}{dt}\right)$$

$$= y_n + \frac{\Delta t}{2}\frac{dy_n}{dt} + \frac{\Delta t}{2}\left(\frac{dy_n}{dt} + \Delta t\frac{d^2 y_n}{dt^2} + \frac{\Delta t^2}{2}\frac{d^3 y_n}{dt^3} + O(\Delta t^4)\right)$$

$$= y_n + \Delta t\frac{dy_n}{dt} + \frac{\Delta t^2}{2}\frac{d^2 y_n}{dt^2} + \frac{\Delta t^3}{4}\frac{d^3 y_n}{dt^3} + \frac{\Delta t}{2}O(\Delta t^4) \tag{C.14}$$

The error caused in going from y_n to y_{n+1} is:

$$\varepsilon_{n+1} = y_{n+1} - \hat{y}_{n+1} = \frac{\Delta t^3}{6}\frac{d^3 y_n}{dt^3} - \frac{\Delta t^3}{4}\frac{d^3 y_n}{dt^3} - \frac{\Delta t}{2}O(\Delta t^4)$$

$$= \frac{-\Delta t^3}{12}\frac{d^3 y_n}{dt^3} - \frac{\Delta t}{2}O(\Delta t^4) = \frac{-\Delta t^3}{12}\frac{d^3 y_\varepsilon}{dt^3}$$

where $t_n \leq \varepsilon \leq t_{n+1}$.

The resulting error arises because the trapezoidal formula represents a truncation of an exact Taylor series expansion, hence the term 'truncation error'.

To illustrate this, consider a simple RC circuit where the voltage across the resistor is of interest, i.e.

$$v_R(t) = RC\frac{dv_C(t)}{dt} = RC\frac{d(v_S(t) - Ri_R(t))}{dt} \tag{C.15}$$

Table C.5 Comparison of numerical integration algorithms ($\Delta T = \tau/10$)

Step	Exact	F. Euler	B. Euler	Trapezoidal	Gear second-order
1	45.2419	45.0000	45.4545	45.2381	–
2	40.9365	40.5000	41.3223	40.9297	40.9273
3	37.0409	36.4500	37.5657	37.0316	37.0211
4	33.5160	32.8050	34.1507	33.5048	33.4866
5	30.3265	29.5245	31.0461	30.3139	30.2891

Table C.6 Comparison of numerical integration algorithms ($\Delta T = \tau$)

Step	Exact	F. Euler	B. Euler	Trapezoidal	Gear second-order
1	18.3940	0.0000	25.0000	16.6667	18.3940
2	6.7668	0.0000	12.5000	5.5556	4.7152
3	2.4894	0.0000	6.2500	1.8519	0.0933
4	0.9158	0.0000	3.1250	0.6173	−0.8684
5	0.3369	0.0000	1.5625	0.2058	−0.7134

If the applied voltage is a step function at $t = 0$ then $dv_S(t)/dt = 0$ and equation C.15 becomes:

$$v_R(t) = RC\frac{dv_R(t)}{dt} = \tau\frac{dv_R(t)}{dt} \tag{C.16}$$

The results for step lengths of $\tau/10$ and τ ($v_s = 50$ V) are shown in Tables C.5 and C.6 respectively. Gear second-order is a two-step method and hence the value at ΔT is required as initial condition.

For the trapezoidal rule the ratio $(1 - \Delta t/(2\tau))/(1 + \Delta t/(2\tau))$ remains less than 1, so that the solution does tend to zero for any time step size. However for $\Delta t > 2\tau$ it does so in an oscillatory manner and convergence may be very slow.

C.3 Stability of integration methods

Truncation error is a localised property (i.e. local to the present time point and time step) whereas stability is a global property related to the growth or decay of errors introduced at each time point and propagated to successive time points. Stability depends on both the method and the problem.

Since general stability analysis is difficult the normal approach is to compare the stability of different methods for a single test equation, such as:

$$\dot{y} = f(y, t) = \lambda y \tag{C.17}$$

Table C.7 Stability region

Integration rule	Formula	Region of stability
Trapezoidal	$v_R(t + \Delta t) = \dfrac{(1 - \Delta t/(2\tau))}{(1 + \Delta t/(2\tau))} v_R(t)$	$0 < \dfrac{\Delta t}{\tau}$
Forward Euler	$v_R(t + \Delta t) = (1 - \Delta t/\tau) v_R(t)$	$0 < \dfrac{\Delta t}{\tau} < 2$
Backward Euler	$v_R(t + \Delta t) = \dfrac{v_R(t)}{(1 + \Delta t/\tau)}$	$\dfrac{\Delta t}{\tau} < -2$ and $0 < \dfrac{\Delta t}{\tau}$
Gear 2nd order	$v_R(t + \Delta t) = \frac{4}{3} v_R(t) - \dfrac{v_R(t - \Delta t)}{3(1 + 2\Delta t/(3\tau))}$	$\dfrac{\Delta t}{\tau} < -4$ and $0 < \dfrac{\Delta t}{\tau}$

An algorithm is said to be A-stable if it results in a stable difference equation approximation to a stable differential equation. Hence the algorithm is A-stable if the numerical approximation of equation C.17 tends to zero for positive step length and eigenvalue, λ, in the left-hand half-plane. In other words for a stable differential equation an A-stable method will converge to the correct solution as time goes to infinity regardless of the step length or the accuracy at intermediate steps.

Table C.7 summarises the stability regions of the test system for the various integration formulae. It should be noted that numerical integration formulae do not possess regions of stability independent of the problem they are applied to. To examine the impact on the numerical stability of a system of equations, the integrator is substituted in and the stability of the resulting difference equations examined.

Where multiple time constants (hence eigenvalues) are present, the stability is determined by the smallest time constant (largest eigenvalue). However the response of interest is often determined by the larger time constants (small eigenvalues) present, thus requiring long simulation times to see it. These conflicting requirements lead to long simulation times with short time steps. Systems where the ratio of the largest to smallest eigenvalue is large are **stiff systems**.

C.4 References

1 CHUA, L. O. and LIN, P. M.: 'Computer aided analysis of electronic circuits: algorithms and computational techniques' (Prentice Hall, Englewood Cliffs, CA, 1975)
2 NAGEL, L. W.: 'SPICE2: a computer program to simulate semiconductor circuits' (Ph.D. thesis, University of California, Berkeley, CA 94720, May 1975, memorandum no. UCB/ERL M520)
3 McCALLA, W. J.: 'Fundamentals of computer-aided circuit simulation' (Kluwer Academic Publishers, Boston, 1988)

4 SMITH, J. M.: 'Mathematical modeling and digital simulation for engineers and scientists' (John Wiley & Sons, New York, 2nd edition, 1987)
5 ALVARADO, F. L., LASSETER, R. H. and SANCHEZ, J. J.: 'Testing of trapezoidal integration with damping for the solution of power transient problems', *IEEE Transactions on Power Apparatus and Systems*, 1983, **102** (12), pp. 3783–90

Appendix D
Test systems data

D.1 CIGRE HVDC benchmark model

The CIGRE benchmark model [1] consists of weak rectifier and inverter a.c. systems resonant at the second harmonic and a d.c. system resonant at the fundamental frequency. Both a.c. systems are balanced and connected in star-ground. The system is shown in Figure D.1 with additional information in Tables D.1 and D.2. The HVDC link is a 12-pulse monopolar configuration with the converter transformers connected star-ground/star and star-ground/delta. Figures D.2–D.4 show the impedance scans of the a.c. and d.c. systems. A phase imbalance is created in the inverter a.c. system by inserting typically a 5.0 per cent resistance into one phase in series with the a.c. system.

D.2 Lower South Island (New Zealand) system

The New Zealand Lower South Island power system, shown in Figure D.5, is a useful test system due to its naturally imbalanced form and to the presence of a large power converter situated at the Tiwai-033 busbar. Apart from the converter installation, the

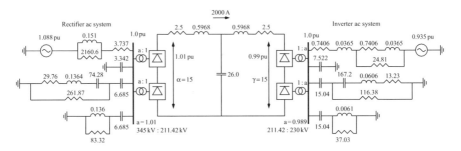

Figure D.1   CIGRE HVDC benchmark test system (all components in Ω, H and μF)

Table D.1 CIGRE model main parameters

Parameter	Rectifier	Inverter
a.c. system voltage	345 kV l–l	230 kV l–l
a.c. system impedance magnitude	119.03 Ω	52.9 Ω
Converter transformer tap (prim. side)	1.01	0.989
Equivalent commutation reactance	27 Ω	27 Ω
d.c. voltage	505 kV	495 kV
d.c. current	2 kA	2 kA
Firing angle	15°	15°
d.c. power	1010 MW	990 MW

Table D.2 CIGRE model extra information

Rectifier a.c. voltage base	345.0 kV
Inverter a.c. voltage base	230 kV
Rectifier voltage source	1.088 ∠22.18°
Inverter voltage source	0.935 ∠−23.14°
Nominal d.c. voltage	500 kV
Transformer power base	598 MVA
Transformer leakage reactance	0.18 pu
Transformer secondary voltage	211.42 kV
Nominal rectifier firing angle	15.0°
Nominal inverter extinction angle	15.0°
Thyristor forward voltage drop	0.001 kV
Thyristor onstate resistance	0.01 Ω
Snubber resistance	5000 Ω
Snubber capacitance	0.05 μF
d.c. current transducer gain	0.5
d.c. current transducer time constant	0.001 s
PI controller proportional gain	1.0989
PI controller time constant	0.0091 s
Type-1 filter	Shunt capacitor
Type-2 filter	Single tuned filter
Type-3 filter	Special CIGRE filter

system includes two major city loads that vary significantly over the daily period. The 220 kV transmission lines are all specified by their geometry and conductor specifications, and consequently are unbalanced and coupled between sequences. Tables D.3–D.9 show all of the necessary information.

Test systems data 361

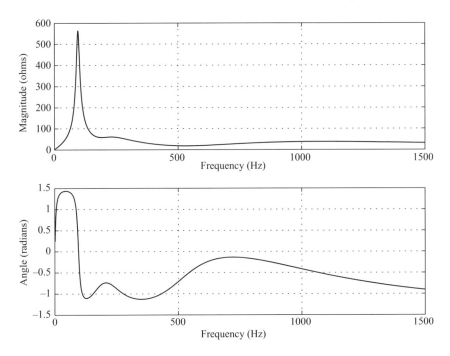

Figure D.2 Frequency scan of the CIGRE rectifier a.c. system impedance

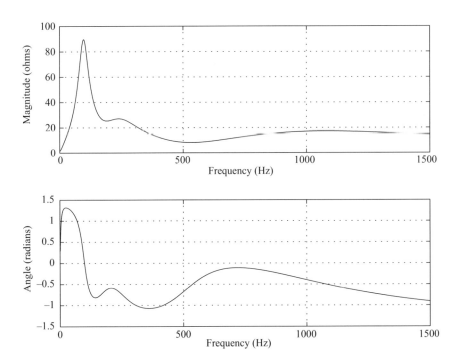

Figure D.3 Frequency scan of the CIGRE inverter a.c. system impedance

Table D.3 Converter information for the Lower South Island test system

	Phase shift	X_1	$V_{pri.}$	$V_{sec.}$	P_{base}
Converter 1	22.5°	0.05 pu	33.0 kV	5.0 kV	100 MVA
Converter 2	7.5°	0.05 pu	33.0 kV	5.0 kV	100 MVA
Converter 3	−7.5°	0.05 pu	33.0 kV	5.0 kV	100 MVA
Converter 4	−22.5°	0.05 pu	33.0 kV	5.0 kV	100 MVA

Table D.4 Transmission line parameters for Lower South Island test system

Busbars	Length	Conductor type	Earth-wire type
Manapouri-220 to Invercargill-220	152.90 km	GOAT (30/3.71 + 7/3.71 ACSR)	(7/3.05 Gehss)
Manapouri-220 to Tiwai-220	175.60 km	GOAT (30/3.71 + 7/3.71 ACSR)	(7/3.05 Gehss)
Invercargill-220 to Tiwai-220	24.30 km	GOAT (30/3.71 + 7/3.71 ACSR)	(7/3.05 Gehss)
Invercargill-220 to Roxburgh-220	129.80 km	ZEBRA (54/3.18 + 7/3.18 ACSR)	(7/3.05 Gehss)
Invercargill-220 to Roxburgh-220	132.20 km	ZEBRA (54/3.18 + 7/3.18 ACSR)	–

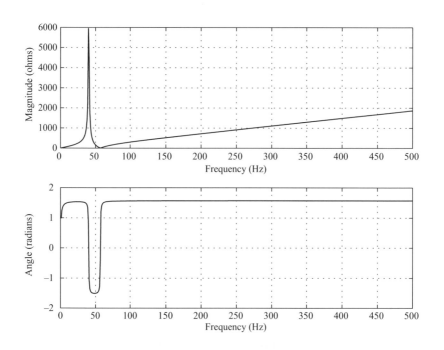

Figure D.4 Frequency scan of the CIGRE d.c. system impedance

Table D.5   Conductor geometry for Lower South Island transmission lines (in metres)

Cx1	Cx2	Cx3	Cy1	Cy2	Cy3	Ex1	Ey1	Bundle sep.	Nc	Ne
4.80	6.34	4.42	12.50	18.00	23.50	0.00	28.94	0.46	2	1
4.80	6.34	4.42	12.50	18.00	23.50	0.00	29.00	0.45	2	1
4.77	6.29	4.41	12.50	17.95	23.41	1.52	28.26	0.46	2	2
0.00	6.47	12.94	12.50	12.50	12.50	4.61	18.41	–	1	2
0.00	7.20	14.40	12.50	12.50	12.50	–	–	–	1	0

Table D.6   Generator information for Lower South Island test system

Busbar	x_d''	V_{set}	P_{set}
Manapouri-1014	0.037	1.0 pu	200.0 MW
Manapouri-2014	0.074	1.0 pu	200.0 MW
Roxburgh-1011	0.062	1.0 pu	Slack

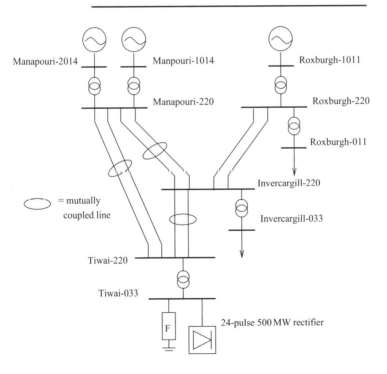

Figure D.5   Lower South Island of New Zealand test system

Table D.7 Transformer information for the Lower South Island test system

Pri. busbar	Sec. busbar	Pri. con.	Sec. con.	R_1(pu)	X_1(pu)	Tap (pu)
Manapouri-220	Manapouri-1014	Star-g	Delta	0.0006	0.02690	0.025 pri.
Manapouri-220	Manapouri-2014	Star-g	Delta	0.0006	0.05360	0.025 pri.
Roxburgh-220	Roxburgh-1011	Star-g	Delta	0.0006	0.03816	0.025 pri.
Invercargill-033	Invercargill-220	Star-g	Delta	0.0006	0.10290	0.025 pri.
Roxburgh-011	Roxburgh-220	Star-g	Delta	0.0006	0.03816	0.025 pri.
Tiwai-220	Tiwai-033	Star-g	Star-g	0.0006	0.02083	–

Table D.8 System loads for Lower South Island test system (MW, MVar)

Busbar	P_a	Q_a	P_b	Q_b	P_c	Q_c
Invercargill-033	45.00	12.00	45.00	12.00	45.00	12.00
Roxburgh-011	30.00	18.00	30.00	18.00	30.00	18.00

Table D.9 Filters at the Tiwai-033 busbar

Connection	R (Ω)	L (mH)	C (μF)
Star-g	0.606	19.30	21.00
Star-g	0.424	9.63	21.50
Star-g	2.340	17.20	3.49
Star-g	1.660	14.40	5.80

In Table D.5 the number of circuits is indicated by (Nc) and the conductor coordinates by (Cx) and (Cy) relative to the origin, which is in the middle of the tower and 12.5 m above the ground. The towers are symmetrical around the vertical axis hence only one side needs specifying. The number of earth-wires is indicated by (Ne) and their coordinates by (Ex) and (Ey). The line to line busbar voltage is given in kV by the last three digits of the busbar name. The power base is 100 MVA and the system frequency is 50 Hz. The filters at the Tiwai-033 busbar consist of three banks of series RLC branches connected in star-ground.

The rating of the rectifier installation at the Tiwai-033 busbar is approximately 480 MW and 130 MVAr. The rectifier has been represented by a 24-pulse installation connected in parallel on the d.c. side, by small linking reactors

($R = 1\,\mu\Omega$, $L = 1\,\mu H$) to an ideal current source (80 kA). The converters are diode rectifiers and the transformer specifications are given in Table D.3.

D.3 Reference

1 SZECHTMAN, M., WESS, T. and THIO, C.V.: 'First benchmark model for HVdc control studies', *ELECTRA*, 1991, **135**, pp. 55–75

Appendix E
Developing difference equations

E.1 Root-matching technique applied to a first order lag function

This example illustrates the use of the root-matching technique to develop a difference equation as described in section 5.3. The first order lag function in the s-domain is expressed as:

$$H(s) = \frac{G}{1 + s\tau}$$

The corresponding z-domain transfer function with pole matched (as $z = e^{s\Delta t}$) is

$$H(z) = \frac{kz}{z - e^{-\Delta t/\tau}}$$

Applying the final value theorem to the s-domain transfer function

$$\mathrm{Lim}_{s \to 0} \{s \cdot H(s)/s\} = G$$

Applying the final value theorem to the z-domain transfer function

$$\mathrm{Lim}_{z \to 1} \left\{ \frac{(z-1)}{z} \cdot H(z) \bigg/ \frac{z}{(z-1)} \right\} = \frac{k}{(1 - e^{-\Delta t/\tau})}$$

Therefore

$$k = G(1 - e^{-\Delta t/\tau})$$

$$\frac{I(z)}{V(z)} = H(z) = \frac{kz}{(z - e^{-\Delta t/\tau})}$$

Rearranging gives:

$$I(z) \cdot (z - e^{-\Delta t/\tau}) = kzV(z)$$

$$I(z) \cdot (1 - z^{-1} \cdot e^{-\Delta t/\tau}) = kV(z)$$

Hence

$$I(z) = kV(z) + e^{-\Delta t/\tau} z^{-1} I(z)$$

or in the time domain the exponential form of difference equation becomes:

$$i(t) = G(1 - e^{-\Delta t/\tau}) \cdot v(t) + e^{-\Delta t/\tau} i(t - \Delta t) \tag{E.1}$$

E.2 Root-matching technique applied to a first order differential pole function

This example illustrates the use of the root-matching technique to develop a difference equation as described in section 5.3, for a first order differential pole function. The s-domain expression for the first order differential pole function is:

$$H(s) = \frac{Gs}{1 + s\tau}$$

The z-domain transfer function with pole and zero matched (using $z = e^{s\Delta t}$) is

$$H(z) = \frac{k(z-1)}{z - e^{-\Delta t/\tau}}$$

Applying the final value theorem to the s-domain transfer function for a unit ramp input:

$$\text{Lim}_{s \to 0} \left\{ s \cdot H(s)/s^2 \right\} = G$$

Applying the final value theorem to the z-domain transfer function:

$$\text{Lim}_{z \to 0} \left\{ \frac{(z-1)}{z} \cdot H(z) \bigg/ \frac{z\Delta t}{(z-1)^2} \right\} = \frac{k\Delta t}{\left(1 - e^{-\Delta t/\tau}\right)}$$

Therefore

$$k = G(1 - e^{-\Delta t/\tau})/\Delta t$$

$$\frac{I(z)}{V(z)} = H(z) = \frac{k(z-1)}{(z - e^{-\Delta t/\tau})}$$

Rearranging gives:

$$I(z)(z - e^{-\Delta t/\tau}) = k(z-1) V(z)$$

$$I(z)(1 - z^{-1} \cdot e^{-\Delta t/\tau}) = k(1 - z^{-1}) V(z)$$

Hence

$$I(z) = kV(z) - k \cdot z^{-1} V(z) + e^{-\Delta t/\tau} z^{-1} I(z)$$

E.3 Difference equation by bilinear transformation for RL series branch

For comparison the bilinear transform is applied to the s-domain rational function for a series RL branch

$$Y(s) = \frac{I(s)}{V(s)} = \frac{1}{R + sL}$$

Applying the bilinear transformation ($s \approx 2(1 - z^{-1})/(\Delta t(1 + z^{-1}))$):

$$\begin{aligned}Y(z) &= \frac{1}{R + 2L(1 - z^{-1})/(\Delta t(1 + z^{-1}))} \\ &= \frac{(1 + z^{-1})}{(R + 2L/\Delta t) + z^{-1} \cdot (R - 2L/\Delta t)} \\ &= \frac{(1 + z^{-1})/(R + 2L/\Delta t)}{1 + z^{-1} \cdot (R - 2L/\Delta t)/(R + 2L/\Delta t)}\end{aligned} \quad (E.2)$$

E.4 Difference equation by numerical integrator substitution for RL series branch

Next numerical integrator substitution is applied to a series RL branch, to show that it does give the same difference equation as using the bilinear transform

$$\frac{di}{dt} = \frac{1}{L}(v - R \cdot i)$$

Hence

$$i_k = i_{k-1} + \int_{t-\Delta t}^{t} \frac{di}{dt} dt$$

Applying the trapezoidal rule gives:

$$i_k = i_{k-1} + \frac{\Delta t}{2}\left(\frac{di_k}{dt} + \frac{di_{k-1}}{dt}\right)$$

Substituting in the branch equation yields:

$$\frac{di}{dt} = \frac{1}{L}(v - R \cdot i)$$

$$i_k = i_{k-1} + \int_{t-\Delta t}^{t} di$$

$$i_k = i_{k-1} + \frac{\Delta t}{2}\left(\frac{di_k}{dt} + \frac{di_{k-1}}{dt}\right)$$

$$i_k = i_{k-1} + \frac{\Delta t}{2}\left(\frac{1}{L}(v_k - R \cdot i_k) + \frac{1}{L}(v_{k-1} - R \cdot i_{k-1})\right)$$

Table E.1 Coefficients of a rational function in the z-domain for admittance

Method	z-domain coefficients
Root-matching method Input type (a)	$a_0 = G(1 - e^{-\Delta t/\tau})$ $a_1 = 0$ $b_0 = 1$ $b_1 = -e^{-\Delta t/\tau}$
Root-matching method Input type (b)	$a_0 = 0$ $a_1 = G(1 - e^{-\Delta t/\tau})$ $b_0 = 1$ $b_1 = -e^{-\Delta t/\tau}$
Root-matching method Input type (c)	$a_0 = G(1 - e^{-\Delta t/\tau})/2$ $a_1 = G(1 - e^{-\Delta t/\tau})/2$ $b_0 = 1$ $b_1 = -e^{-\Delta t/\tau}$
Root-matching method Input type (d)	$a_0 = G\left(-e^{-\Delta t/\tau} + \dfrac{\tau}{\Delta t}(1 - e^{-\Delta t/\tau})\right)$ $a_1 = G\left(1 - \dfrac{\tau}{\Delta t}(1 - e^{-\Delta t/\tau})\right)$ $b_0 = 1$ $b_1 = -e^{-\Delta t/\tau}$
Recursive convolution (second order)	$a_0 = \lambda \qquad a_1 = \mu \qquad a_2 = \nu$ $b_0 = 1 \qquad b_1 = -e^{-\Delta t/\tau}$ $\lambda = G\left(\left(\dfrac{\tau}{\Delta t}\right)^2 (1 - e^{-\Delta t/\tau}) - \dfrac{\tau}{2\Delta t}(3 - e^{-\Delta t/\tau}) + 1\right)$ $\mu = G\left(-2\left(\dfrac{\tau}{\Delta t}\right)^2 (1 - e^{-\Delta t/\tau}) + \dfrac{2\tau}{\Delta t}(1 - e^{-\Delta t/\tau}) + \dfrac{2\tau}{\Delta t} - e^{-\Delta t/\tau}\right)$ $\nu = G\left(\left(\dfrac{\tau}{\Delta t}\right)^2 (1 - e^{-\Delta t/\tau}) - \dfrac{\tau}{2\Delta t}(1 + e^{-\Delta t/\tau})\right)$
Trapezoidal integrator Substitution	$a_0 = G/(1 + 2\tau/\Delta t)$ $a_1 = G/(1 + 2\tau/\Delta t)$ $b_0 = 1$ $b_1 = (1 - 2\tau/\Delta t)/(1 + 2\tau/\Delta t)$

Table E.2   Coefficients of a rational function in the z-domain for impedance

Method	z-domain coefficients
Root-matching method Input type (a)	$a_0 = 1/G(1 - e^{-\Delta t/\tau})$ $a_1 = -e^{-\Delta t/\tau}/G(1 - e^{-\Delta t/\tau})$ $b_0 = 1 \quad b_1 = 0$
Root-matching method Input type (b)	$a_0 = 1/G(1 - e^{-\Delta t/\tau})$ $a_1 = -e^{-\Delta t/\tau}/G(1 - e^{-\Delta t/\tau})$ $b_0 = 0 \quad b_1 = 1$
Root-matching method Input type (c)	$a_0 = 2/G(1 - e^{-\Delta t/\tau})$ $a_1 = -2e^{-\Delta t/\tau}/G(1 - e^{-\Delta t/\tau})$ $b_0 = 1 \quad b_1 = 1$
Root-matching method Input type (d)	$a_0 = 1/G\left(1 - \frac{\tau}{\Delta t}(1 - e^{-\Delta t/\tau})\right)$ $a_1 = -e^{-\Delta t/\tau}/G\left(1 - \frac{\tau}{\Delta t}(1 - e^{-\Delta t/\tau})\right)$ $b_0 = 1$ $b_1 = \left(-e^{-\Delta t/\tau} + \frac{\tau}{\Delta t}(1 - e^{-\Delta t/\tau})\right) \bigg/ \left(1 - \frac{\tau}{\Delta t}(1 - e^{-\Delta t/\tau})\right)$
Recursive convolution (second order)	$a_0 = 1/\lambda \quad a_1 = -e^{-\Delta t/\tau}/\lambda$ $b_0 = 1 \quad b_1 = \mu/\lambda \quad b_2 = \nu/\lambda$ $\lambda = G\left(\left(\frac{\tau}{\Delta t}\right)^2(1 - e^{-\Delta t/\tau}) - \frac{\tau}{2\Delta t}(3 - e^{-\Delta t/\tau}) + 1\right)$ $\mu = G\left(-2\left(\frac{\tau}{\Delta t}\right)^2(1 - e^{-\Delta t/\tau}) + \frac{2\tau}{\Delta t}(1 - e^{-\Delta t/\tau}) + \frac{2\tau}{\Delta t} - e^{-\Delta t/\tau}\right)$ $\nu = G\left(\left(\frac{\tau}{\Delta t}\right)^2(1 - e^{-\Delta t/\tau}) - \frac{\tau}{2\Delta t}(1 + e^{-\Delta t/\tau})\right)$
Trapezoidal integrator Substitution	$a_0 = (1 + 2\tau/\Delta t)/G$ $a_1 = (1 - 2\tau/\Delta t)/G$ $b_0 = 1 \quad b_1 = 1$

Table E.3 Summary of difference equations

Method	Difference equations
Root-matching method Input type (a)	$y_t = e^{-\Delta t/\tau} y_{t-\Delta t} + G(1 - e^{-\Delta t/\tau}) u_t$
Root-matching method Input type (b)	$y_t = e^{-\Delta t/\tau} y_{t-\Delta t} + G(1 - e^{-\Delta t/\tau}) \dfrac{(u_t + u_{t-\Delta t})}{2}$
Root-matching method Input type (c)	$y_t = e^{-\Delta t/\tau} y_{t-\Delta t} + G(1 - e^{-\Delta t/\tau}) u_{t-\Delta t}$
Root-matching method Input type (d)	$y_t = e^{-\Delta t/\tau} y_{t-\Delta t} + G\left(-e^{-\Delta t/\tau} + \dfrac{\tau}{\Delta t}(1 - e^{-\Delta t/\tau})\right) u_{t-\Delta t}$ $+ G\left(1 - \dfrac{\tau}{\Delta t}(1 - e^{-\Delta t/\tau})\right) u_t$
Trapezoidal integrator substitution	$y_t = \dfrac{(1 - \Delta t/(2\tau))}{(1 + \Delta t/(2\tau))} y_{t-\Delta t} + \dfrac{G\Delta t/(2\tau)}{(1 + \Delta t/(2\tau))} (u_t + u_{t-\Delta t})$

Rearranging this gives:

$$i_k \left(1 + \frac{\Delta t R}{2L}\right) = i_{k-1}\left(1 - \frac{\Delta t R}{2L}\right) + \frac{\Delta t}{2L}(v_k + v_{k-1})$$

or

$$i_k = \frac{(1 - \Delta t R/(2L))}{(1 + \Delta t R/(2L))} i_{k-1} + \frac{(\Delta t/(2L))}{(1 + \Delta t R/(2L))} (v_k + v_{k-1})$$

Taking the z-transform:

$$\left(1 - z^{-1}\frac{(1 - \Delta t R/(2L))}{(1 + \Delta t R/(2L))}\right) I(z) = \frac{(\Delta t/(2L))}{(1 + \Delta t R/(2L))}(1 + z^{-1}) V(z)$$

Rearranging gives:

$$\frac{I(z)}{V(z)} = \frac{(1 + z^{-1})/(R + 2L/\Delta t)}{1 + z^{-1}(R - 2L/\Delta t)/(R + 2L/\Delta t)}$$

This is identical to the equation obtained using the bilinear transform (equation E.2).

Tables E.1 and E.2 show the first order z-domain rational functions associated with admittance and impedance respectively, for a first order lag function, for each of the exponential forms described in section 5.5. The rational functions have been converted to the form $(a_0' + a_1' z^{-1})/(1 + b_1' z^{-1})$ if b_0 is non-zero, otherwise left in the form $(a_0 + a_1 z^{-1})/(b_0 + b_1 z^{-1})$. Table E.3 displays the associated difference equation for each of the exponential forms.

Appendix F

MATLAB code examples

F.1   Voltage step on *RL* branch

In this example a voltage step (produced by a switch closed on to a d.c. voltage source) is applied to an *RL* load. The results are shown in section 4.4.2. The *RL* load is modelled by one difference equation rather than each component separately.

```
% EMT_StepRL.m
clear all

% Initialize Variables
R = 1.00000;
L = 0.05E-3;
Tau= L/R;         % load Time-constant
Delt = 250.0E-6; % Time-step
Finish_Time = 4.0E-3;
V_mag= 100.0;
V_ang= 0.0;

l=1;
i(1)    = 0.0;
time(1)= 0.0;
v(1)    = 0.0;

K_i = (1-Delt*R/(2.0*L))/(1+Delt*R/(2.0*L));
K_v = (Delt/(2.0*L))/(1+Delt*R/(2.0*L));
G_eff= K_v;

% Main Time-step loop
for k=Delt:Delt:Finish_Time
    l=l+1;
    time(l)=k;

    if time(l)>=0.001
       v(l) = 100.0;
```

```
    else
        v(1)=0.0;
    end;
    I_history = K_i*i(l-1) + K_v*v(l-1);
    I_inst = v(l)*G_eff;
    i(l) = I_inst+I_history;
end;

% Plot results
figure(1)
plot(time,v,'-k',time,i,':k');
legend('Voltage','Current');
xlabel('Time (Seconds)');
```

F.2 Diode fed *RL* branch

This simple demonstration program is used to show the numerical oscillation that occurs at turn-off by modelling an *RL* load fed from an a.c. source through a diode. The results are shown in section 9.4. The *RL* load is modelled by one difference equation rather than each component separately.

```
% Diode fed RL load
% A small demonstration program to demonstrate numerical noise
%%%%%%%%%%%%%%%%%%%%%%%%%%%%%%%%%%%%%%%%%%%%%%
clear all

f =50.0;                  % Source Frequency
Finish_Time = 60.0E-3;
R= 100.0;
L= 500.0E-3;
Tau= L/R;                 % Load time-constant
Delt = 50.0E-6;           % Time-step
V_mag= 230.0*sqrt(2.);    % Peak source magnitude
V_ang= 0.0;
R_ON = 1.0E-10;  % Diode ON Resistance
R_OFF= 1.0E10;   % Diode OFF Resistance

% Initial State
ON=1;
R_switch = R_ON;

l=1;
i(1)=0.0;
time(1)=0.0;
v(1) = V_mag*sin(V_ang*pi/180.0);
v_load(1)=v(1);
ON = 1;

K_i=(1-Delt*R/(2.0*L))/(1+Delt*R/(2.0*L));
K_v=(Delt/(2.0*L))/(1+Delt*R/(2.0*L));
G_eff= K_v;
G_switch = 1.0/R_switch ;
```

```
for k=Delt:Delt:Finish_Time
    % Advance Time
    l=l+1;
    time(l)=k;

    % Check Switch positions
    if i(l-1) <= 0.0 & ON==1 & k > 5*Delt
      % fprintf('Turn OFF time = %12.1f usecs \n',time(l-1)*1.0E6);
      % fprintf('i(l) = %12.6f  \n',i(l-1));
           fprintf('v_load(l) = %12.6f \n',v_load(l-1));
       ON=0;
       Time_Off=time(l);
       R_switch=R_OFF;
       G_switch = 1.0/R_switch;
       i(l-1)=0.0;
    end;
    if v(l-1)-v_load(l-1) > 1.0 & ON==0
      % fprintf('Turn ON time = %12.1f usecs\n',time(l)*1.0E6);
       ON=1;
       R_switch=R_ON;
       G_switch = 1.0/R_switch;
    end;

    % Update History Term
    I_history = K_i*i(l-1) + K_v*v_load(l-1);

    % Update Voltage Sources
    v(l) = V_mag*sin(2.0*pi*50.0*time(l) + V_ang*pi/180.0);

    % Solve for V and I
    v_load(l)= (-I_history+v(l)*G_switch)/(G_eff+G_switch);
    I_inst = v_load(l)*G_eff;
    i(l) = I_inst + I_history;
   %fprintf( '%12.5f %12.5f %12.5f %12.5f \n',time(l)*1.0E3,v(l),
            v_load(l),i(l));
end;

figure(1);
clf;
subplot(211);

plot(time,v_load,'k');
legend('V_{Load}');

ylabel('Voltage (V)')'
xlabel('Time (S)')'
grid;
title('Diode fed RL Load');

subplot(212);
plot(time,i,'-k');
legend('I_{Load}');
```

```
ylabel('Current (A)')'
grid;
axis([0.0 0.06 0.0000 2.5])
xlabel('Time (S)')'
```

F.3 General version of example F.2

This program models the same case as the program of section F.2, i.e. it shows the numerical oscillation that occurs at turn-off by modelling an RL load fed from an a.c. source through a diode. However it is now structured in a general manner, where each component is subject to numerical integrator substitution (NIS) and the conductance matrix is built up. Moreover, rather than modelling the switch as a variable resistor, matrix partitioning is applied (see section 4.4.1), which enables the use of ideal switches.

```
% General EMT Program
%%%%%%%%%%%%%%%%%%%%%%%%%%%%%%%%%%%%%%%%%
clear all
global Branch
global No_Brn
global Source
global No_Source
global No_Nodes
global i_RL
global t
global v_load
global v_s
global V_n
global V_K
global ShowExtra
%%%%%%%%%%%%%%%%%%%%%%%%%%%%%%%%%%%%%%%%%%%%
% Initialize
%%%%%%%%%%%%%%%%%%%%%%%%%%%%%%%%%%%%%%%%%%%%
format long

ShowExtra=0;
TheTime = 0.0;
Finish_Time = 60.0E-3;
DeltaT = 50.0E-6;
%%%%%%%%%%%%%%%%%%%%%%%%%%%%%%%%%%%%%%%%%%%%
% Specify System
%%%%%%%%%%%%%%%%%%%%%%%%%%%%%%%%%%%%%%%%%%%%
No_Brn=3;
No_Nodes=3;
No_UnkownNodes=2;
No_Source=1;

Branch(1).Type = 'R';
Branch(1).Value= 100.0;
Branch(1).Node1= 2;
Branch(1).Node2= 1;
Branch(1).V_last = 0.0;
```

```
Branch(1).I_last = 0.0;
Branch(1).I_history = 0.0;

Branch(2).Type = 'L';
Branch(2).Value= 500.0E-3;
Branch(2).Node1= 1;
Branch(2).Node2= 0;
Branch(2).V_last = 0.0;
Branch(2).I_last = 0.0;
Branch(2).I_history = 0.0;

Branch(3).Type = 'S';
Branch(3).Node1= 3;
Branch(3).Node2= 2;
Branch(3).R_ON  = 1.0E-10;
Branch(3).R_OFF = 1.0E+10;
Branch(3).V_last = 0.0;
Branch(3).I_last = 0.0;
Branch(3).I_history = 0.0;

Branch(3).Value= Branch(3).R_ON;
Branch(3).State= 1;

Source(1).Type = 'V';
Source(1).Node1= 3;
Source(1).Node2= 0;
Source(1).Magnitude= 230.0*sqrt(2.);
Source(1).Angle= 0.0;
Source(1).Frequency= 50.0;

% Initialize
% ----------
for k=1:No_Nodes
   BusCnt(k,1) = k;
end;
for k=1:No_Brn
   Branch(k).V_last  = 0.0;
   Branch(k).V_last2 = 0.0;
   Branch(k).I_last  = 0.0;
   Branch(k).I_last2 = 0.0;
end;
%%%%%%%%%%%%%%%%%%%%%%%%%%%%%%%%%%%%%%%%%%%%%%%%%%%%%%%
% Form Conductance Matrix
%%%%%%%%%%%%%%%%%%%%%%%%%%%%%%%%%%%%%%%%%%%%%%%%%%%%%%%
[G] = Form_G(DeltaT,ShowExtra);

% Initial Source Voltage
v_source = V_Source(0.0);
% Initial Branch Voltage
Branch(2).V_last = v_source;
%%%%%%%%%%%%%%%%%%%%%%%%%%%%%%%%%%%%%%%%%%%%%%%%%%%%%%%
% Main TheTime loop
%%%%%%%%%%%%%%%%%%%%%%%%%%%%%%%%%%%%%%%%%%%%%%%%%%%%%%%
```

```
l = 1;
while TheTime <= Finish_Time
   %%%%%%%%%%%%%%%%%%%%
   % Advance TheTime %
   %%%%%%%%%%%%%%%%%%%%
   TheTime = TheTime+DeltaT;
   l = l+1;
   t(l) = TheTime;
   %fprintf('\n ------------ %12.6f -- l=%d ---------- \n',TheTime,l);

   %%%%%%%%%%%%%%%%%%%%%%%%%%%%
   % Check Switch positions   %
   %%%%%%%%%%%%%%%%%%%%%%%%%%%%
   [Reform_G] = Check_Switch(DeltaT,TheTime,v_source,l);
   if Reform_G == 1
      ShowExtra = 0;
      [G] = Form_G(DeltaT,ShowExtra);
   end;
   %%%%%%%%%%%%%%%%%%%%%%%%%%%%%%%%%%%%%%%%%%%
   % Update Sources
   %%%%%%%%%%%%%%%%%%%%%%%%%%%%%%%%%%%%%%%%%%%
   v_source = V_Source(TheTime);
   %%%%%%%%%%%%%%%%%%%%%%%%%%%%%%%%%%%%%%%%%%%
   % Calculate History Terms
   %%%%%%%%%%%%%%%%%%%%%%%%%%%%%%%%%%%%%%%%%%%
   CalculateBrn_I_history;
   %%%%%%%%%%%%%%%%%%%%%%%%%%%%%%%%%%%%%%%%%%%
   % Calculate Injection Current Vector
   %%%%%%%%%%%%%%%%%%%%%%%%%%%%%%%%%%%%%%%%%%%
   [I_history] = Calculate_I_history;
   %%%%%%%%%%%%%%%%%%%%%%%%%%%%%%%%%%%%%%%%%%%%%%%%
   % Partition Conductance Matrix and Injection Current Vector
   %%%%%%%%%%%%%%%%%%%%%%%%%%%%%%%%%%%%%%%%%%%%%%%%
   G_UU = G(1:No_UnkownNodes,1:No_UnkownNodes);
   G_UK = G(1:No_UnkownNodes,No_UnkownNodes+1:No_Nodes);
   I_U  = I_history(1:No_UnkownNodes)';
   % Known Node Voltage
   % -------------------
   V_K(1:1) = v_source;
   %%%%%%%%%%%%%%%%%%%%%%%%%%%%%%%%%%%%%%%%%%%
   % Modified Injection Current Vector
   %%%%%%%%%%%%%%%%%%%%%%%%%%%%%%%%%%%%%%%%%%%
   I_d_history = I_U - G_UK*V_K;
   %%%%%%%%%%%%%%%%%%%%%%%%%%%%%%%%%%%%%%%%%%%
   % Solve for Unknown Node Voltages
   %%%%%%%%%%%%%%%%%%%%%%%%%%%%%%%%%%%%%%%%%%%
   V_U = G_UU\I_d_history;
   %%%%%%%%%%%%%%%%%%%%%%%%%%%%%%%%%%%%%%%%%%%
   % Rebuild Node Voltage Vector
   %%%%%%%%%%%%%%%%%%%%%%%%%%%%%%%%%%%%%%%%%%%
   V_n(1:No_UnkownNodes,1) = V_U;
   V_n(No_UnkownNodes+1:No_Nodes,1)=V_K;
   %%%%%%%%%%%%%%%%%%%%%%%%%%%%%%%%%%%%%%%%%%%
```

MATLAB code examples 379

```
      % Calculate Branch Voltage
      %%%%%%%%%%%%%%%%%%%%%%%%%%%%%%%%%%%%%%%%%%%
      V_Branch(V_n);
      if(ShowExtra==1)
         for k=1:No_Brn
            fprintf('V_Branch(%d)= %12.6f  \n',k,Branch(k).V_last);
         end;
      end; %if
      %%%%%%%%%%%%%%%%%%%%%%%%%%%%%%%%%%%%%%%%%%%
      % Calculate Branch Current
      %%%%%%%%%%%%%%%%%%%%%%%%%%%%%%%%%%%%%%%%%%%
      I_Branch;
      %%%%%%%%%%%%%%%%%%%%%%%%%%%%%%%%%%%%%%%%%%%
      % Load information to be plotted
      %%%%%%%%%%%%%%%%%%%%%%%%%%%%%%%%%%%%%%%%%%%
      i_Brn2 = Branch(2).I_last;
      LoadforPlotting(1,TheTime,i_Brn2);
end;
%%%%%%%%%%%%%%%%%%%%
% Generate Plots %
%%%%%%%%%%%%%%%%%%%%
figure(1);
clf;
subplot(211);
plot(t,v_load,'k');
legend('V_{Load}');
ylabel('Voltage (V)')'
xlabel('Time (S)')'
grid;
title('Diode fed RL Load');

subplot(212);
plot(t,i_RL,'-k');
legend('I_{Load}');
ylabel('Current (A)')'
grid;
axis([0.0 0.06 0.0000 2.5])
xlabel('Time (S)')'
% =============== END of emtn.m ================
function [Reform_G] = Check_Switch (DeltaT,time,v_source,1)

global Branch
global No_Brn
global i_RL
global ShowExtra

Reform_G = 0;
for k=1:No_Brn
   Type = Branch(k).Type;
   From = Branch(k).Node1;
   To   = Branch(k).Node2;

   if Type == 'S'
```

```
         i_last = Branch(k).I_last;
         v_last = Branch(k).V_last;
         ON_or_OFF = Branch(k).State;

         % Check if Switch needs to be turned OFF
         if i_last <= 0.0 & ON_or_OFF==1 & time > 5*DeltaT
            if(ShowExtra==1)
               fprintf('Turn OFF (Brn %d) time = %12.1f usecs \n',...
                    k,time*1.0E6-DeltaT);
               fprintf('i_Switch(1) = %12.6f   \n',i_last);
               fprintf('v_load(1)   = %12.6f   \n',v_last);
            end;
            Branch(k).State=0;
            Branch(k).Value = Branch(k).R_OFF;
            Branch(1).I_last=0.0;
            Branch(2).I_last=0.0;
            Branch(3).I_last=0.0;
            i_RL(1-1) = 0.0;
            Branch(2).V_last = v_source;
            Reform_G = 1;
         end;
         % Check if Switch needs to be turned ON
         if v_last > 1.0 & ON_or_OFF==0
            Branch(k).Value = Branch(k).R_ON;
            Branch(k).Reff  = Branch(k).Value;
            Branch(k).State=1;
            Reform_G = 1;
            if(ShowExtra==1)
               fprintf('Turn ON time = %12.1f usecs\n',time*1.0E6);
               fprintf('v_last = %20.14f \n',v_last);
            end;
         end;
      end; % if
end; % for
if(ShowExtra==1) fprintf('Reform_G = %d   \n',Reform_G); end;

return
% =============== END of Check_Switch.m ================
function [G] = Form_G(DeltaT,Debug);
global Branch
global No_Brn
global No_Nodes
global ShowExtra

% Initialize [G] matrix to zero
for k=1:No_Nodes;
   for kk=1:No_Nodes;
      G(kk,k) = 0.0;
   end;
end;

for k-1:No_Brn
   Type = Branch(k).Type;
```

```
    From = Branch(k).Node1;
    To   = Branch(k).Node2;

    Series = 1;
    if To == 0
       To = From;
       Series = 0;
    end;
    if From == 0
       From = To;
       Series = 0;
    end;
    if To==0
       disp('*** Both Nodes Zero ***');
       exit;
    end;
    Branch(k).Series = Series;

    if Type == 'R'
       R_eff = Branch(k).Value;
    elseif Type == 'S'
        R_eff = Branch(k).Value;
    elseif Type == 'L'
       L = Branch(k).Value;
       R_eff = (2*L)/DeltaT;
    elseif Type == 'C'
       C = Branch(k).Value;
       R_eff = DeltaT/(2*C);
    else
       disp('*** Invalid Branch Type ***');
       exit;
    end;
    Branch(k).Reff = R_eff;

    if(ShowExtra==1)
       fprintf('Branch %d From %d to %d has Reff=  %20.14f Ohms \n',
              k,From,To,R_eff);
    end;

    if Series==1
       G(To,To)     = G(To,To) + 1/R_eff;
       G(From,From) = G(From,From) + 1/R_eff;
       G(From,To)   = G(From,To) - 1/R_eff;
       G(To,From)   = G(To,From) - 1/R_eff;
    else
       G(To,To)     = G(To,To) + 1/R_eff;
    end;
end;
if(ShowExtra==1)
  G
  pause;
end;
return
```

```
% =============== END of Form_G.m ================
function [] = CalculateBrn_I_history ()
% Calculate Branch History Term
%----------------------------------------
global Branch
global No_Brn

for k=1:No_Brn
   Type = Branch(k).Type;
   if Type == 'R'
     Branch(k).I_history =0.0;
   elseif Type == 'S'
     Branch(k).I_history =0.0;
   elseif Type == 'L'
     Branch(k).I_history = Branch(k).I_last + Branch(k).V_last/Branch(k).Reff;
   elseif Type == 'C'
     Branch(k).I_history = -Branch(k).I_last - Branch(k).V_last/Branch(k).Reff;
   end;
end;

return
% =============== END of CalculateBrn_I_history.m ================
function [I_History] = Calculate_I_history;
% Calculate Current Injection Vector from Branch History Terms
% ------------------------------------------------------------
global Branch
global No_Brn
global No_Nodes

for k=1:No_Nodes
   I_History(k)=0.0;
end;

for k=1:No_Brn
   Type = Branch(k).Type;
   From = Branch(k).Node1;
   To   = Branch(k).Node2;
   Brn_I_History = Branch(k).I_history;

   if Branch(k).Series==1
     % Series Component
     I_History(To)   =  I_History(To)   + Brn_I_History;
     I_History(From) =  I_History(From) - Brn_I_History;
   else
     if To~= 0
        I_History(To)   = I_History(To)   + Brn_I_History;
     elseif From~= 0
        I_History(From) = I_History(From) - Brn_I_History;
     else
        disp('*** Error:  Both Nodes Zero ***');
        exit;
     end;
   end; %if
```

```
end; % for
% ============== END of Calculate_I_history.m ================
function [] = I_Branch;
% Calculate Branch Current
% ------------------------
global Branch
global No_Brn

for k=1:No_Brn
    Type    = Branch(k).Type;
    From    = Branch(k).Node1;
    To      = Branch(k).Node2;
    V       = Branch(k).V_last;

    % Save last value to another variable before reassigning last.
    % Extra Past terms added so Interpolation can be added
    Branch(k).I_last3 = Branch(k).I_last2;
    Branch(k).I_last2 = Branch(k).I_last;

    if Type == 'R'
        Branch(k).I_last = V/Branch(k).Reff;
    elseif Type == 'S'
        Branch(k).I_last = V/Branch(k).Reff;
    elseif Type == 'L'
        Branch(k).I_last = V/Branch(k).Reff + Branch(k).I_history;
    elseif Type == 'C'
        Branch(k).I_last = V/Branch(k).Reff + Branch(k).I_history;
    end;
end; % for
return;
% ============== END of I_Branch.m ================
function [] = V_Branch(V);
% Calculates the Branch Voltage
% -----------------------------
global Branch
global No_Brn

for k=1:No_Brn
    From    = Branch(k).Node1;
    To      = Branch(k).Node2;
    % Save last value to another variable before reassigning last
    Branch(k).V_last3 = Branch(k).V_last2;
    Branch(k).V_last2 = Branch(k).V_last;

    if Branch(k).Series==1
        % Series Component
        Branch(k).V_last = V(From)-V(To);
    else
        if To~= 0
            Branch(k).V_last= -V(To);
        elseif From~= 0
            Branch(k).V_last= V(From);
        else
```

```
            disp('*** Error:  Both Nodes Zero ***');
            exit;
        end;
    end; %if
end; % for
return
% =============== END of V_Branch.m ================
function [V_instantaneous] = V_Source(TheTime)
% Calculates Source Voltage at TheTime
% --------------------------------
global Source
global No_Source

for k=1:No_Source
    V_mag = Source(k).Magnitude;
    V_ang = Source(k).Angle;
    freq  = Source(k).Frequency;

    V_instantaneous(k) = V_mag*sin(2.0*pi*50.0*TheTime + V_ang*pi/180.0);
end;

return
% =============== END of V_Source.m ================
function LoadforPlotting(l,TheTime,i_check);
% Load Variables for Plotting
global v_load
global v_s
global V_n
global V_K
global i_RL
global t

t(1)       = TheTime;
v_load(1)  = V_n(2);
v_s(1)     = V_K(1);
i_RL(1)    = i_check;
return;
% =============== END of LoadforPlotting.m ================
```

F.4 Frequency response of difference equations

This MATLAB procedure generates the frequency response of different discretisation methods by evaluating the frequency response of the resulting rational function in z that characterises an RL circuit. The results are shown in section 5.5.3.

```
% Compare Methods for Continuous to discrete conversion %
%%%%%%%%%%%%%%%%%%%%%%%%%%%%%%%%%%%%%%%%%%%%%%%%%
% Initialise the variable space                 %
%%%%%%%%%%%%%%%%%%%%%%%%%%%%%%%%%%%%%%%%%%%%%%%%%
clear;
format long;

StepLength = input('Enter Step-length (usecs) > ');
```

```
deltaT = StepLength * 1.0E-6;
%%%%%%%%%%%%%%%%%%%%%%%%%%%%%%%%%%%%%%%%%%%%%%%%%%%%%%%%%%%%%
% Intitalize variables and set up some intermediate results %
%%%%%%%%%%%%%%%%%%%%%%%%%%%%%%%%%%%%%%%%%%%%%%%%%%%%%%%%%%%%%
  fmax = 1001;
  R =1.0;
  L = 50.0E-6;
  Tau = L/R;
  G = 1/R;
  expterm = exp(-deltaT/Tau);
  Gequ = G*(1.0-exp(-deltaT/Tau));

%% Recursive Convolution
  a_RC = R/L;
  ah = a_RC*deltaT;
  Gequ_RC2 = G*((1.-expterm)/(ah*ah)-(3.-expterm)/(2*ah)+1);
  mu = G*(-2.*(1.-expterm)/(ah*ah) + 2.0/ah - expterm);
  vg = G*((1.-expterm)/(ah*ah)-(1.+expterm)/(2.*ah));

  for fr = 1:fmax
    f(fr) = (fr-1)*5.0; % 5 Hz increment in Data Points
    Theory(fr) = R+j*2.*pi*f(fr)*L;
  end;

  for i=1:6,
    NumOrder = 1;
    DenOrder = 1;

    if i==1 %           Root-Matching (Type a)
        a(1) = 1./Gequ;                         %a0
        a(2) = -expterm/Gequ;                   %a1
        b(1) = 1.;                              %b0
        b(2) = 0.;                              %b1
    elseif i==2 %       Root-Matching (Type c) Average
        a(1) = 2./Gequ;                         %a0
        a(2) = -2.*expterm/Gequ;                %a1
        b(1) = 1.;                              %b0
        b(2) = 1.;                              %b1
    elseif i==3 %       Root-Matching (Type b)  Z-1
        a(1) = 1./Gequ;                         %a0
        a(2) = -expterm/Gequ;                   %a1
        b(1) = 0.;                              %b0
        b(2) = 1.;                              %b1
    elseif i==4 %       Root-Matching (Type d) (or 1st order
                                      Recursive Convolution)
        Gequ_RC1 = G*(1.-(1.-expterm)/ah);
        Cterm = G*(-expterm + (1.-expterm)/ah);

        a(1) = 1./Gequ_RC1;                     %a0
        a(2) = -expterm/Gequ_RC1;               %a1
        b(1) = 1.;                              %b0
        b(2) = Cterm/Gequ_RC1;                  %b1
    elseif i==5 %       Trapezoidal integration
```

```
            kk = (2*L/deltaT + R);

            a(1) = 1/kk;
            a(2) = a(1);
            b(1) = 1.0;
            b(2) = (R-2*L/deltaT)/kk;
        elseif i==6 %         2nd order Recursive Convolution
            NumOrder = 1;
            DenOrder = 2;

            a(1) = 1./Gequ_RC2;                        %a0
            a(2) = -expterm/Gequ_RC2;                  %a1
            a(3) = 0.0;
            b(1) = 1.;                      %b0
            b(2) = mu/Gequ_RC2;                        %b1
            b(3) = vg/Gequ_RC2;
        end;

        for fr = 1:fmax,
            w = 2*pi*f(fr);
            den(fr)=0;
            num(fr)=0;
%           Calculate Denominator polynomial
            for h=1:DenOrder+1,
                den(fr) = den(fr) + b(h)*exp(-j*w*(h-1)*deltaT);
            end;
%           Calculate Numerator polynomial
            for v=1:NumOrder+1,
                num(fr) = num(fr) + a(v)*exp(-j*w*(v-1)*deltaT);
            end;
        end;
%       Calculate Rational Function
        if i==1
          Gt1 = num./den;
        elseif i==2
          Gt2 = num./den;
        elseif i==3
          Gt3 = num./den;
        elseif i==4
          Gt4 = num./den;
        elseif i==5
          Gt5 = num./den;
        elseif i==6
          Gt6 = num./den;
        end;
    end;
%%%%%%%%%%%%%%%%%%%%%%%%%%%%%%%%%%%% PLOT 1 %%%%%%%%%%%%%%%%%%%%%%%%%%%%%
figure(1);
clf;
subplot(211);

plot(f,abs(Gt1),'r-.',f,abs(Gt2),'y:',f,abs(Gt3),'b:',f,abs(Gt4),...
     'g-.',f,abs(1./Gt5),'c:',f,abs(Gt6),'m--',f,abs(Theory),'k-');
```

```
ylabel('Magnitude');
legend('RM','RM-Average','RM Z-1','RC','Trap. Int.','RC 2','Theoretical');
grid;

subplot(212);
aGt1= (180./pi)*angle(Gt1);
aGt2= (180./pi)*angle(Gt2);
aGt3= (180./pi)*angle(Gt3);
aGt4= (180./pi)*angle(Gt4);
aGt5= (180./pi)*angle(1./Gt5);
aGt6= (180./pi)*angle(Gt6);
aTheory = (180./pi)*angle(Theory);
plot(f,aGt1,'r-.',f,aGt2,'y:',f,aGt3,'b:',f,aGt4,'g-.',f,aGt5,'c:',
     f,aGt6,'m--',f,aTheory,'k-');
xlabel('Frequency - Hz');
ylabel('Phase - (Degrees)');
legend('RM','RM-Average','RM Z-1','RC','Trap. Int.','
       RC 2','Theoretical');
grid;
```

Appendix G
FORTRAN code for state variable analysis

G.1 State variable analysis program

This program demonstrates the state variable analysis technique for simulating the dynamics of a network. The results of this program are presented in section 3.6.

```
!****************************************************************
!
!       PROGRAM StateVariableAnalysis
!
!****************************************************************
!
! This program demonstrates state space analysis
!
!                 .
!              x = Ax + Bu
!              y = Cx + Du
!
!       Where   x in the state variable vector.
!               y is the output vector.
!               u the input excitation vector.
!
!                .
!               x = (dx/dt)
!
! The program is set up to solve a second order RLC circuit at
! present, and plot the results. The results can then be compared
! with the analytic solution given by SECORD_ORD.
!
!-----------------------------------------------------------------
! Version 1.0    (25 April 1986) converted to FORTRAN90 2001
!****************************************************************
      IMPLICIT NONE
! Definition of variables.
!     ----------------------
      INTEGER, PARAMETER:: RealKind_DP = SELECTED_REAL_KIND(15,307)
      REAL (Kind = RealKind_DP), PARAMETER :: pi = 3.141592653589793233D0
```

```
      INTEGER :: Iter_Count         ! Iteration Counter
      INTEGER :: Step_CHG_Count     ! Counts the number of step changes that
                                    ! have occured in this time-step
      REAL (Kind = RealKind_DP) :: Xt(2)        ! State variable vector
                                                ! at Time=t
      REAL (Kind = RealKind_DP) :: Xth(2)       ! State variable vector
                                                ! at Time=t+h
      REAL (Kind = RealKind_DP) :: XtDot(2)     ! Derivative of state
                                                ! variables at Time=t
      REAL (Kind = RealKind_DP) :: XthDot(2)    ! Derivative of state
                                                ! variables at Time=t+h
      REAL (Kind = RealKind_DP) :: XthO(2)      ! Previous iterations
                                                ! estimate for state
                                                ! variables at t+h
      REAL (Kind = RealKind_DP) :: XthDotO(2)   ! Previous iterations
                                                ! estimate for derivative
                                                ! of state variables at t+h
      REAL (Kind = RealKind_DP) :: h            ! Step length
      REAL (Kind = RealKind_DP) :: Time         ! Current Time
      REAL (Kind = RealKind_DP) :: Switch_Time  ! Switching Time
      REAL (Kind = RealKind_DP) :: Time_Left    ! Time left until switching
      REAL (Kind = RealKind_DP) :: Finish_Time  ! Simulation Finish Time
      REAL (Kind = RealKind_DP) :: StepWidth    ! Step Width
      REAL (Kind = RealKind_DP) :: StepWidthNom ! Step Width
      REAL (Kind = RealKind_DP) :: EPS          ! Convergence tolerance
      REAL (Kind = RealKind_DP) :: R,L,C        ! Circuit Parameters
      REAL (Kind = RealKind_DP) :: E            ! Source Voltage
      REAL (Kind = RealKind_DP) :: f_res        ! Resonant Frequency

      CHARACTER*1 CHARB
      CHARACTER*10 CHARB2

      LOGICAL :: STEP_CHG      ! Step Change
      LOGICAL :: CONVG         ! Converged
      LOGICAL :: Check_SVDot   ! Check Derivative of State Variable
      LOGICAL :: OptimizeStep  ! Optimize Step length

      COMMON/COMPONENTS/R,L,C,E

!     Initialize variables.
!     --------------------
      Time = 0.0
      Xt(1) = 0.0
      Xt(2) = 0.0
      E=0.0
      Switch_Time =0.1D-3 ! Seconds

      Check_SVDot=.FALSE.
      OptimizeStep=.FALSE.

      OPEN(Unit=1,STATUS='OLD',file='SV_RLC.DAT')
      READ(1,*) R,L,C
      READ(1,*) EPS,Check_SVDot,OptimizeStep
```

```fortran
      READ(1,*) Finish_Time
      CLOSE(1)

      f_res=1.0/(2.0*pi*sqrt(L*C))
      WRITE(6,*) R,L,C,f_res,1./f_res
      WRITE(6,*) 'Tolerance =',EPS
      WRITE(6,*) 'Check State Variable Derivatives ',Check_SVDot
      WRITE(6,*) 'Optimize step-length ',OptimizeStep
      WRITE(6,*) 'Finish Time ',Finish_Time

!     Put header on file.
!     ------------------
      OPEN(UNIT=98,status='unknown',file='SVanalysis.out')
      WRITE(98,9860)R,L,C,E
9860  FORMAT(1X,'%R = ',F8.3,' Ohms L = ',F8.5,' Henries C = ',G16.5,'
     Farads E = ',F8.3)   WRITE(98,9870)
9870  FORMAT(1X,'% Time',7X,'X(1)',7X,'X(2)',6X,'STEP W',
     2X,'No. iter.',&
&              ' XDot(1)',' XDot(2)')

      StepWidth = 0.05D0
8     WRITE(6,'(/X,A,F8.6,A)')'Default = ',STEPWIDTH,' msec.'
      WRITE(6,'(1X,A,$)')'Enter Nominal stepwidth. (msec.) : '
      READ(5,'(A)')CHARB2
      IF(CHARB2.NE.' ') READ(CHARB2,'(BN,F8.4)',ERR=8) StepWidth
      write(6,*)'StepWidth=',StepWidth,' msec.'
      pause
      StepWidth = StepWidth/1000.0D0

      CALL XDot (Xt,XtDot)
      WRITE(98,9880)TIME,Xt(1),Xt(2),H,Iter_Count,XtDot(1),XtDot(2)

      DO WHILE(TIME .LE. Finish_Time)
         h = StepWidth

!        Limit step to fall on required switching instant
!        ------------------------------------------------
         Time_Left= Switch_Time-Time
         IF(Time_Left .GT.0.0D0 .AND. Time_Left.LE.h) THEN
            h = Time_Left
         END IF

         XthDot(1) = XtDot(1)
         XthDot(2) = XtDot(2)
         CONVG=.FALSE.

         Step_CHG_Count=0

         Xth0(1) = Xth(1)
         Xth0(2) = Xth(2)
         XthDot0(1) = XthDot(1)
         XthDot0(2) = XthDot(2)
```

```
          DO WHILE((.NOT.CONVG).AND.(Step_CHG_Count.LT.10))

!         Trapezoidal Integration (as XthDot=XtDot)
!         ------------------------------------------
          Xth(1) = Xt(1)+XtDot(1)*h
          Xth(2) = Xt(2)+XtDot(1)*h

          Step_CHG = .FALSE.
          Iter_Count=0

          DO WHILE((.NOT.CONVG).AND.(.NOT.Step_CHG))

           CALL XDot(Xth,XthDot)

!          Trapezoidal Integration
!          -----------------------
           Xth(1) = Xt(1) + (XthDot(1)+XtDot(1))*h/2.0
           Xth(2) = Xt(2) + (XthDot(2)+XtDot(2))*h/2.0

           Iter_Count = Iter_Count+1

           IF((ABS(Xth(1)-XthO(1)).LE.EPS).AND.             &
     &        (ABS(Xth(2)-XthO(2)).LE.EPS)) CONVG=.TRUE.

           IF((CONVG).AND.(Check_SVDot)) THEN

             IF((ABS(XthDot(1)-XthDotO(1)).GT.EPS).OR.      &
     &          (ABS(XthDot(2)-XthDotO(2)).GT.EPS)) CONVG=.FALSE.
           END IF

           XthO(1) = Xth(1)
           XthO(2) = Xth(2)
           XthDotO(1) = XthDot(1)
           XthDotO(2) = XthDot(2)

           IF(Iter_Count.GE.25) THEN
!              If reached 25 iteration half step-length regardless of
               convergence FLAG status
               Step_CHG=.TRUE.
               CONVG=.FALSE.
               h = h/2.0D0

               WRITE(6,986)Time*1000.0,h*1000.0
 986           FORMAT(1X,'Time(msec.)=',F6.3,' *** Step Halved ***',1X, &
     &                  'New Step Size (msec.)= ',F9.6)
               Step_CHG_Count = Step_CHG_Count+1
           END IF
          END DO
         END DO

         TIME = TIME+h
         Xt(1)= Xth(1)
         Xt(2)= Xth(2)
```

```
      XtDot(1) = XthDot(1)
      XtDot(2) = XthDot(2)

      IF(Step_CHG_Count.GE.10) THEN
         WRITE(6,'(/X,A)')'FAILED TO CONVERGE'
         WRITE(98,'(/X,A)')'FAILED TO CONVERGE'
         STOP
      ELSE IF(OptimizeStep) THEN
         ! Optimize Step Length based on step length and number of
         ! iterations.
         IF(Iter_Count.LE.5) THEN
            StepWidth = h*1.10
            TYPE *,'Step-length increased by 10%'
         ELSE IF(Iter_Count.GE.15) THEN
            StepWidth = h*0.9
            TYPE *,'Step-length decreased by 10%'
         END IF
      END IF
      IF (Iter_Count.EQ.25) THEN
         TYPE *,'How come?'
         TYPE *,'Time (msec.)=',TIME*1000.0
         TYPE *,'CONVG=',CONVG
         pause
      END IF

      WRITE(98,9880)TIME*1000.0,Xt(1),Xt(2),h*1000.0,Iter_Count,
     XtDot(1),XtDot(2)
9880  FORMAT(2X,F8.6,3X,F10.6,3X,F10.6,5X,F10.7,2X,I2,2X,F16.5,
     2X,F12.5)

      IF(abs(Switch_Time-Time).LE.1.0E-10) THEN
         ! State Variable can not change instantaneously hence are
         ! the same
         ! but dependent variables need updating. i.e. the derivative
         ! of state variables
         ! as well as those that are functions of state variables
         ! or their derivative.
         ! now two time points for the same time.
         E = 1.0
         CALL XDot(Xt,XtDot)
         WRITE(98,9880)TIME*1000.0,Xt(1),Xt(2),h*1000.0,Iter_Count,
     XtDot(1),XtDot(2)
      END IF
      END DO
      CLOSE(98)
      TYPE *,' *** THE END ***'
      END
```

```
!***********************************************************
         SUBROUTINE XDot(SVV,DSVV)
!       .
!       X   = [A]X+[B]U
!***********************************************************

! SVV = State variable vector
! DSVV = Derivative of state variable vector
! X(1) = Capacitor Voltage
! X(2) = Inductor Current

         IMPLICIT NONE

         INTEGER, PARAMETER:: RealKind_DP = SELECTED_REAL_KIND(15,307)

         REAL (Kind = RealKind_DP) ::  SVV(2),DSVV(2)
         REAL (Kind = RealKind_DP) ::  R,L,C,E
         COMMON/COMPONENTS/R,L,C,E

         DSVV(1) = SVV(2)/C
         DSVV(2) =-SVV(1)/L-SVV(2)*R/L+E/L

         RETURN
         END
```

Appendix H
FORTRAN code for EMT simulation

H.1 DC source, switch and RL load

In this example a voltage step (produced by a switch closed on to a d.c. voltage source) is applied to an RL Load. Results are shown in section 4.4.2. The RL load is modelled by one difference equation rather than each component separately.

```fortran
!======================================================================
   PROGRAM EMT_Switch_RL
!======================================================================
   IMPLICIT NONE
   INTEGER, PARAMETER:: RealKind_DP = SELECTED_REAL_KIND(15,307)
   INTEGER, PARAMETER:: Max_Steps = 5000

   REAL (Kind = RealKind_DP), PARAMETER :: pi = 3.141592653589793233D0
   REAL (Kind = RealKind_DP) :: R,L,Tau
   REAL (Kind = RealKind_DP) :: DeltaT,Time_Sec
   REAL (Kind = RealKind_DP) :: K_i,K_v
   REAL (Kind = RealKind_DP) :: R_switch,G_switch
   REAL (Kind = RealKind_DP) :: V_source
   REAL (Kind = RealKind_DP) :: I_inst,I_history
   REAL (Kind = RealKind_DP) :: i(Max_Steps),v(Max_Steps), &
                                v_load(Max_Steps)
   REAL (Kind = RealKind_DP) :: R_ON,R_OFF
   REAL (Kind = RealKind_DP) :: G_eff,Finish_Time

   INTEGER :: k,m,ON,No_Steps

!  Initalize Variables
!  ------------------
   Finish_Time = 1.0D-3   ! Seconds
   R = 1.0D0              ! Ohms
   L = 50.00D-6           ! Henries
   V_source = 100.0       ! Volts
   Tau = L/R              ! Seconds
   DeltaT = 50.0D-6       ! Seconds
```

```
      R_ON    = 1.0D-10        ! Ohms
      R_OFF = 1.0D10           ! Ohms
      R_Switch = R_OFF         ! Ohms
      ON = 0                   ! Initially switch is open

      m = 1
      i(m) = 0.0
      Time_Sec = 0.0
      v(m) =100.0
      v_load(m) = 0.0

      K_i = (1-DeltaT*R/(2*L))/(1+DeltaT*R/(2*L))
      K_v = (DeltaT/(2*L))/(1+DeltaT*R/(2*L))
      G_eff = K_v
      G_switch = 1.0/R_switch

      OPEN (unit=10,status='unknown',file='SwitchRL1.out')

      No_Steps= Finish_Time/DeltaT
      IF(Max_Steps<No_Steps) THEN
          STOP '*** Too Many Steps ***'
      END IF
      MainLoop: DO k=1,No_Steps,1
         m=m+1
         Time_Sec = k*DeltaT

!     Check Switch position
!     --------------------
      IF(k==3) THEN
          ON = 1
          R_switch = R_ON
          G_switch = 1.0/R_switch
      END IF

!     Update History term
!     ------------------
      I_history = k_i*i(m-1) + k_v*v_load(m-1)

!     Update Voltage Sources
!     ---------------------
      v(m) = V_source

!     Solve for V and I
!     ----------------
      v_load(m) = (-I_history + v(m)* G_switch)/(G_eff+G_Switch)
      I_inst    = v_load(m)*G_eff
      i(m)      = I_inst + I_history
      write(10,*) Time_Sec,v_load(m),i(m)
      END DO MainLoop
      CLOSE(10)
      PRINT *,' Execution Finished'

END PROGRAM EMT_Switch_RL
```

H.2 General EMT program for d.c. source, switch and *RL* load

The same case as in section H.1 is modelled here, however, the program is now structured in a general manner, where each component is subjected to numerical integrator substitution (NIS) and the conductance matrix is built up. Moreover, rather than modelling the switch as a variable resistor, matrix partitioning is applied (see section 4.4.1), which enables the use of ideal switches.

```fortran
!===============================
!
   PROGRAM EMT_Switch_RL
!
! Checked and correct 4 May 2001
!===============================
   IMPLICIT NONE
   ! INTEGER SELECT_REAL_KIND
   !INTEGER ,PARAMETER:: Real_18_4931 =SELECT_REAL_KIND(P=18,R=4931)

   REAL*8 :: R,L,Tc
   REAL*8 :: DeltaT
   REAL*8 :: i_L,i_R,i_Source,i_Source2
   REAL*8 :: R_switch,G_switch
   REAL*8 :: G(3,3),v(3),I_Vector(2)
   REAL*8 :: G_L,G_R
   REAL*8 :: V_source
   REAL*8 :: I_L_history
   REAL*8 :: Multiplier
   REAL*8 :: R_ON,R_OFF

   INTEGER k,n,NoTimeSteps
   INTEGER NoColumns

   open(unit=11,status='unknown',file='vi.out')

   NoTimeSteps = 6
   NoColumns = 3
   R_ON  = 1.0D-10   ! Ohms
   R_OFF = 1.0D+10   ! Ohms

   DeltaT = 50.0D-6 ! Seconds

   R = 1.0D0         ! Ohms
   L = 50.00D-6      ! Henries
   V_source=100      ! Volts
   Tc = L/R
   PRINT *,'Time Constant=',Tc
   R_switch = R_OFF
   G_switch = 1/R_switch
   G_R = 1/R
   G_L = DeltaT/(2*L)   ! G_L_eff
```

```
!   Initialize Variables
!   -------------------
    I_Vector(1) =0.0D0
    I_Vector(2) =0.0D0
    i_L =0.0D0
    DO k=1,3
      v(k) = 0.000D0
    END DO

!   Form System Conductance Matrix
!   ------------------------------
    CALL Form_G(G_R,G_L,G_switch,G)

!   Forward Reduction
!   -----------------
    CALL Forward_Reduction_G(NoColumns,G,Multiplier)

!   Enter Main Time-stepping loop
!   -----------------------------
    DO n=1,NoTimeSteps

       IF(n==1)THEN
          write(10, *) ' Switch Turned ON'
          R_switch = R_ON
          G_switch = 1/R_switch

          CALL Form_G(G_R,G_L,G_switch,G)
          CALL Forward_Reduction_G(NoColumns,G,Multiplier)
       END IF

!      Calculate Past History Terms
!      ----------------------------
       I_L_history =i_L + v(3)*G_L
       I_Vector(1) = 0.0
       I_Vector(2) = -I_L_history

!      Update Source Values
!      --------------------
       V_source = 100.0
       V(1) = V_source

!      Forward Reduction of Current Vector
!      -----------------------------------
       I_Vector(2) = I_Vector(2) - Multiplier*I_Vector(1)

!      Move Known Voltage to RHS (I_current Vector)
!      --------------------------------------------
       I_Vector(1) = I_Vector(1)- G(1,3) * V(1)
       I_Vector(2) = I_Vector(2)- G(2,3) * V(1)

!      Back-substitution
!      -----------------
       v(3) = I_Vector(2)/G(2,2)
       v(2) = (I_Vector(1)-G(1,2)*v(3))/G(1,1)
```

```fortran
!      Calculate Branch Current
!      -----------------------
       i_R = (v(2) - v(3))/R
       i_L = v(3)*G_L + I_L_history
       i_Source2 = (v(1)-v(2))*G_switch
       i_Source = G(3,1)*v(2) + G(3,2)*v(3) + G(3,3)*v(1)

       WRITE(11,1160) n*DeltaT,v(1),v(2),v(3),i_Source,i_R,i_L

   END DO
   close(11)
 1160 FORMAT(1X,F8.6,1X,6(G16.10,1X))
   STOP
   END
!=====================================================
   SUBROUTINE Form_G(G_R,G_L,G_switch,G)
!=====================================================
   IMPLICIT NONE
   REAL*8 :: G(3,3)
   REAL*8 :: G_L,G_R,G_switch

   G(1,1) = G_switch + G_R
   G(2,1) = - G_R
   G(1,2) = - G_R

   G(2,2) = G_L + G_R
   G(1,3) = -G_switch
   G(2,3) = 0.0D0

   G(3,1) =   -G_switch
   G(3,2) =    0.0D0
   G(3,3) =    G_switch

   CALL Show_G(G)
   RETURN
   END
!=====================================================
   SUBROUTINE Forward_Reduction_G(NoColumns,G,Multiplier)
!=====================================================
   IMPLICIT NONE
   REAL*8 :: G(3,3)
   REAL*8 :: Multiplier
   INTEGER :: k,NoColumns

   Multiplier = G(2,1)/G(1,1)
   PRINT *,' Multiplier= ',Multiplier

   DO k=1,NoColumns
      G(2,k) = G(2,k) - Multiplier*G(1,k)
   END DO

   RETURN
   END
```

```
!========================================================
      SUBROUTINE Show_G(G)
!========================================================
      IMPLICIT NONE
      REAL*8 :: G(3,3)

      WRITE(10,*) ' Matrix'
      WRITE(10,2000) G(1,1:3)
      WRITE(10,2000) G(2,1:3)
 2000 FORMAT(1X,'[',G16.10,' ',G16.10,' ',G16.10,']')
      RETURN
      END
```

H.3 AC source diode and RL load

This program is used to demonstrate the numerical oscillation that occurs at turn-off, by modelling an RL load fed from an a.c. source through a diode. The RL load is modelled by one difference equation rather than each component separately. The results are given in section 9.4.

```
!========================================================================
      PROGRAM EMT_DIODE_RL1
!========================================================================
      IMPLICIT NONE
      INTEGER, PARAMETER:: RealKind_DP = SELECTED_REAL_KIND(15,307)
      INTEGER, PARAMETER:: Max_Steps = 5000
      REAL (Kind = RealKind_DP), PARAMETER :: pi = 3.141592653589793233D0

      REAL (Kind = RealKind_DP) :: R,L,Tau,f
      REAL (Kind = RealKind_DP) :: DeltaT,Time_Sec
      REAL (Kind = RealKind_DP) :: K_i,K_v
      REAL (Kind = RealKind_DP) :: R_switch,G_switch
      REAL (Kind = RealKind_DP) :: V_mag,V_ang
      REAL (Kind = RealKind_DP) :: I_inst,I_history
      REAL (Kind = RealKind_DP) :: i(Max_Steps),v(Max_Steps),v_load(Max_Steps)
      REAL (Kind = RealKind_DP) :: R_ON,R_OFF
      REAL (Kind = RealKind_DP) :: G_eff,Finish_Time

      INTEGER :: k,m,ON,No_Steps

! Initalize Variables
! ------------------
      f=50.0
      Finish_Time = 60.0D-3
      R = 100.0
      L = 500D-3
      Tau = L/R
      DeltaT = 50.0D-6
      V_mag = 230.0*sqrt(2.)
      V_ang = 0.0
```

```fortran
      R_ON  = 1.0D-10
      R_OFF = 1.0D10
      R_Switch = R_ON

      m=1
      i(m) = 0.0
      Time_Sec = 0.0
      v(m) = V_mag*sin(V_ang*pi/180)
      v_load(m)=v(m)
      ON=1
      K_i = (1-DeltaT*R/(2*L))/(1+DeltaT*R/(2*L))
      K_v = (DeltaT/(2*L))/(1+DeltaT*R/(2*L))
      G_eff = K_v
      G_switch = 1.0/R_switch

      OPEN (unit=10,status='unknown',file='DiodeRL1.out')

      No_Steps= Finish_Time/DeltaT
      IF(Max_Steps<No_Steps) THEN
          STOP '*** Too Many Steps ***'
      END IF

      MainLoop: DO k=1,No_Steps,1
         m=m+1
         Time_Sec = k*DeltaT

!     Check Switch position
!     --------------------
         IF (i(m-1)<= 0.0 .and. ON==1 .and. k >5*DeltaT) THEN
            ON = 0
            R_switch = R_OFF
            G_switch = 1.0/R_switch
            i(m-1)   = 0.0
         END IF
         IF (v(m-1)-v_load(m-1) > 1.0 .and. ON==0) THEN
            ON = 1
            R_switch = R_ON
            G_switch = 1.0/R_switch
         END IF

!     Update History term
!     -------------------
         I_history = k_i*i(m-1) + k_v*v_load(m-1)

!     Update Voltage Sources
!     ---------------------
         v(m) =   V_mag*sin(2*pi*f*Time_Sec + V_ang*pi/180)

!     Solve for V and I
!     -----------------
         v_load(m) = (-I_history + v(m)* G_switch)/(G_eff+G_Switch)
         I_inst    = v_load(m)*G_eff
```

```
            i(m)          = I_inst + I_history
            write(10,*) Time_Sec,v_load(m),i(m)

      END DO MainLoop
      CLOSE(10)

      PRINT *,' Execution Finished'

END PROGRAM EMT_DIODE_RL1
```

H.4 Simple lossless transmission line

This program evaluates the step response of a simple lossless transmission line, as shown in section 6.6.

```
!=============================================================
    PROGRAM Lossless_TL
!
! A simple lossless travelling wave transmission line
!=============================================================
    IMPLICIT NONE
    INTEGER, PARAMETER:: RealKind_DP = SELECTED_REAL_KIND(15,307)
    INTEGER, PARAMETER:: TL_BufferSize = 100

! Transmission Line Buffer
! -----------------------
    REAL (Kind=RealKind_DP) :: Vsend(TL_BufferSize)
    REAL (Kind=RealKind_DP) :: Vrecv(TL_BufferSize)
    REAL (Kind=RealKind_DP) :: Isend_Hist(TL_BufferSize)
    REAL (Kind=RealKind_DP) :: Irecv_Hist(TL_BufferSize)

    REAL (Kind=RealKind_DP) :: L_dash,C_dash,Length
    REAL (Kind=RealKind_DP) :: DeltaT
    REAL (Kind=RealKind_DP) :: Time
    REAL (Kind=RealKind_DP) :: R_Source,R_Load
    REAL (Kind=RealKind_DP) :: V_Source,I_Source
    REAL (Kind=RealKind_DP) :: Gsend,Grecv,Rsend,Rrecv
    REAL (Kind=RealKind_DP) :: Zc, Gamma
    REAL (Kind=RealKind_DP) :: Finish_Time,Step_Time
    REAL (Kind=RealKind_DP) :: i_send ! Sending to Receiving end current
    REAL (Kind=RealKind_DP) :: i_recv ! Receiving to Sending end current

    INTEGER Position
    INTEGER PreviousHistoryPSN
    INTEGER k,NumberSteps,Step_No
    INTEGER No_Steps_Delay

    OPEN(UNIT=10,file='TL.out',status="UNKNOWN")

! Default Line Parameters
! -----------------------
    L_dash = 400D-9
```

```fortran
      C_dash = 40D-12
      Length = 2.0D5
      DeltaT = 50D-6
      R_Source = 0.1
      R_Load =  100.0
      Finish_Time = 10.0D-3
      Step_Time=5*DeltaT

      CALL ReadTLData(L_dash,C_Dash,Length,DeltaT,R_Source,R_Load,
     +     Step_Time,Finish_Time)

      Zc = sqrt(L_dash/C_dash)
      Gamma = sqrt(L_dash*C_dash)
      No_Steps_Delay = Length*Gamma/DeltaT

!     Write File Header Information
!     ----------------------------
      WRITE(10,10) L_dash,C_dash,Length,Gamma,Zc
      WRITE(10,11) R_Source,R_Load,DeltaT,Step_Time
 10   FORMAT(1X,'% L =',G16.6,' C =',G16.6,' Length=',F12.2,'
     + Propagation Constant=',G16.6,' Zc=',G16.6)
 11   FORMAT(1X,'% R_Source =',G16.6,' R_Load =',G16.6,' DeltaT=',F12.6,'
     + Step_Time=',F12.6)

      Gsend = 1.0D0/ R_Source + 1.0D0/Zc
      Rsend = 1.0D0/Gsend
      Grecv = 1.0D0/ R_Load + 1.0D0/Zc
      Rrecv = 1.0D0/Grecv

!     Initialize Buffers
      DO k=1,TL_BufferSize
         Vsend(k) = 0.0D0
         Vrecv(k) = 0.0D0
         Isend_Hist(k) = 0.0D0
         Irecv_Hist(k) = 0.0D0
      END DO

      Position = 0
      NumberSteps = NINT(Finish_Time/DeltaT)

!     DO Time = DeltaT,Finish_Time,DeltaT (Note REAL DO loop variables
!     removed in FORTRAN95)
      DO Step_No = 1,NumberSteps,1
         Time = DeltaT*Step_No
         Position = Position+1

!        Make sure index the correct values in Ring Buffer
!        ------------------------------------------------
         PreviousHistoryPSN = Position - No_Steps_Delay
         IF(PreviousHistoryPSN>TL_BufferSize) THEN
            PreviousHistoryPSN = PreviousHistoryPSN-TL_BufferSize
         ELSE IF(PreviousHistoryPSN<1) THEN
            PreviousHistoryPSN = PreviousHistoryPSN+TL_BufferSize
         END IF
```

```fortran
            IF(Position>TL_BufferSize) THEN
                Position = Position-TL_BufferSize
            END IF


!       Update Sources
!       --------------
            IF(Time< 5*DeltaT) THEN
                V_Source = 0.0
            ELSE
                V_Source = 100.0
            END IF
            I_Source = V_Source/R_Source


!       Solve for Nodal Voltages
!       ------------------------
            Vsend(Position) = (I_Source-Isend_Hist(PreviousHistoryPSN))*Rsend
            Vrecv(Position) = (          -Irecv_Hist(PreviousHistoryPSN))*Rrecv

!       Solve for Terminal Current
!       --------------------------
            i_send = Vsend(Position)/Zc + Isend_Hist(PreviousHistoryPSN)
            i_recv = Vrecv(Position)/Zc + Irecv_Hist(PreviousHistoryPSN)

!       Calculate History Term (Current Source at tau later).
!       -----------------------------------------------------
            Irecv_Hist(Position) = (-1.0/Zc)*Vsend(Position) - i_send
            Isend_Hist(Position) = (-1.0/Zc)*Vrecv(Position) - i_recv

        WRITE(10,1000) Time,Vsend(Position),Vrecv(Position),i_send,i_recv,
            Isend_Hist(Position),Irecv_Hist(Position)
    END DO
1000 FORMAT(1X,7(G16.6,1X))

    CLOSE(10)

    PRINT *,' Successful Completion'
    END
```

H.5 Bergeron transmission line

In this example the step response of a simple transmission line with lumped losses (Bergeron model) is evaluated (see section 6.6).

```fortran
!=============================================================
    PROGRAM TL_Bergeron
! Bergeron Line Model (Lumped representation of Losses)
!=============================================================
    IMPLICIT NONE
    INTEGER, PARAMETER:: RealKind_DP = SELECTED_REAL_KIND(15,307)
    INTEGER, PARAMETER:: TL_BufferSize = 100
```

```fortran
!   Transmission Line Buffer
!   ------------------------
    REAL (Kind=RealKind_DP) :: Vsend(TL_BufferSize)
    REAL (Kind=RealKind_DP) :: Vrecv(TL_BufferSize)
    REAL (Kind=RealKind_DP) :: Isend_Hist(TL_BufferSize)
    REAL (Kind=RealKind_DP) :: Irecv_Hist(TL_BufferSize)

    REAL (Kind=RealKind_DP) :: R_dash,L_dash,C_dash,Length
    REAL (Kind=RealKind_DP) :: DeltaT,Time
    REAL (Kind=RealKind_DP) :: R,R_Source,R_Load
    REAL (Kind=RealKind_DP) :: V_Source,I_Source
    REAL (Kind=RealKind_DP) :: Gsend,Grecv,Rsend,Rrecv
    REAL (Kind=RealKind_DP) :: Zc, Zc_Plus_R4,Gamma
    REAL (Kind=RealKind_DP) :: Finish_Time,Step_Time
    REAL (Kind=RealKind_DP) :: i_send ! Sending to Receiving end current
    REAL (Kind=RealKind_DP) :: i_recv ! Receiving to Sending end current

    INTEGER Position
    INTEGER PreviousHistoryPSN
    INTEGER k,NumberSteps,Step_No
    INTEGER No_Steps_Delay

    OPEN(UNIT=10,file='TL.out',status="UNKNOWN")

    R_dash = 100D-6
    L_dash = 400D-9
    C_dash = 40D-12
    Length = 2.0D5
    DeltaT = 50D-6
    R_Source = 0.1
    R_Load =  100.0
    Finish_Time = 1.0D-4
    Step_Time=5*DeltaT

    CALL ReadTLData(R_dash,L_dash,C_Dash,Length,DeltaT,R_Source, &
         R_Load,Step_Time,Finish_Time)

    R=R_Dash*Length

    Zc = sqrt(L_dash/C_dash)
    Gamma = sqrt(L_dash*C_dash)
    No_Steps_Delay = Length*Gamma/DeltaT
    Zc_Plus_R4 = Zc+R/4.0

!   Write File Header Information
!   -----------------------------
    WRITE(10,10) R_dash,L_dash,C_dash,Length,Gamma,Zc
    WRITE(10,11) R_Source,R_Load,DeltaT,Step_Time
10  FORMAT(1X,'% R =',G16.6,' L =',G16.6,' C =',G16.6,' Length=',F12.2,' &
    Propagation Constant=',G16.6,' Zc=',G16.6)
11  FORMAT(1X,'% R_Source =',G16.6,' R_Load =',G16.6,' DeltaT=',F12.6,' &
    Step_Time=',F12.6)
```

```
      Gsend = 1.0D0/ R_Source + 1.0D0/Zc_Plus_R4
      Rsend = 1.0D0/Gsend
      Grecv = 1.0D0/ R_Load + 1.0D0/Zc_Plus_R4
      Rrecv = 1.0D0/Grecv

      DO k=1,TL_BufferSize
         Vsend(k) = 0.0D0
         Vrecv(k) = 0.0D0
         Isend_Hist(k) = 0.0D0
         Irecv_Hist(k) = 0.0D0
      END DO
      Position = 0

      Position = 0
      NumberSteps = NINT(Finish_Time/DeltaT)
      DO Step_No = 1,NumberSteps,1
         Time = DeltaT*Step_No
         Position = Position+1

!        Make sure index the correct values in Ring Buffer
!        -------------------------------------------------
         PreviousHistoryPSN = Position - No_Steps_Delay
         IF(PreviousHistoryPSN>TL_BufferSize) THEN
            PreviousHistoryPSN = PreviousHistoryPSN-TL_BufferSize
         ELSE IF(PreviousHistoryPSN<1) THEN
            PreviousHistoryPSN = PreviousHistoryPSN+TL_BufferSize
         END IF

         IF(Position>TL_BufferSize) THEN
            Position = Position-TL_BufferSize
         END IF

!        Update Sources
!        --------------
         IF(Time< 5*DeltaT) THEN
            V_Source = 0.0
         ELSE
            V_Source = 100.0
         END IF
         I_Source = V_Source/R_Source


!        Solve for Nodal Voltages
!        ------------------------
         Vsend(Position) = (I_Source-Isend_Hist(PreviousHistoryPSN))*Rsend
         Vrecv(Position) = (        -Irecv_Hist(PreviousHistoryPSN))*Rrecv

         WRITE(12,1200)Time,Position,PreviousHistoryPSN,
     &       Isend_Hist(PreviousHistoryPSN),Irecv_Hist(PreviousHistoryPSN)
 1200 FORMAT(1X,G16.6,1X,I5,1X,I5,2(G16.6,1X))
```

```fortran
!   Solve for Terminal Current
!   -------------------------
    i_send = Vsend(Position)/Zc_Plus_R4 + Isend_Hist(PreviousHistoryPSN)
    i_recv = Vrecv(Position)/Zc_Plus_R4 + Irecv_Hist(PreviousHistoryPSN)

!   Calculate History Term (Current Source at tau later).
!   -----------------------------------------------------
    Irecv_Hist(Position) = (-Zc/(Zc_Plus_R4**2))*(Vsend(Position)
                            +(Zc-R/4.0)*i_send)  &
                           +((-R/4.0)/(Zc_Plus_R4**2))
                            *(Vrecv(Position)+(Zc-R/4.0)*i_recv)

    Isend_Hist(Position) = (-Zc/(Zc_Plus_R4**2))*(Vrecv(Position)
                            +(Zc-R/4.0)*i_recv)  &
                           +((-R/4.0)/(Zc_Plus_R4**2))
                            *(Vsend(Position)+(Zc-R/4.0)*i_send)

    WRITE(10,1000) Time,Vsend(Position),Vrecv(Position),
        i_send,i_recv,Isend_Hist(Position),Irecv_Hist(Position)
    END DO
1000 FORMAT(1X,7(G16.6,1X))

    CLOSE(10)
    PRINT *,' Successful Completion'
    END
```

H.6 Frequency-dependent transmission line

This program demonstrates the implementation of a full frequency-dependent transmission line and allows the step response to be determined. This is an *s*-domain implementation using recursive convolution. For simplicity interpolation of buffer values is not included. Results are illustrated in section 6.6.

```
!============================================================
   PROGRAM TL_FDP_s
!
! Simple Program to demonstrate the implementation of a
! Frequency-Dependent Transmission Line using s-domain representation.
!
!         Isend(w)                        Irecv(w)
!        ------>----                     ------<-----
!           |                               |
!         --------                        --------
!         |      |                        |      |
!        _|_    /                        /      _|_
!        | |   / ^                      / ^     | |
! Vs(w)  |Yc|  | /                      | /     |Yc|   Vr(w)
!        |_|    /                        /      |_|
!         |    | Ih_s                    | Ih_r |
!         --------                        --------
!           |                               |
!        -----------                     -----------
!
!============================================================
```

```fortran
      IMPLICIT NONE
      INTEGER, PARAMETER:: RealKind_DP = SELECTED_REAL_KIND(15,307)
      INTEGER, PARAMETER:: TL_BufferSize = 100
      INTEGER, PARAMETER:: TL_MaxPoles = 5

!     Transmission Line Buffer
!     ------------------------
      REAL (Kind=RealKind_DP) :: Vsend(TL_BufferSize)
      REAL (Kind=RealKind_DP) :: Vrecv(TL_BufferSize)
      REAL (Kind=RealKind_DP) :: i_send(TL_BufferSize)
              ! Sending to Receiving end current
      REAL (Kind=RealKind_DP) :: i_recv(TL_BufferSize)
              ! Receiving to Sending end current

      REAL (Kind=RealKind_DP) :: ApYcVr_Ir(TL_BufferSize,TL_MaxPoles)
      REAL (Kind=RealKind_DP) :: ApYcVs_Is(TL_BufferSize,TL_MaxPoles)

      REAL (Kind=RealKind_DP) :: I_Yc_Send(TL_BufferSize,TL_MaxPoles)
                ! Current in each pole
      REAL (Kind=RealKind_DP) :: I_Yc_Recv(TL_BufferSize,TL_MaxPoles)
                ! Current in each pole
      REAL (Kind=RealKind_DP) :: I_A_Send(TL_BufferSize,TL_MaxPoles)
                ! Current in each pole
      REAL (Kind=RealKind_DP) :: I_A_Recv(TL_BufferSize,TL_MaxPoles)
                ! Current in each pole
      REAL (Kind=RealKind_DP) :: YcVr_Ir(TL_BufferSize)
      REAL (Kind=RealKind_DP) :: YcVs_Is(TL_BufferSize)
      REAL (Kind=RealKind_DP) :: Time,DeltaT
      REAL (Kind=RealKind_DP) :: R_Source,R_Load
      REAL (Kind=RealKind_DP) :: V_Source,I_Source
      REAL (Kind=RealKind_DP) :: Gsend,Grecv,Rsend,Rrecv
      REAL (Kind=RealKind_DP) :: Y_TL              ! Total TL Admittance
      REAL (Kind=RealKind_DP) :: Finish_Time,Step_Time

      REAL (Kind=RealKind_DP) :: H_Yc,K_Yc,Pole_Yc(TL_MaxPoles),
              Residue_Yc(TL_MaxPoles)
      REAL (Kind=RealKind_DP) :: Alpha_Yc(TL_MaxPoles),
              Lambda_Yc(TL_MaxPoles),mu_Yc(TL_MaxPoles)

      REAL (Kind=RealKind_DP) :: H_Ap,Pole_A(TL_MaxPoles),
              Residue_A(TL_MaxPoles)
      REAL (Kind=RealKind_DP) :: Alpha_A(TL_MaxPoles) ,Lambda_A(TL_MaxPoles),
              mu_A(TL_MaxPoles)
      REAL (Kind=RealKind_DP) :: I_s_Yc_History(TL_MaxPoles)
      REAL (Kind=RealKind_DP) :: I_r_Yc_History(TL_MaxPoles)
      REAL (Kind=RealKind_DP) :: I_s_Yc,I_r_Yc
      REAL (Kind=RealKind_DP) :: I_s_Ap,I_r_Ap
      REAL (Kind=RealKind_DP) :: ah
      REAL (Kind=RealKind_DP) :: I_Send_History, I_Recv_History
      REAL (Kind=RealKind_DP) :: YcVs_I_Total,YcVr_I_Total

      INTEGER :: Position
      INTEGER :: Last_Position
```

```
      INTEGER :: t_tau,t_Tau_1,t_Tau_2
      INTEGER :: k,m
      INTEGER :: No_Steps_Delay
      INTEGER :: RecursiveConvType
      INTEGER :: No_Poles_Yc, No_Poles_A

      OPEN(UNIT=10,file='TL.out',status="UNKNOWN")

      DeltaT = 50D-6       ! Time-step
      R_Source = 0.1D0     ! Source resistance
      R_Load =   100.0D0
      Finish_Time = 1.0D-2
      Step_Time   = 5*DeltaT
      RecursiveConvType = 1
      No_Steps_Delay = 7

!     Partial Fraction Expansion of Yc and Ap
!     ---------------------------------------
      H_Yc =0.00214018572698*0.91690065830247
      No_Poles_Yc = 3
      K_Yc = 1.0
      Pole_Yc(1)  = -1.00514000000000D5
      Pole_Yc(2)  = -0.00625032000000D5
      Pole_Yc(3)  = -0.00028960740000D5

      Residue_Yc(1) =   -19.72605872772154D0
      Residue_Yc(2) =    -0.14043511946635D0
      Residue_Yc(3) =    -0.00657234249032D0

      Y_TL = H_Yc*K_Yc
      DO k=1,No_Poles_Yc
         Residue_Yc(k) =     H_Yc * Residue_Yc(k)
      END DO
!     -------------- Ap ----------------------
      No_Poles_A=4
      H_Ap = 0.995
      Residue_A(1) =    2.13779561263148D6
      Residue_A(2) =   -2.18582740962054D6
      Residue_A(3) =    0.04688271799632D6
      Residue_A(4) =    0.00114907899276D6

      Pole_A(1) = -5.58224599999997D5
      Pole_A(2) = -5.46982800000003D5
      Pole_A(3) = -0.47617630000000D5
      Pole_A(4) = -0.06485341000000D5

      DO k=1,No_Poles_A
         Residue_A(k) =    H_Ap * Residue_A(k)
      END DO

!     Initialize variables to zero
!     ----------------------------
      Time = 0.0D0
```

```
DO k=1,TL_BufferSize
   Vsend(k)    = 0.0D0
   Vrecv(k)    = 0.0D0
   i_send(k)   = 0.0D0
   i_recv(k)   = 0.0D0
   YcVs_Is(k)  = 0.0D0
   YcVr_Ir(k)  = 0.0D0

   DO m=1,No_Poles_Yc
      I_Yc_send(k,m) = 0.0D0
      I_Yc_recv(k,m) = 0.0D0
   END DO
   DO m=1,No_Poles_A
      I_A_send(k,m) = 0.0D0
      I_A_recv(k,m) = 0.0D0
   END DO
END DO

IF (RecursiveConvType.EQ.0) THEN
   DO k=1,No_Poles_Yc
      Alpha_Yc (k)= exp(Pole_Yc(k)*DeltaT)
      Lambda_Yc(k)= (Residue_Yc(k)/(-Pole_Yc(k)))*(1.0-Alpha_Yc(k))
      mu_Yc  (k)   = 0.0
      Y_TL = Y_TL+ Lambda_Yc(k)
   END DO
   DO k=1,No_Poles_A
      Alpha_A (k) = exp(Pole_A(k)*DeltaT)
      Lambda_A(k) = (Residue_A(k)/(-Pole_A(k)))*(1.0-Alpha_A(k))
      mu_A  (k)   = 0.0
   END DO
ELSE IF (RecursiveConvType.EQ.1) THEN
   DO k=1,No_Poles_Yc
      ah = -Pole_Yc(k)*DeltaT
      Alpha_Yc (k) = exp(Pole_Yc(k)*DeltaT)
      Lambda_Yc(k) = (Residue_Yc(k)/(-Pole_Yc(k)))
                     *(1.0 - (1.0-Alpha_Yc(k))/ah)
      mu_Yc  (k)   = (Residue_Yc(k)/(-Pole_Yc(k)))
                     *(((1.0-Alpha_Yc(k))/ah)-Alpha_Yc(k) )
      Y_TL = Y_TL+ Lambda_Yc(k)

   END DO
   DO k=1,No_Poles_A
      ah = -Pole_A(k)*DeltaT
      Alpha_A (k) = exp(Pole_A(k)*DeltaT)
      Lambda_A(k) = (Residue_A(k)/(-Pole_A(k)))
                     *(1.0 - (1.0-Alpha_A(k))/(ah))
      mu_A  (k)   = (Residue_A(k)/(-Pole_A(k)))
                     *(((1.0-Alpha_A(k))/(ah))-Alpha_A(k))

   END DO
END IF
```

```fortran
!    Add transmisison line admittance to system Addmittance
!    ------------------------------------------------------
     Gsend = 1.0D0/ R_Source + Y_TL
     Rsend = 1.0D0/Gsend
     Grecv = 1.0D0/ R_Load + Y_TL
     Rrecv = 1.0D0/Grecv

!    Enter Main time-step loop
!    -------------------------
     Position = 0
     DO Time = DeltaT,Finish_Time,DeltaT
         Last_Position = Position
         Position = Position+1

!        Make sure index the correct values in Ring Buffer
!        -------------------------------------------------
         t_Tau   = Position - No_Steps_Delay
         IF(t_Tau >TL_BufferSize) THEN
             t_Tau = t_Tau -TL_BufferSize
         ELSE IF(t_Tau <1) THEN
                 t_Tau  = t_Tau +TL_BufferSize
         END IF

         IF(Position>TL_BufferSize) THEN
             Position = Position-TL_BufferSize
         END IF
         IF(Last_Position==0) THEN
             Last_Position = TL_BufferSize
         END IF
         t_Tau_1 = t_Tau-1
         IF(t_Tau_1==0) THEN
             t_Tau_1 = TL_BufferSize
         END IF
         t_Tau_2 = t_Tau_1-1
         IF(t_Tau_2==0) THEN
             t_Tau_2 = TL_BufferSize
         END IF

!        Update Sources
!        --------------
         IF(Time< 5*DeltaT) THEN
             V_Source = 0.0
         ELSE
             V_Source = 100.0
         END IF
         I_Source = V_Source/R_Source

!        Yc(t)*Vs(t) and Yc(t)*Vr(t)
!        This calculates the history terms
!        (instantaneous term comes from  admittance added to
!         system equation)
!        -------------------------------------------------------------
```

412 *Power systems electromagnetic transients simulation*

```
            I_s_Yc = 0.0D0
            I_r_Yc = 0.0D0
            DO k=1,No_Poles_Yc
               I_s_Yc_History(k) = Alpha_Yc (k)*I_Yc_send(Last_Position,k)
                                 + mu_Yc(k)*Vsend(Last_Position)
               I_s_Yc = I_s_Yc +  I_s_Yc_History(k)

               I_r_Yc_History(k) = Alpha_Yc (k)*I_Yc_recv(Last_Position,k)
                                 + mu_Yc(k)*Vrecv(Last_Position)
               I_r_Yc = I_r_Yc + I_r_Yc_History(k)
            END DO
!
!           Calculate Ap*(Yc(t-Tau)*Vs(t-Tau)+Is(t-Tau)) and Ap*(Yc(t-Tau)
!            *Vr(t-Tau)+Ir(t-Tau))
!           As these are using delayed terms i.e. (Yc(t-Tau)*Vs(t-Tau)
!            +Is(t-Tau)) then the complete
!           convolution can be achieved (nothing depends on present
!           time-step values).
!           ----------------------------------------------------------------
            I_s_Ap = 0.0D0
            I_r_Ap = 0.0D0
            DO k=1,No_Poles_A
               ! Sending End
               ApYcVr_Ir(Position,k) = Alpha_A(k)*ApYcVr_Ir(Last_Position,k)&
     &                               + Lambda_A(k)*(YcVr_Ir(t_Tau))         &
     &                               + mu_A(k) *  (YcVr_Ir(t_Tau_2) )
               I_s_Ap = I_s_Ap + ApYcVr_Ir(Position,k)

               ! Receiving End
               ApYcVs_Is(Position,k) = Alpha_A(k)*ApYcVs_Is(Last_Position,k)&
     &                               + Lambda_A(k)*(YcVs_Is(t_Tau))         &
     &                               + mu_A(k)* (YcVs_Is(t_Tau_2) )
               I_r_Ap = I_r_Ap + ApYcVs_Is(Position,k)

            END DO

!           Sum all the current source contributions from Characteristic
!           Admittance and Propagation term.

!           ----------------------------------------
            I_Send_History = I_s_Yc - I_s_Ap
            I_Recv_History = I_r_Yc - I_r_Ap

!           Solve for Nodal Voltages
!           ------------------------
            Vsend(Position) = (I_Source-I_Send_History)*Rsend
            Vrecv(Position) = (         -I_Recv_History)*Rrecv

!           Solve for Terminal Current
!           --------------------------
            i_send(Position) = Vsend(Position)*Y_TL + I_Send_History
            i_recv(Position) = Vrecv(Position)*Y_TL + I_Recv_History
```

```fortran
!       Calculate current contribution from each block
!       ----------------------------------------------
        YcVs_I_Total = H_Yc*K_Yc*Vsend(Position)
        YcVr_I_Total = H_Yc*K_Yc*Vrecv(Position)
        DO k=1,No_Poles_Yc
           I_Yc_send(Position,k) = Lambda_Yc(k)*Vsend(Position)
                                 + I_s_Yc_History(k)
           I_Yc_recv(Position,k) = Lambda_Yc(k)*Vrecv(Position)
                                 + I_r_Yc_History(k)

           YcVs_I_Total = YcVs_I_Total +  I_Yc_send(Position,k)
           YcVr_I_Total = YcVr_I_Total +  I_Yc_recv(Position,k)
        END DO

!       Calculate (Yc(t)*Vs(t)+Is(t)) and (Yc(t)*Vr(t)+Ir(t)) and store
!       Travelling time (delay) is represented by accessing values that
!       are No_Steps_Delay old.  This gives a travelling
!       time of No_Steps_Delay*DeltaT (7*50=350 micro-seconds)
!       ----------------------------------------------------------------
        YcVs_Is(Position) =  YcVs_I_Total + i_send(Position)
        YcVr_Ir(Position) =  YcVr_I_Total + i_recv(Position)

        WRITE(10,1000) Time,Vsend(Position),Vrecv(Position),
            i_send(Position),i_recv(Position)

     END DO

     CLOSE(10)
     PRINT *,' Successful Completion'

! Format Statements
 1000 FORMAT(1X,8(G16.6,1X))
     END
```

H.7 Utility subroutines for transmission line programs

```fortran
!========================================================================
    SUBROUTINE ReadTLData(R_Dash,L_dash,C_Dash,Length,DeltaT,R_Source,
    R_Load,Step_Time,Finish_Time)
!========================================================================
    IMPLICIT NONE
    INTEGER, PARAMETER:: RealKind_DP = SELECTED_REAL_KIND(15,307)

    REAL (Kind=RealKind_DP) :: R_dash,L_dash,C_dash,Length
    REAL (Kind=RealKind_DP) :: DeltaT
    REAL (Kind=RealKind_DP) :: R_Source,R_Load
    REAL (Kind=RealKind_DP) :: Step_Time
    REAL (Kind=RealKind_DP) :: Finish_Time

    INTEGER :: Counter
    INTEGER :: Size
```

```fortran
      INTEGER :: Psn
      CHARACTER(80) Line,String

      OPEN(UNIT=11,file='TLdata.dat',status='OLD',err=90)
      Counter=0

      DO WHILE (Counter<10)
         Counter = Counter+1
         READ(11,'(A80)',end=80) Line
         CALL STR_UPPERCASE(Line)   ! Convert to Uppercase
         String = ADJUSTL(Line)     ! Left hand justify
         Line   = TRIM(String)      ! Trim trailing blanks
         Size   = LEN_TRIM(Line)    ! Size excluding trailing blanks
         Psn    = INDEX(Line,'=')

         IF(Psn < 2) CYCLE
         IF(Psn >= Size) CYCLE
         IF(Line(1:1)=='!') CYCLE
         IF(Line(1:1)=='%') CYCLE

         IF(Line(1:6)=='R_DASH')THEN
            READ(Line(Psn+1:Size),*,err=91) R_Dash
            PRINT *,'R_Dash set to ',R_Dash
         ELSE IF(Line(1:6)=='L_DASH')THEN
            READ(Line(Psn+1:Size),*,err=91) L_Dash
            PRINT *,'L_Dash set to ',L_Dash
         ELSE IF(Line(1:6)=='C_DASH')THEN
            READ(Line(Psn+1:Size),*,err=91) C_Dash
            PRINT *,'C_Dash set to ',C_Dash
         ELSE IF(Line(1:6)=='LENGTH')THEN
            READ(Line(Psn+1:Size),*,err=91) Length
            PRINT *,'Length set to ',Length
         ELSE IF(Line(1:8)=='R_SOURCE')THEN
            READ(Line(Psn+1:Size),*,err=91) R_source
            PRINT *,'R_source set to ',R_source
         ELSE IF(Line(1:6)=='R_LOAD')THEN
            READ(Line(Psn+1:Size),*,err=91) R_Load
            PRINT *,'R_Load set to ',R_Load
         ELSE IF(Line(1:6)=='DELTAT')THEN
            READ(Line(Psn+1:Size),*,err=91) DeltaT
            PRINT *,'DeltaT set to ',DeltaT
         ELSE IF(Line(1:9)=='STEP_TIME')THEN
            READ(Line(Psn+1:Size),*,err=91) Step_Time
            PRINT *,'Step_Time set to ',Step_Time
         ELSE IF(Line(1:11)=='FINISH_TIME')THEN
            READ(Line(Psn+1:Size),*,err=91) Finish_Time
            PRINT *,'Finish_Time set to ',Finish_Time
         ELSE IF(Line(1:8)=='END_DATA')THEN

         END IF
      END DO
 80   CLOSE(11)
      RETURN
```

```
   90 PRINT *,' *** UNABLE to OPEN FILE TLdata.dat'
      STOP
   91 PRINT *,' *** Error reading file TLdata.dat'
      STOP
      END
!==================================================
!
      SUBROUTINE STR_UPPERCASE(CHAR_STR)
! Convert Character String to Upper Case
!==================================================
      IMPLICIT NONE
      CHARACTER*(*) CHAR_STR
      INTEGER :: SIZE,I,INTEG

      Size  = LEN_TRIM(CHAR_STR)
      IF (SIZE.GE.1)THEN
         DO I=1,SIZE
            IF((CHAR_STR(I:I).GE.'a').AND.(CHAR_STR(I:I).LE.'z')) THEN
               INTEG = ICHAR(CHAR_STR(I:I))
               CHAR_STR(I:I)=CHAR(IAND(INTEG,223))
            END IF
         END DO
      END IF
      RETURN
      END
```

Index

A-stable 357
active power (real power) 307, 308
admittance matrix 98, 170, 174, 257
analogue computer, electronic 4
arc resistance 210, 292
ARENE 329
ATOSEC 8, 37
ATP (alternative transient program) 6, 8, 206
attenuation of travelling waves 126, 128, 132, 134, 148, 155, 295
auto regressive moving average (ARMA) 30, 148, 267

backward wave 75, 128, 129
Bergeron line model 5, 9, 124, 126, 149, 150, 157
bilinear transform 6, 28, 100, 369

cable 6, 92, 123, 142, 144, 230, 252, 270, 333, 340
capacitance 1, 45, 70, 74, 109, 126, 138, 142, 176, 208, 218, 297
Carson's technique 123, 137, 139, 156
characteristic equations 76
characteristic impedance 75, 130, 136, 137, 146, 153
chatter 82, 97, 217, 220, 222, 227
CIGRE HVdc benchmark model 359
circuit breaker 3, 9, 54, 194, 210, 230, 321, 326, 334

Clarke transformation 128, 157, 239
commutation 222, 236, 248, 287, 289
commutation reactance 360
companion circuit 6, 69, 78
compensation method 89, 212
computer systems
  graphical interface 7
  languages 195, 325
  memory 95, 220
  software 118, 205, 233, 234, 322, 323, 325, 333
conductance matrix 69, 76, 83, 91, 93, 95, 106, 185, 213, 219–221, 224, 225, 230, 340, 376
constant current control 55
continuous systems 5, 11, 22
convergence 21, 43, 44, 60, 244, 248, 279, 285, 286, 356, 357
converter 7, 35–37, 44, 49, 53, 55, 94, 194, 217–219, 231–236, 241–248, 278–290, 296–304, 313–322, 359–365
convolution 21, 114, 130, 132, 134, 251, 347, 370
corona losses 140
cubic spline interpolation 253
current chopping 97, 109, 220, 221, 227, 272, 274
curve fitting 254, 313, 348

DFT (Discrete Fourier Transform) 203, 255, 260

difference equation 99–120, 367–372
  exponential form 99–120
digital TNA 321, 322, 327
Discrete Fourier Transform (DFT) 203, 260
discrete systems 11, 30, 34, 100
distributed parameters 3, 5, 9
Dommel's method 5, 6, 9, 67, 73, 98, 105–118
dq transformation 239

earth impedance 3, 139, 142
earth return 144, 157
eigenvalues 18, 21, 127, 357
eigenvectors 21, 127
electromagnetic transients
  EMTP 5–9, 25, 52, 67, 68, 98, 105, 123, 155, 171, 177, 185, 189, 194, 206–208, 211, 217, 219, 277, 284, 285, 290, 297, 329, 333
  EMTDC 7, 8, 14, 24, 68, 94, 95, 118, 126, 136, 139, 140, 159, 166, 171, 177, 185, 190, 195–205, 217–219, 222–224, 230, 232, 235, 238–250, 255, 256, 278, 290–296, 303–307, 311–318, 333–338
  NETOMAC 8, 54, 225, 249, 303
  PSCAD/EMTDC program 7, 8, 14, 24, 80, 95, 118, 127, 139, 140, 155
  real time digital simulation 8, 80, 205, 290, 321–330
  root matching 6, 99–120
  state variables 35–64
  subsystems 219, 220, 230, 244, 248, 322, 325, 340
  synchronous machines 89, 176–190
  transformers 159–176
  transmission lines and cables 123–156
electromechanical transients 1, 303, 304

electronic analogue computer 4
EMTDC *see* electromagnetic transients
EMTP *see* electromagnetic transients
equivalent circuits
  induction motors 190, 290, 297
  Norton 6, 31, 69, 71–84, 102, 104, 105, 132, 166, 169, 174, 218, 238, 245, 264, 282, 353
  subsystems *see* electromagnetic transients
  synchronous machines 176–190
  Thevenin 84, 94, 132, 238, 245, 309, 312, 316
equivalent pi 123
Euler's method 21, 72, 100, 101, 225, 228, 351–357
extinction angle control 49, 57, 232, 248, 360

FACTS 219, 233, 304, 319, 324
fast transients 9, 176
Fast Fourier Transform (FFT) 52, 203, 281, 286, 292, 313
flexible a.c. transmission systems *see* FACTS
Ferranti effect 131
Ferroresonance 9, 164, 208
finite impulse response (FIR) 30
fitting of model parameters 251, 262
forward Euler 21, 100, 101, 351–357
forward wave 128, 131
Fourier Transform 282
frequency-dependent model 6, 44, 45, 127, 129, 130, 139, 176, 213, 251–275
frequency domain 126, 130, 132, 251, 253, 257, 277, 278, 279, 281, 295, 341–345
frequency response 67, 117, 217, 251, 253, 258, 260, 261–267, 299, 341–345, 384

Gaussian elimination 37, 73, 84, 259
ground impedance *see* earth impedance
graphical interface 7

Index   419

graph method 40
GTO 80, 204, 222, 224, 233, 241

harmonics 58, 277, 279, 282, 284, 293, 297
HVdc simulator 4
high voltage direct current transmission (HVdc) 230–233, 359
  a.c.-d.c. converter 230, 313
  CIGRE benchmark model 234–236, 359
  simulator 322
history term 26, 31, 69, 75–84, 103, 125, 134, 224
homogeneous solution 21, 105
hybrid solution 244, 245, 286, 303–308
hysteresis 54, 91, 176, 206, 208, 240

ideal switch 115, 219, 339, 376
ill-conditioning 70, 162
imbalance *see* unbalance
implicit integration 22, 44, 71, 100, 101, 351
impulse response 21, 23, 30, 94, 133
induction machines 190, 290, 297
infinite impulse response (IIR) 30
inrush current 164
insulation co-ordination 1, 3, 9, 211
instability 6, 56, 89, 116, 185, 308
integration
  accuracy 44, 67, 100
  Adam-Bashforth 352, 353
  backward Euler 72, 100, 101, 225, 228, 351, 353–357
  forward Euler 100, 101, 225, 228, 351, 353–357
  Gear-2nd order 72, 353–357
  implicit 100, 101
  predictor-corrector methods 5, 22
  Runge-Kutta 352
  stability 356
  step length 44, 51, 53–64, 85–88, 111, 114, 202, 218, 220, 238, 243, 303, 314, 315, 356

  trapezoidal 72, 100, 101, 225, 353–357
instantaneous term 69, 79, 89, 103, 162
interpolation 53, 59, 80, 91, 198, 212, 220–227, 241, 253, 323
iterative methods 12, 22, 44, 89, 171, 207, 248, 260, 265, 278

Jacobian matrix 279, 282, 285
Jury table 266, 276

Krean 8
Kron's reduction 35, 79

Laplace Transform 11, 17, 20, 33, 133, 134
LDU factorisation 230
leakage reactance 159, 162, 186, 190, 239, 360
lead-lag control 27, 29
lower south island (New Zealand) 359
LSE (least square error) 253
lightning transient 1, 3, 159
linear transformation 37
loss-free transmission line 73, 76, 123, 124, 147, 148
losses 4, 164, 176, 263
LTE (local truncation error) 36, 351
lumped parameters 4, 245, 289
lumped resistance 124

magnetising current 159, 162, 165, 171, 284
mapping 100, 220, 285
MATLAB 8, 37, 84, 118, 198, 222, 337
method of companion circuits 6
MicroTran 8
modal analysis 11, 21, 123, 126, 131, 137, 340
multi-conductor lines 126
mutual inductance 160, 178

NETOMAC 8, 54, 55
NIS (numerical integration substitution) 67–99
nodal analysis 47, 332
nodal conductance 6, 185
non-linearities 3, 4, 5, 36, 42, 54, 164, 208, 252, 277, 281
  compensation method 6
  current source representation 36, 47, 69, 76, 78, 89–92, 115, 125, 164, 185, 193, 218, 245, 267, 279, 295, 308
  piecewise linear representation 89, 91, 92, 97, 206, 219
non-linear resistance 212
Norton equivalent *see* equivalent circuits
numerical integrator substitution *see* NIS
numerical oscillations 5, 44, 67, 99, 105, 200
numerical stability 357
Nyquist frequency 42, 264, 346

optimal ordering 95

Park's transformation 177
partial fraction expansion 18, 133, 153, 155, 267
per unit system 45, 184
phase-locked oscillator (PLO) 56, 58, 65, 231
PI section model 123, 124
piecewise linear representation *see* non-linearities
poles 6, 18, 23, 32, 155, 156, 268, 368
Pollaczek's equations 157
power electronic devices 5, 109, 193, 217, 243, 279, 284, 288, 319
PowerFactory 8
PSCAD (power system computer aided design) *see* electromagnetic transients
predictor corrector methods *see* integration
prony analysis 262, 346

propagation constant 130, 151, 252
propagation function 134, 135

rational function 31, 263, 268, 269, 307, 308
reactive power 193, 239, 307, 308, 317
real time digital simulation (RTDS) 8, 80, 205, 290, 321–330
recovery voltage 278, 291
recursive formula 26, 114, 130, 133, 148, 205, 313, 348
recursive least squares 313, 348
relays 208, 209, 210
resonance 109–111, 176, 184, 208, 253, 286, 297
RLC branch 1, 60, 74, 262
r.m.s. power 313, 314
root matching 99–121
Routh-Hurwitz stability criteria 266
row echelon form 40, 41
RTDS *see* real time digital simulation
Runge-Kutta method 352

sample data 341, 343, 352
saturation 3, 44, 54, 88, 159, 164, 190, 208, 237
s-domain 25, 32, 103, 112, 117, 136, 153, 155, 264, 367
sequence components 310, 314
short circuit impedance 44, 162
short circuit level 251, 292, 297
shunt capacitance 297
snubber 217, 225, 230, 231, 237, 360
sparsity 48, 95
s-plane (s-domain) 16, 22, 23, 99, 103, 136, 210, 266
stability
  hybrid program 244, 245, 286, 303–308
  transient 1, 9, 301, 303–320
standing wave 252
STATCOM 241, 242
state space analysis 36, 243
state variable 5, 35–64
  choice 35
  formulation 13–22

Index 421

state variable (contd.)
  valve switching 51
static VAR compensator 233, 236–241
step function 105, 356
step length 44, 51, 53, 59, 64, 85, 111, 203, 218, 220, 238, 243, 303, 314, 315, 356, 357
stiffness 357
subsynchronous resonance 9, 184
subsystems *see* electromagnetic transients
subtransient reactance 186, 289
surge arrester 89, 97, 194, 211, 225
surge impedance 124, 130, 152
swing curves 317
switch representation 79
switching
  chatter 82, 97, 217, 220, 222, 227
  discontinuities 53, 54, 97, 220, 243, 333
Synchronous machine 89, 176–190
  excitation 194
  impedance 310

TACS (transient analysis of control systems) 6, 25, 194, 208–213
Taylor's series 67, 99, 351, 354
TCR (thyristor controlled reactor) 233, 236, 240
TCS (transient converter simulation) 44–55
  automatic time step adjustment 53
  converter control 55
  valve switching 51
Thevenin equivalent circuit *see* equivalent circuits
three-phase fault 236, 290
time constants 3, 22, 49, 84, 105, 136, 184, 278, 303, 357, 360
time domain 20, 132, 255, 281, 345
time step (step length) *see* integration
TNA *see* transient network analyser
transfer function 13, 18, 24, 55, 100, 102, 104, 195, 267, 367
transformers 159–176

single phase model 166–171
three phase model 172–175
transient network analyser 4, 9, 275, 300
transient stability 303–319
  hybrid program 303–308
  test system 317
transmission lines 123–142
  Bergeron model 124–126
  equivalent pi 123
  frequency dependent 130–137
  multi-conductor 126–129
trapezoidal integration *see* integration
travelling waves 129, 131
  attenuation 132–134
  velocity of propagation 75
triangular factorization 77, 80, 98
truncation errors 6, 36, 97, 99, 351, 354, 356
TS *see* transient stability
TS/EMTDC interface 307–311
  equivalent circuit
    component 308–310
  interface variables derivation 311
  location 316
Tustin method *see* bilinear transform

UMEC (unified magnetic equivalent circuit) 165–172
unbalance 242, 277, 360
underground cables 142

valve
  extinction 49, 53, 232, 248
  group 56, 230, 232
VAR compensator *see* static VAR compensator
velocity of wave propagation 75
voltage sag 278, 288–292, 300

WLS (weighted least squares) 265

zeros 30, 99, 112, 136, 155, 267, 268
zero sequence 130, 171, 186, 239
z-plane (z-domain) 22, 32, 99, 101, 116, 199, 266
z-transform 31, 276, 345, 346